Springer Optimization and Its Applications

VOLUME 89

Managing Editor
Panos M. Pardalos (University of Florida)

Editor–Combinatorial Optimization
Ding-Zhu Du (University of Texas at Dallas)

Advisory Board
J. Birge (University of Chicago)
C.A. Floudas (Princeton University)
F. Giannessi (University of Pisa)
H.D. Sherali (Virginia Polytechnic and State University)
T. Terlaky (McMaster University)
Y. Ye (Stanford University)

Aims and Scope
Optimization has been expanding in all directions at an astonishing rate during the last few decades. New algorithmic and theoretical techniques have been developed, the diffusion into other disciplines has proceeded at a rapid pace, and our knowledge of all aspects of the field has grown even more profound. At the same time, one of the most striking trends in optimization is the constantly increasing emphasis on the interdisciplinary nature of the field. Optimization has been a basic tool in all areas of applied mathematics, engineering, medicine, economics, and other sciences.

The series *Springer Optimization and Its Applications* publishes undergraduate and graduate textbooks, monographs and state-of-the-art expository work that focus on algorithms for solving optimization problems and also study applications involving such problems. Some of the topics covered include nonlinear optimization (convex and nonconvex), network flow problems, stochastic optimization, optimal control, discrete optimization, multi-objective programming, description of software packages, approximation techniques and heuristic approaches.

For further volumes:
http://www.springer.com/series/7393

Marie E. Schmidt

Integrating Routing Decisions in Public Transportation Problems

 Springer

Marie E. Schmidt
Institute for Numerical and Applied Mathematics
Göttingen, Germany

ISSN 1931-6828   ISSN 1931-6836 (electronic)
ISBN 978-1-4614-9565-9   ISBN 978-1-4614-9566-6 (eBook)
DOI 10.1007/978-1-4614-9566-6
Springer New York Heidelberg Dordrecht London

Library of Congress Control Number: 2013957935

Mathematics Subject Classification (2010): 90-02

© Springer Science+Business Media New York 2014
This work is subject to copyright. All rights are reserved by the Publisher, whether the whole or part of the material is concerned, specifically the rights of translation, reprinting, reuse of illustrations, recitation, broadcasting, reproduction on microfilms or in any other physical way, and transmission or information storage and retrieval, electronic adaptation, computer software, or by similar or dissimilar methodology now known or hereafter developed. Exempted from this legal reservation are brief excerpts in connection with reviews or scholarly analysis or material supplied specifically for the purpose of being entered and executed on a computer system, for exclusive use by the purchaser of the work. Duplication of this publication or parts thereof is permitted only under the provisions of the Copyright Law of the Publisher's location, in its current version, and permission for use must always be obtained from Springer. Permissions for use may be obtained through RightsLink at the Copyright Clearance Center. Violations are liable to prosecution under the respective Copyright Law.
The use of general descriptive names, registered names, trademarks, service marks, etc. in this publication does not imply, even in the absence of a specific statement, that such names are exempt from the relevant protective laws and regulations and therefore free for general use.
While the advice and information in this book are believed to be true and accurate at the date of publication, neither the authors nor the editors nor the publisher can accept any legal responsibility for any errors or omissions that may be made. The publisher makes no warranty, express or implied, with respect to the material contained herein.

Printed on acid-free paper

Springer is part of Springer Science+Business Media (www.springer.com)

Preface

To model and solve optimization problems arising in public transportation planning, data about the passengers is necessary and has to be included in the models in any phase of the planning process. In particular, when aiming at minimizing the passengers' travel time, information about passengers' travel routes is indispensable. For example, if many passengers travel between two stations using the same route, it may be beneficial to establish a high-speed line on this route; a good timetable should offer low waiting times for connections with high passenger numbers.

For this reason, many solution approaches assume a two-step procedure: in the first step, passenger routes are determined using shortest path, flow, or traffic assignment procedures. In the second step, the actual planning of lines, timetables, etc., is done. This approach ignores that for most passengers there are many possible ways to reach their destinations in the public transportation network; thus, the actual connections the passengers will take depend strongly on the decisions made during the planning phase.

In this book we aim at integrating the route determination procedure in the optimization process of public transportation problems. We consider three planning problems arising in public transportation planning: *line planning* specifies the paths and frequencies of the train lines to be established based on data about the infrastructure; *timetabling* determines arrival and departure times of trains at stations; and the task of *delay management* is to decide whether trains should wait for feeder trains or depart on time in case of delays. In order to maximize the passengers' benefit, our objective is the minimization of the passengers' overall travel time.

After formulating suitable network models for the considered problems with integrated routing, we investigate the computational complexity of the integrated problems. We prove NP-hardness results and derive polynomial algorithms for special cases, furthermore we investigate the approximability of some of the considered problems. For solving the integrated problems, we present exact integer programming approaches as well as a heuristic approach which alternates route determination and optimization steps.

Göttingen, Germany   Marie E. Schmidt

Contents

1	**Introduction**	1
	1.1 Overview	1
	1.2 Outline	4
	1.3 Basic Concepts Used in This Book	5
2	**Line Planning**	9
	2.1 Introduction to Line Planning	9
	2.1.1 Line Planning Problems	9
	2.1.2 Literature Overview	12
	2.1.3 The Line Planning Problems Considered in This Book	14
	2.1.4 Outline of the Line Planning Chapter	20
	2.2 Modeling Line Planning with Routing	20
	2.2.1 The Change-and-Go Network	21
	2.2.2 Line Planning in the Change-and-Go Network	23
	2.2.3 The Line Network	25
	2.2.4 Uncapacitated Line Planning in the Line Network	28
	2.3 Solving Line Planning When Part of the Solution Is Fixed	30
	2.3.1 Solving Uncapacitated Line Planning When Part of the Solution Is Fixed	30
	2.3.2 Solving Capacitated Line Planning When Part of the Solution Is Fixed	31
	2.4 Classification of Line Planning Problems with Routing	33
	2.5 Uncapacitated Line Planning with Routing	35
	2.5.1 Uncapacitated Line Planning in General Networks	35
	2.5.2 Uncapacitated Line Planning in Linear Networks	44
	2.6 Capacitated Line Planning with Routing	60
	2.6.1 Complexity of Capacitated Line Planning with Routing	60
	2.6.2 Integer Programming Formulations	64

3 Timetabling 73
- 3.1 Introduction to Timetabling 73
  - 3.1.1 Timetabling Problems 73
  - 3.1.2 Literature Overview 76
  - 3.1.3 The Timetabling Problem Considered in This Book 79
  - 3.1.4 Outline of the Timetabling Chapter 81
- 3.2 Modeling Timetabling with Routing Using Event-Activity Networks 82
- 3.3 Solving Timetabling When Part of the Solution Is Fixed 85
- 3.4 Complexity of Timetabling with Routing 88
  - 3.4.1 NP-Hardness of Timetabling with Routing 88
  - 3.4.2 Inapproximability of Timetabling with Routing 92
- 3.5 Timetabling with Routing Between Events 97
- 3.6 An Exact Algorithm for Timetabling with Routing 99
- 3.7 Integer Programming Formulations 101
  - 3.7.1 Flow-Based Formulation 102
  - 3.7.2 Virtual-Activity-Based Formulation 108

4 Delay Management 113
- 4.1 Introduction to Delay Management 113
  - 4.1.1 Delay Management Problems 113
  - 4.1.2 Literature Overview 117
  - 4.1.3 The Delay Management Problem Considered in This Book 120
  - 4.1.4 Outline of the Delay Management Chapter 122
- 4.2 Modeling Delay Management with Routing Using Event-Activity Networks 123
- 4.3 Solving Delay Management When Part of the Solution Is Fixed 126
- 4.4 Complexity of Delay Management with Routing 130
  - 4.4.1 NP-Hardness of Delay Management with Routing for One OD-Pair 131
  - 4.4.2 An Algorithm for Delay Management with Routing with One OD-Pair 136
  - 4.4.3 NP-Hardness of Delay Management with Routing for Instances with NTT Property 151
  - 4.4.4 Inapproximability of Delay Management with Routing 155
- 4.5 Integer Programming Formulation 162

5 An Iterative Solution Approach for General Network Problems with Routing 167
- 5.1 Classification of Network Problems with Routing 167
- 5.2 An Iterative Heuristic 173
- 5.3 The Price of Sequentiality 175
- 5.4 Analysis of the Iterative Heuristic for Timetabling with Routing 176

	5.5	Analysis of the Iterative Heuristic for Arc Speed-Up	181
		5.5.1 Definition of the Problem Arc Speed-Up	181
		5.5.2 The Iterative Heuristic for Arc Speed-Up	183
		5.5.3 Solving Arc Speed-Up Exactly	185
		5.5.4 The Price of Sequentiality for Arc Speed-Up	186
6	**Conclusions and Outlook**		207

Frequently Used Notation ... 213

References ... 217

Index ... 225

Chapter 1
Introduction

1.1 Overview

This book treats three optimization problems arising in public railway transportation, namely

- the problem of *line planning* which, given the infrastructural data, i.e., information about stations, tracks and track lengths, consists of finding a *line concept* that specifies the paths and frequencies of the train lines to be established;
- the problem of *timetabling*, i.e., to specify arrival and departure times of trains at stations in a *timetable*; and
- the problem of *delay management*: given an operating public transportation system and a set of delays occurring in the daily operations, the task of delay management is to decide whether trains should wait for delayed feeder trains or depart on time and to update the existing timetable to incorporate these changes. The updated timetable is called a *disposition timetable*.

Our objective in the above-mentioned problems is to minimize the overall travel time of the passengers who use the public transportation system. Certainly, the routes that the passengers take influence the travel time. Hence, a good solution algorithm for the above-described problems takes passenger routes into account. That is, if many passengers travel between two stations using the same route, it might be beneficial to establish a high-speed line on this route. Or, to give an example with regard to the problems of timetabling and delay management, if many passengers change at a station from one train to another, the passengers would favor a timetable that does not make them wait too long for the second train, while in case of a delay on the incoming train, a good *disposition timetable* would ensure that the passengers do not miss the connecting train. Hence, in order to find good line concepts, timetables, or disposition timetables, it is important to know the passenger routes.

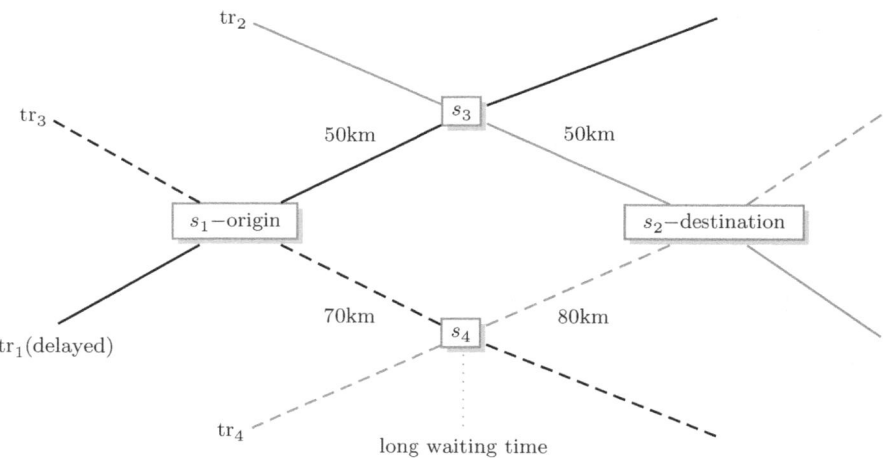

Fig. 1.1 The public transportation network, consisting of stations and tracks, described in the example. The *dashed trains* are high-speed trains, the other trains are regional trains

For this reason, in the literature it is often assumed that a good *routing*, i.e., a collection of a route for every passenger, is part of the input of the above-described problems.

Thereby, many approaches lose sight of the fact that the line concepts, timetables, or disposition timetables themselves considerably influence which routes are favorable and which are not. Hence, approaches that estimate passenger routes without any knowledge of the transportation system to be established on the infrastructure are likely to produce suboptimal solutions.

This fact is illustrated in the following example which is sketched in Fig. 1.1:

Consider a passenger p who wants to travel from station s_1 to station s_2. There are two possible ways to do that: a shorter connection with a length of 100 km passing a station s_3 and another connection which makes a slight detour via station s_4 and thus has length 150 km. Without further information, one would suggest that p take the shorter connection.

However, due to constraints on the line concept or preferences of other passengers, it may turn out that a line concept is established in which the shorter route via s_3 is served by very slow regional trains tr_1 and tr_2, making many intermediate stops at regional stations, while on the other hand the tracks (s_1, s_4) and (s_4, s_2) are served by high-speed trains tr_3 and tr_4. In view of our objective of minimizing the travel time, the passenger would choose the high-speed connection. Thus, our initial 'guess' for an optimal route, taking only the lengths of the tracks between the stations into account was misleading and did not find an optimal route with regard to the line concept.

1.1 Overview

Given additionally the knowledge of the timetable, it may turn out that the train tr_4 in fact departs much later in s_4 than the arrival of tr_3 takes place. In this setting it turns out to be advantageous for p to choose the regional trains instead of waiting a long time at station s_4. Again, our guess given only the input information of the timetabling problem, consisting in this case of the network and the line concept assumed above, leads to a suboptimal path for passenger p.

To round the example off with a quick glance on delay management we remark that for the given timetable in case of a delay on tr_1, train tr_2 might or might not wait for train tr_1 at station s_3, depending on the number of passengers that want to transfer at s_3 and the number of passengers which already sit in tr_2 and hope to continue their journey without delays. If tr_2 does not wait, p would have better chosen the path via station s_4, preferring a long waiting time to the prospect of being not able to continue the journey at station s_3.

We conclude that the passenger routes are not at all fixed, i.e., contained in the *input* of our optimization problems, but depend on the chosen line concept, timetable, or disposition timetable, respectively, and should hence be considered as part of the *solution* instead.

The task of this book is to conceptually integrate the determination of passenger routes in the optimization step of line planning, timetabling, and delay management. In other words, our goal is to determine a line concept/timetable/disposition timetable which is optimal in the sense that it allows a routing with travel time shorter than or equal to the travel time of routings that are possible for other line concepts (or timetables/disposition timetables, respectively).

An important step when approaching an optimization problem, in particular if real-world instances can be expected to be large, is to investigate the complexity of (solving or approximating) the problem, since it allows to assess which solution methods may be possible to solve the problems and which running times can be hoped for. For this reason, a large part of this book concentrates on an analysis of the border between easy and hard problems in line planning, timetabling, and delay management, which can later on be used as a basis for developing solution methods.

In particular, we are interested in the following questions:

- How can existing models for line planning, timetabling, and delay management be extended to allow a free choice of the routing?
- What is the computational complexity of determining a line concept/timetable/ disposition timetable and a routing, simultaneously?
  Can we identify classes of instances for which this can be done in polynomial or pseudopolynomial time? Where is the borderline between easily solvable and NP-hard problems?
- To what extent can existing solution methods for line planning, timetabling, and delay management support the simultaneous calculation of both a line concept/timetable/disposition timetable and a routing?
  Can existing integer programming approaches be modified to solve our problems? Is it possible to approximate optimal solutions calculating line concepts/timetables/disposition timetables and routings iteratively?

1.2 Outline

Since public transportation problems are often very complex and hard to solve, most works concentrate only on one problem at a time, although there exist some attempts to combine multiple problems identified in public transportation planning into one large model (see, e.g., [LM07a] and references therein). Following this approach, we regard the problems of *line planning with routing, (aperiodic) timetabling with routing*, and *delay management with routing* as independent problems and dedicate a chapter to each of these problems.

An introduction to the respective problem and an overview over models, complexity results, and solution techniques is given at the beginning of each of the three chapters. To obtain an overview about the field of public transportation in general, we refer the interested reader to [BLZ97, GJP+04, HKLV05, DH07, GH08, CKT10], and references therein.

Chapter 2 treats line planning with routing. Based on a result implying strong NP-hardness for general line planning problems with routing, we concentrate on identifying restrictions on the input parameter of the problems which allow polynomial- or pseudopolynomial-time algorithms for line planning problems without capacity restrictions.

When taking into account capacity restrictions, line concepts that represent a system optimum may force some passengers to take long detours which do not represent the individually optimal route choices for these passengers. We formulate a model for capacitated line planning where every passenger can take a route that is individually optimal for him/her without provoking capacity conflicts and present an integer programming formulation to solve our problem.

In Chap. 3, after proving NP-hardness of aperiodic timetabling with routing in general, we utilize the fact that for a fixed routing timetabling can be solved by linear programming to define a timetabling problem with limited route choices which is still solvable in polynomial time. This motivates an exact algorithm for timetabling with routing as well as an integer programming formulation which differs from the straightforward approach of modeling passenger routes using multi-commodity flow constraints.

Surprisingly, in Chap. 4 delay management with routing turns out to be NP-hard even if there is only one passenger traveling. Hence, our main focus in this chapter is to develop a polynomial-time approximation algorithm for instances of delay management with routing considering only one passenger. We identify a restriction under which this algorithm provides optimal results and investigate whether these results can be extended to instances with more passengers. Although this turns out not to be the case, the algorithm can still be used in the last section of this chapter to enhance out the integer programming formulation which we develop based on a formulation for a delay management problem with fixed routing.

Motivated by the structural similarities between the considered problems, in Chap. 5 we develop a heuristic which can be applied to a whole class of network problems with routing. This heuristic alternatingly calculates a routing and a line concept/timetable/disposition timetable, relying on shortest path algorithms and existing optimization algorithms for finding an optimal line concept/timetable/disposition timetable in presence of a fixed routing. Furthermore, we establish an index measuring the worst-case quality of the heuristic.

Before we start the discussion of the problems, we define some concepts from graph theory needed in the following chapters and other prerequisites and notions used in this book in Sect. 1.3.

The problems of line planning, timetabling, and delay management are treated as independent problems in Chaps. 2–4. Hence, to some extent it is possible to read individual chapters of this book independently from the others. However, since the problems of line planning, timetabling, and delay management are subsequent steps of the public transportation planning process, definitions and notations in timetabling and delay management rely on precedent chapters and are not repeated but assumed to be known in later chapters. The iterative heuristic developed in Chap. 5 is stated for a general problem class, hence Chap. 5 could also stand alone, but examples for the illustration of the notions and the evaluation of the heuristic are taken from the preceding chapters.

1.3 Basic Concepts Used in This Book

In this book, we use some basic concepts from graph theory, complexity theory, and combinatorial and linear integer optimization.

For the theory of NP-completeness we refer to [GJ79]. Note that since all decision problems considered in this book are in NP, as can be easily verified, in our proofs of NP-completeness, we only specify a polynomial transformation from a problem Q' that is known to be NP-complete to our problem Q. If all numbers occurring in the constructed instances of Q are polynomially bounded in the size of the corresponding instances of Q', strong NP-completeness follows. (Strong) NP-hardness for an optimization problem follows from (strong) NP-completeness of the corresponding decision problem.

Models and concepts from linear integer and combinatorial optimization can be found in, e.g., [NW88]. We use the abbreviations *LP* for *linear program* and *IP* for *linear integer program*. When considering optimization problems, we use the term solution in the sense of *feasible solution*.

To avoid confusion we define all concepts from graph theory needed in this book below, following [Die06, AMO93] for the most part. Furthermore, at the end of this section we introduce the *public transportation network* (PTN) which models the infrastructure for public transportation problems and is hence used in the definitions of line planning, timetabling, and delay management.

Definition 1.3.1. An *(undirected) graph* $G = (V, E)$ is a pair of disjoint sets V and $E \subset V \times V$. The elements in V are called the *vertices* or *nodes* of G, the elements in E are called (undirected) *edges*. We denote an edge by $e = \{i, j\}$.

Most problems in public transportation are represented using directed graphs.

Definition 1.3.2. A *directed graph* or *digraph* $G = (V, A)$ is a pair of disjoint sets V and A together with two mappings α and ω from A to V. We denote a *directed edge* or *arc* of G by $e = (\alpha(e), \omega(e))$.

Note that we can easily 'forget' that a digraph is directed replacing the arc set A by the edge set $E := \{\{i, j\} : (i, j) \in A\}$. Since $V \cap E = \emptyset$, we sometimes slightly abuse notation and write $i \in G$ or $e \in G$ for a vertex i or an edge e in a (di)graph $G = (V, E)$.

Let $G = (V, A)$ be a digraph. For $v \in V$, let $\delta^-(v) := \{a : \alpha(a) = v\} \subset A$ denote the set of arcs starting in v and $\delta^+(v) := \{a : \omega(a) = v\} \subset A$ denote the set of arcs ending in v.

In our application, all considered graphs are *finite*, e.g., they have a finite number of nodes and they do not have *loops*, e.g., no edges $\{i, i\}$ (or (i, i), in the directed case), or *multiple edges*. Thus, from now on, when talking about graphs we refer to finite graphs.

Since the graphs we use for modeling are also used for distance calculations, in many cases we assign labels to the edges or arcs representing their lengths.

Definition 1.3.3. An (undirected) graph $G = (V, E)$ with edge length L_e for every edge $e \in E$ is called an *(undirected) network*, analogously we call a directed graph $G = (V, A)$ with arc length L_a for every arc $a \in A$ a *(directed) network*.

Definition 1.3.4. A *linear network* is an undirected network $G = (V, E)$ with node set $V = \{v_1, v_2, \ldots, v_n\}$ and edge set $E = \{\{v_1, v_2\}, \{v_2, v_3\}, \ldots, \{v_{n-1}, v_n\}\}$.

Let $G = (V, E)$ be a graph/undirected network. We refer to $G' = (V', E')$ with $V' \subset V$ and $E' \subset (V' \times V') \cap E$ as a *subgraph/subnetwork* of G and write $G' \subseteq G$. Let V' be a subset of V and $E' := \{\{i, j\} \in E : i, j \in V'\}$. Then $G' = (V', E') \subseteq G$ is called the *subgraph/subnetwork induced by V'*.

In a directed graph/directed network $G = (V, A)$, we call $G' = (V', A')$ with $V' \subset V$ and $A' \subset A$ a *subgraph/subnetwork* of G and write $G' \subseteq G$. For a subset V' of V and $A' := \{(i, j) \in A : i, j \in V'\}$, we call $G' = (V', A') \subseteq G$ the *subgraph/subnetwork induced by V'*.

Definition 1.3.5.

- Let $G = (V, E)$ be an undirected graph. A *path* in G is a subgraph $P = (V_P, E_P)$ defined by a sequence of nodes $i_1, \ldots, i_n$ that are linked by edges $\{i_j, i_{j+1}\}$ for $i = 1, \ldots, n - 1$.
- Let $G = (V, A)$ be a directed graph. A *directed path* in G is a subgraph $P = (V_P, A_P)$ given by a sequence of nodes $i_1, \ldots, i_n$ that are linked by arcs (i_j, i_{j+1}) for $i = 1, \ldots, n - 1$. In contrast, an *undirected path* in G is given as a sequence of nodes $i_1, \ldots, i_n$ such that for every $j = 1, \ldots, n - 1$ either $(i_j, i_{j+1}) \in A$ or $(i_{j+1}, i_j) \in A$.

Note that nodes can repeat in a path according to this definition. This is often referred to as a *walk* in the literature. A path $P = (i_1, i_2, \ldots, i_n)$ is called *simple* if $i_j \neq i_k$ for all $j \neq k$. In directed graphs, we often refer to directed paths simply as *paths*. If we refer to undirected paths, we always emphasize that.

1.3 Basic Concepts Used in This Book

Note that a path is also uniquely determined by its edge or arc sequence, hence, we sometimes write $P = (i_1, e_1, i_2, e_2, \ldots, e_{n-1}, i_n)$, $P = (i_1, i_2, \ldots, i_n)$, or $P = (e_1, e_2, \ldots, e_{n-1})$ instead of $P = (\{i_1, i_2, \ldots, i_n\}, \{e_1, e_2, \ldots, e_{n-1}\})$.

Definition 1.3.6. The *length* of a (directed or undirected) path $P = (V_P, A_P)$ is given as $l(P) = \sum_{a \in A_P} L_a$.

The *shortest path problem* is the task to find a path P with first node s and last node t of minimum length in a (directed or undirected network) with two specified nodes s and t.

Let $G = (V, E)$ be a directed or undirected graph with nonnegative edge lengths. For a given start node s, *Dijkstra's algorithm* (see, e.g., [NW88]) calculates a set of shortest paths $\{P_t : t \in V\}$ to every other node t in the node set V; thus, in particular it solves the shortest path problem defined above. If this algorithm is implemented using *Fibonacci Heaps*, a running time of $O(m + n \log(n))$ can be achieved (see [FT87]), where m is the number of edges/arcs and n is the number of nodes in the considered graph. Note that in directed acyclic graphs, it is possible to find shortest paths in $O(m)$ (see, e.g., [AMO93]).

The variant where shortest paths between all pairs of nodes have to be found is called the *all-pairs shortest path problem*. It can be solved, e.g., using the *Floyd–Warshall algorithm* in $O(n^3)$.

Notation 1.3.7. Let $G = (V, E)$ be a (directed or undirected) network and let $\{(s_i, t_i) : i \in I\} \subset V \times V$ denote a set of pairs of nodes. We call the set of paths $\{P_i : i \in I\}$ where P_i is a shortest path from s_i to t_i in G a shortest-path routing.

Note that, in case of positive edge or arc lengths, every shortest path is simple.

Definition 1.3.8. In an undirected graph, a path C defined by a node sequence $i_1, \ldots, i_n$ with $i_1 = i_n$ is called a *cycle*. If a graph does not contain any cycle as a subgraph, it is called *acyclic*.

Let $G = (V, A)$ be a directed graph and $i, j \in V$.

- j is called a *successor* of a node i if there exists a directed path from i to j in G.
- j is called a *predecessor* of i if there exists a directed path from j to i in G.
- If G is acyclic and i is a predecessor of j in G, we write $i \leq_G j$ and $j \geq_G i$. If $i \neq j$, we can also write $i <_G j$ and $j >_G i$. $\leq_G, \geq_G, <_G,$ and $>_G$ define a partial order on the vertices of G.
- j is the *direct successor* of i on a path P if $(i, j) \in P$. i is then called *direct predecessor* of j on P.

Definition 1.3.9. Let $G = (V, E)$ be a non-empty directed or undirected graph.

- G is called *connected* if for every pair $i, j \in V$; it contains a path from i to j as a subgraph.
- A connected subgraph H of G is called a *connected component* of G if there is no connected subgraph H' of G such that $H \subsetneq H'$.

Definition 1.3.10. A *tree* is an undirected, connected, acyclic graph.

The basic network, common to all three problems studied in this book, is the so-called PTN, modeling the underlying infrastructure of stations and tracks.

Definition 1.3.11. A PTN is an undirected network

$$\mathcal{PTN} = (S, E).$$

In railway transportation, the vertices S represent train stations, the (undirected) edges E represent the physical connections between the vertices, i.e., the tracks. The length of a track e is included in the model as edge length $L_e \in \mathbb{N}_0$.

Chapter 2
Line Planning

2.1 Introduction to Line Planning

2.1.1 Line Planning Problems

Given the public transportation network, i.e., information about the location of the stations, the tracks connecting the stations, and the lengths of the tracks, *line planning* aims at determining the *lines*, i.e., the routes served regularly by a train. Furthermore, in many line planning approaches, not only the routes which should be served are considered, but also the *frequencies* of the services are planned.

Lines

A *line* is a path in the PTN which is served by *trains*. In the line planning problems investigated in this book we are given a predefined set of potential lines, the *line pool*, among which we choose the lines we want to establish. It is also possible to define the set of potential lines only implicitly as a path from a predefined origin to a predefined destination [BGP08, BN10]. Another approach is to construct lines by combining small line parts, see, e.g., [LS67, SBP74, Son79, Fuh08].

Trains and Frequencies

The *frequency* of a line indicates how many trains of the line are operated in a certain time period. In the literature, both the problems of finding a line concept and the problem of finding a line concept *together* with a corresponding frequency assignment are referred to as line planning, while the problem of assigning a frequency to a given line concept is called *frequency setting* (see, e.g., [CF95, GSS04, dMI06]). There are various reasons why frequencies are an important feature in line planning.

When a passenger arrives at a station by train, his/her waiting time for the next connecting train depends on the frequency of the connecting train; hence in practice, line frequencies have a significant impact on travel times. However, there is a second reason to consider line frequencies: the capacity of the transportation system. If there are 500 passengers who want to take the same line on the same edge of the PTN during a certain time period and the capacity of a train is only 200, we need at least three trains serving the line.

In this book we ignore the impact of line frequencies on waiting times and refer the interested reader to [CF95, GSS04, dMI06] and references therein. Instead, we focus on *uncapacitated line planning*, where frequencies are not taken into account, and *capacitated line planning*, where frequencies are introduced due to capacity reasons.

In capacitated line planning we make use of the definition of *trains* as single realizations of a line which can transport only a certain number of passengers Cap at a time. In uncapacitated line planning capacity restrictions are not considered, i.e., Cap is set to ∞.

Properties of Lines

Every line l is assigned a monotonously increasing function $\alpha_l : \mathbb{R}_0^+ \to \mathbb{R}_0^+$, called *line driving time function*, which is often assumed to be the identity. However, like in [SS10b, SS12a, Sch12] in this book we consider the general case where line driving time functions can differ from the identity.

Furthermore, every line is assigned a line cost β_l which specifies the cost caused by establishing line l with frequency f_l. Often, β_l consists of a fixed cost $\kappa_l \geq 0$ and a per-train cost $b_l > 0$, i.e., $\beta_l(f_l) = b_l f_l + \kappa_l$ (see, e.g., [TTBP11]). In the problems considered in this book, we only consider per-train costs b_l, i.e., $\kappa_l = 0$. This simplification is common in the literature, it is, e.g., used in [GYW06, BN10, TTBP11]. More realistic models apportion costs according to rolling stock utilization [Bus98, CvDZ98, GvHK04, GvHK06].

Some models additionally assume upper bounds on the frequency of a line [TTBP11] or an edge of the PTN [Bus98, CvDZ98, GvHK04, GvHK06].

Transportation Modes

Some publications consider the simultaneous optimization of several transportation modes or train types which may differ, for example, in halting patterns, speed, and capacity (see, e.g., [CvDZ98, GvHK06, BGP07, BGP08, TTBP11, BK12]), while others assume that all considered vehicles are uniform and that different transportation modes or types of vehicles can be dealt with separately [BKZ97, Sch05, SS06]. The approach followed in this book allows different line driving time functions and can hence be classified somewhere in between these approaches.

2.1 Introduction to Line Planning   11

Demand

In cost-oriented models (see, e.g., [BLL04, TTBP11, GvHK04, CvDZ98]) and some earlier models in passenger-oriented transportation (e.g., [BKZ97]) passenger demand is modeled by defining minimal frequencies on the edges of the PTN and has to be covered by the line concept. The minimal frequencies may be a direct input to the problem [Bus98, BLL04, TTBP11] or pre-estimated by a shortest-path routing in the PTN [CvDZ98, GvHK04, GvHK06].

However, when trying to maximize passengers' convenience, predefined passenger routes do not seem to be the right modeling approach since passengers choose routes according to the travel time and/or the numbers of transfers observed in the present line concept. Consequently, most line planning models assume that we are given every passenger's origin and his/her destinations as a set of *origin-destination pairs (OD-pairs)*.

Budget Considerations

Without budget considerations, a solution that maximizes passengers' convenience would lead to many lines which are rarely used. Therefore, some passenger-oriented line planning models minimize a weighted sum of the chosen passenger-oriented objective and the costs of the line concepts (see, e.g., [GYW06, BGP07, BGP08, NJ08, BN10]). Another option is to impose an overall *budget B* on the line costs [Sch05, SS06, SS10b, SS12a, Sch12]. Additionally, setting up too many lines can be avoided by *upper bounds* on frequencies, as done, e.g., in [Bus98, SS06, BGP08, NJ08].

Objective Functions

The aim of *cost-oriented line planning* [CvDZ98, BLL04, GvHK04, TTBP11] is to minimize the costs of the established line concept subject to covering the passengers' demand. To cover passengers' demand, minimal frequencies are imposed on the edges of the PTN.

In contrast to cost-oriented line planning models, passenger-oriented line planning aims at maximizing the quality of the public transportation system from the passengers' point of view. There are two main approaches to passenger-oriented line planning: the *direct traveler approach* and the *minimization of travel time*.

The *direct traveler approach* maximizes passengers' travel quality by maximizing the number of travelers that do not have to change lines during their journey. It is introduced in [Die78] and developed further in [BKZ97, Bus98]. In [BKZ97] the routes for the OD-pairs are predefined and only direct travelers on these routes are counted. In [Die78, Bus98] in contrast, for every OD-pair a set of the so-called *pleasant paths* in the PTN is given. Any of these pleasant paths can be chosen for the passenger, if it allows a direct connection from origin to destination.

Also the *travel time* is an important criterion to measure passengers' convenience. It consists of the *driving time* between stations and the *transfer time* between trains. Since in line planning no timetable is known, the transfer time is estimated by *transfer penalties* $p_s^{ll'}$ for transfers from line l to line l' at station s. To simplify the model, sometimes it is assumed that the transfer penalties are equal for all transfers (see, e.g., [Sch05, SS06]) or even set to 0 (see, e.g., [PB06, BGP07, BGP08]), which implies that only the driving time is minimized in the corresponding models. Borndörfer and Neumann [BN10] and Borndörfer and Karbstein [BK12] use a model that estimates transfer penalties to allow faster computation times with respect to exact models such as [SS06]. Guan et al. [GYW06] minimize a weighted sum of costs, travel time, and transfers.

Apart from the cited papers, the objective of minimizing travel time has already been formulated, e.g., in [LS67, SBP74, Son79, Man80]. However, solution approaches in these early papers are purely heuristic.

Another approach of passenger-oriented line planning with integrated passenger routing has been treated in [NJ08] under the objective of maximizing the *travel quality* which is defined as the time a passenger can save compared to the maximal accepted travel time.

2.1.2 Literature Overview

In Sect. 2.1.1 we presented different modeling paradigms in line planning. In this section we give a broad overview on complexity results and exact solution techniques for solving line planning problems. Since there is a multitude of different line planning models and results, a more detailed overview is beyond the scope of this book. We refer the interested reader to the surveys about line planning given in [Chu84, IC95, Sch11], the more general survey papers listed in Sect. 1.2 and references therein.

Computational Complexity

In most more recent papers on line planning, the use of sophisticated solution techniques is justified with proofs of computational hardness for the considered problems.

In [Bus98] it is shown that in presence of lower and upper frequency bounds, choosing a feasible line concept with frequencies from a given line pool is (strongly) NP-complete by reduction from Hamiltonian Path. Given a passenger routing, in [CvDZ98] it is shown that covering the demand in a cost-optimal way is (strongly) NP-hard. In [TTBP11], the computational complexity of a cost-oriented line planning problem on linear networks and tree networks is investigated.

2.1 Introduction to Line Planning

In [BNP09], Borndörfer et al. define the *line connectivity problem*: given an undirected graph, a set of OD-pairs and a set of lines with costs, the task is to find a set of lines with minimal costs such that for every OD-pair there exists a path in the graph from origin to destination which is completely covered by lines. For the corresponding decision problem, the authors show that it is in P if there is only one OD-pair, but strongly NP-complete in general, even if $\mathcal{OD} = S \times S$, using a reduction from Set Cover.

Line planning with travel-time objective is (strongly) NP-complete if lines are constructed from scratch, even if only the number of transfers is counted in the objective function [Sch05]. In [PB06], Pfetsch and Borndörfer show that line planning with a weighted objective functions consisting of travel time and costs is NP-hard in case of free route choices as well as for the restrictions to unsplittable routes and shortest paths in the PTN. Furthermore, they investigate the gap between solutions to the line planning problems under different restrictions on the routings. In [Sch05, SS06] line planning with travel-time objective is considered on a linear network. The problem is shown to be (strongly) NP-hard even if only transfers are counted in the objective function and the line costs are equal for all lines. Here, the proof takes advantage of different halting patterns of the trains. [SS10b, SS12a] give an extensive overview about the complexity of different uncapacitated line planning problems on linear networks where all trains have the same halting patterns and extend their results to line planning problems with capacities. The results are presented in Sects. 2.5 and 2.6.1 of this book.

IP Formulations and Solution Techniques

The different line planning problems in the literature are often modeled as integer programs. In cost-oriented models, the variables represent the frequencies of lines in the line pool, subject to capacity and coverage constraints that vary with the model [Bus98, GvHK04, GvHK06, TTBP11]. Sometimes, other decisions like the number of coaches [CvDZ98, Bus98, BLL04] are included in the formulation. Solution techniques include branch-and-bound [Bus98, CvDZ98], branch-and-cut [GvHK04], and variable fixing heuristics [BLL04].

In passenger-oriented approaches, passengers' demand is given in form of OD-pairs that have to be routed through the network. When using integer programming methods, passenger routes are modeled as paths or flows in the network.

In [BGP08], Borndörfer et al. propose an IP model to minimize a weighted sum of driving time and costs modeling passengers' route choices as flows in a directed version of the PTN. Instead of using a predefined line pool, also the lines are modeled as flows between line origins and line destinations.

To account for transfers, many flow-based formulations [Sch05, SS06, SS10b, SS12a, Sch12] make use of an extension of the PTN, the *change-and-go network (CGN)* (see Sect. 2.2.1), which consists of a path for each line and transfer arcs. For every arc of this network and every OD-pair, a flow variable is introduced.

To solve the LP-relaxation of a line planning problem with travel-time objective, in [Sch05, SS06] Scholl and Schöbel make use of the (almost) block-diagonal structure of the constraint matrix and apply Dantzig–Wolfe decomposition.

In contrast to flow-based formulations, path-based formulation introduce a variable for every possible path from origin to destination of an OD-pair.

Guan et al. [GYW06] propose an integer program which minimizes a weighted sum of costs, travel time, and transfers using variables for assigning passengers to routes and to lines covering the routes. To keep the number of variables small, the number of possible routes is restricted in the problem input.

Borndörfer et al. [BGP07, BGP08] propose an IP model using path variables for minimizing a weighted sum of driving time and costs. Like in the flow-based model proposed in [BGP08], they do not assume a fixed line pool, but allow all possible lines between a set of line origins and line destinations. However, in contrast to the flow-based model, the lines are not modeled by flows but, like the passengers' routes, by path variables. Borndörfer et al. [BGP07, BGP08] solve a relaxation of the model using column generation. While the pricing problem for the passengers' routes is a shortest path problem, the pricing problem for the lines turns out to be a longest path problem and strongly NP-hard. However, if the lengths of the lines are restricted, dynamic programming techniques can be used to solve the pricing problem efficiently [BGP07]. A similar approach is followed in [NJ08].

A disadvantage of the models proposed by [BGP07, BGP08] is that they do not account for passengers' transfers, i.e., only the driving time is considered in the objective function. The models proposed in [BN10, BK12] try to overcome this disadvantage by estimating the number of direct travelers [BN10] or the number of passengers' transfers [BK12]. Also here, column generation is used to solve a relaxation of the problem. Borndörfer and Neumann compare the obtained results to the results obtained using the model of [BN10].

Path-based IP models for the direct traveler approach are given in [Die78, BKZ97, Bus98].

2.1.3 The Line Planning Problems Considered in This Book

We now describe the line planning problems considered in this book in more detail and introduce definitions and notations used throughout the book. We consider three different passenger-oriented line planning problems with the objective of minimizing the overall travel time: uncapacitated line planning with routing (uLPwR), simplified capacitated line planning with routing (scLPwR), and capacitated line planning with routing (cLPwR).

2.1 Introduction to Line Planning

Lines, Line Concepts, and Budget Restriction

Let $\mathcal{PTN} = (S, E)$ be a PTN. We define a line l to be a path in $\mathcal{PTN}$ specifying also a direction:

Definition 2.1.1.

- A *one-directional line* l is given by a sequence

$$(s_1, e_1, s_2, e_2, \ldots, e_{k-1}, s_k)$$

of vertices $s_i \in S$ and edges $e_i \in E$ such that $e_i = \{s_i, s_{i+1}\}$ which contains every node of the PTN at most once, unless $s_1 = s_k$.
- A sequence $(s_1, e_1, s_2, e_2, \ldots, e_{k-1}, s_k, e_{k-1}, e_2, s_2, e_1, s_1)$ with $s_i \neq s_j$ for all i, j is called a *bi-directional line*.
- Both one- and bi-directional lines are summarized under the name of *lines*.

Since a line $(s_1, e_1, s_2, e_2, \ldots, e_{k-1}, s_k)$ is uniquely represented by its node sequence $(s_1, s_2, \ldots, s_k)$ or its edge sequence $(e_1, e_2, \ldots, e_{k-1})$, we sometimes represent it using one of these shorter notations. We denote by $S(l) := \{s \in S : s \in l\}$ and $E(l) := \{e \in E : e \in l\}$ the stations and edges visited by line l, and by $\mathcal{L}(s) := \{l : l \ni s\}$ and $\mathcal{L}(e) := \{l : l \ni e\}$ the lines visiting a given station s or edge e. To emphasize that s_{i+1} is a direct successor of s_i on l (in the order defined by the line), we write $(s_i, s_{i+1}) \in l$. Note that for bi-directional lines, we have both $(s_i, s_{i+1}) \in l$ and $(s_{i+1}, s_i) \in l$.

We denote by $\mathcal{L}$ a given line pool which can contain both one-directional and bi-directional lines l with associated line driving time function α_l. Line costs are given by a multiplicative factor b_l, i.e., $\beta_l(f_l) := b_l f_l$ for every $l \in \mathcal{L}$.

In capacitated line planning, we call a single realization of a line a *train* and assume that every train has a *capacity* Cap, which does not depend on the train or the line.

Definition 2.1.2. For a line $l \in \mathcal{L}$, the *frequency* $f_l \in \mathbb{N}$ states how many trains of line l are operated in a certain time period.

In capacitated line planning, a *line concept* is a set $\mathcal{L}' \subset \mathcal{L}$ together with a frequency $f_l \in \mathbb{N}$ for every $l \in \mathcal{L}'$. To indicate that a line l is not contained in $\mathcal{L}'$, we sometimes write $f_l = 0$.

Since in uncapacitated line planning the frequencies are either 0 (the line is not contained in the line concept) or 1 (the line is contained in the line concept), we often omit stating the frequencies explicitly and call $\mathcal{L}'$ a line concept.

We assume an overall budget B on the costs of a line concept. That is, a line concept $\mathcal{L}' \subset \mathcal{L}$ is *feasible* if $\sum_{l \in \mathcal{L}'} b_l f_l \leq B$. Due to the $\{0, 1\}$-frequencies, in uncapacitated line planning this can be written as $\sum_{l \in \mathcal{L}'} b_l \leq B$.

Passenger Routes

Assuming that the passengers' demand for traveling between the stations is known, we use the notion of *OD-pairs* to define demand on the network.

Definition 2.1.3. A set of *OD-pairs* is a subset of pairs of stations $\mathcal{OD} \subset S \times S$. For every OD-pair $(u, v) \in \mathcal{OD}$, the *number of passengers* or *passenger weight* w_{uv} is specified.

A passenger's journey can now be described in the following way:

Definition 2.1.4. A *line-route* P_{uv} for a passenger of OD-pair $(u, v) \in \mathcal{OD}$ specifies

- a path $P = (s_1, e_1, s_2, e_2, \ldots, e_{j-1}, s_j)$ with $s_1 := u$ and $s_j := v$ and
- for each edge $e_i \in P$ a line $l \in \mathcal{L}(e_i)$ which is chosen for traveling on e_i.

P is called the *path corresponding to P_{uv} in the PTN*.

P_{uv} does not only specify the path in the PTN, but also the lines used for traveling on the path and the stations where transfers between lines take place. We say that a line-route P_{uv} is *feasible* for a line concept $\mathcal{L}'$ if P_{uv} uses only lines contained in $\mathcal{L}'$.

In order to estimate travel times, we make use of the line driving time function α_l associated to every line l. Let L_e denote the edge length for $e \in E$ and let α_l be the line driving time function of a line $l \in \mathcal{L}(e)$. The *driving time* of line l on edge $e \in E$ is calculated as $L_e^l := \alpha_l(L_e)$.

Transfers from one line to another are penalized using *transfer penalties* $p_s^{ll'} \in \mathbb{N}_0$ which arise when a passenger transfers from line l to line l' at station s. Given driving times and transfer penalties, we can estimate the travel time along a line-route:

Let $\mathcal{L}'$ be a line concept and P be a feasible line-route for $\mathcal{L}'$.

- The *driving time along P*, $d(\mathcal{L}', P)$, is calculated summing up the driving times $L_e^l = \alpha_l(L_e)$ for every edge $e \in P$ and the line l used on this edge.
- The *transfer time along P*, $t(\mathcal{L}', P)$, is calculated summing up the transfer penalties $p_s^{ll'}$ whenever the line-route suggests the transfer from one line l to another line l' at a station s.
- The *travel time along P* is given as $c(\mathcal{L}', P) := d(\mathcal{L}', P) + t(\mathcal{L}', P)$.
- The *cost along P_{uv}*, $b(\mathcal{L}', P_{uv})$, is calculated as the sum over the costs of all lines used by P_{uv}.

Since a line-route P_{uv} does not only specify a path in the PTN but also the lines taken on the path, the line concept decides the feasibility of P_{uv}. However, for any two line concepts $\mathcal{L}'_1, \mathcal{L}'_2$ for which P_{uv} is feasible, the driving, transfer and travel times, and the cost along P_{uv} coincide. Even so, to keep notation consistent to later chapters, we decide against the abbreviated notation $d(P_{uv}) := d(\mathcal{L}', P_{uv})$, etc.

2.1 Introduction to Line Planning

Line-Routings

A *line-routing for OD-pair* (u, v)

$$\mathcal{R}_{uv} := \{P : \text{line-route } P \text{ is used by } (u, v)\}$$

specifies a line-route for every passenger in (u, v). We denote by w_{uv}^P the *number of passengers of OD-pair* (u, v) *using line-route* P and require that $\sum_{P \in \mathcal{R}_{uv}} w_{uv}^P = w_{uv}$. A *line-routing* $\mathcal{R}$ is a set which contains a line-routing for every OD-pair $(u, v) \in \mathcal{OD}$.

In uncapacitated line planning we assume that all passengers of an OD-pair take the same line-route. Hence, in this case a line-routing is just a collection of exactly one path P_{uv} for every OD-pair (u, v). When the set of OD-pairs is indexed, e.g., $\mathcal{OD} = \{(u_k, v_k) : k \in K\}$ for an index set K, for the sake of simplicity, we also use the notation $w_k := w_{u_k v_k}$ and $P_k := P_{u_k v_k}$.

In capacitated line planning, passengers of the same OD-pair may take different line-routes.

A line-routing $\mathcal{R}$ is *feasible for a given line concept* $\mathcal{L}'$ if every line-route in $\mathcal{R}$ is feasible for $\mathcal{L}'$ and (in case of capacitated line planning) capacity constraints are not violated, i.e., for every edge e and every $l \in \mathcal{L}(e)$, the number of passengers traveling in l on e is at most $f_l \cdot \text{Cap}$.

On the other hand, given a line-routing we can also demand that the line concept should be chosen in a way to make the line-routing feasible: A line concept $\mathcal{L}'$ is *feasible for a line-routing* $\mathcal{R}$ if it is a feasible line concept, if all lines used by $\mathcal{R}$ are contained in $\mathcal{L}'$, and if (in case of capacity restrictions) the frequency f_l of each line l on each edge $e \in E(l)$ is at least the number of passengers using l on e divided by Cap.

Let $\mathcal{L}'$ be a line concept and $\mathcal{R} = \{P_{uv} : (u, v) \in \mathcal{OD}\}$ a line-routing. The overall *travel time* is defined as

$$c(\mathcal{L}', \mathcal{R}) := \sum_{(u,v) \in \mathcal{OD}} \sum_{P \in \mathcal{R}_{uv}} w_{uv}^P c(\mathcal{L}', P).$$

Uncapacitated Line Planning and Simplified Capacitated Line Planning

We are now ready to define the first two line planning problems considered in this book.

Definition 2.1.5.

- *Uncapacitated line planning with routing*
  An instance $(\mathcal{PTN}, \mathcal{L}, \mathcal{OD}, B)$ of *uncapacitated line planning with routing* (uLPwR) consists of a public transportation network $\mathcal{PTN}$, a line pool $\mathcal{L}$, a set of OD-pairs $\mathcal{OD}$, and a budget B.

Fig. 2.1 Example for an instance where the solution found by the conventional frequency model is unrealistic. Travel demand is indicated by the *dotted arrows*

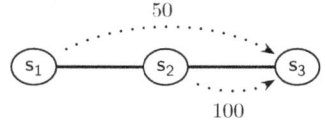

$$l_1 \xrightarrow{\alpha_{l_1}(x) = 2x, \quad b_{l_1} = 2}$$

$$l_2 \xrightarrow{\alpha_{l_2}(x) = x, \quad b_{l_2} = 3}$$

$$l_3 \xrightarrow{\alpha_{l_3}(x) = 2.5x, \quad b_{l_3} = 1}$$

The task is to choose a feasible line concept $\mathcal{L}'$ and a feasible (in the uncapacitated sense) line-routing $\mathcal{R}$ for $\mathcal{L}'$, such that the overall travel time $c(\mathcal{L}', \mathcal{R})$ of the passengers is minimized.

- *Simple capacitated line planning with routing*
  An instance $(N, \mathcal{L}, \mathcal{OD}, B, \text{Cap})$ of *simple capacitated line planning with routing (scLPwR)* consists of a public transportation network $\mathcal{PTN}$, a line pool $\mathcal{L}$, a set of OD-pairs $\mathcal{OD}$, a budget B, and a restriction on the capacity of the trains Cap.
  The task is to choose a feasible line concept $\mathcal{L}'$ together with frequencies $f_l \in \mathbb{N}$ for all $l \in \mathcal{L}'$ and a feasible (in the capacitated sense) routing $\mathcal{R}$, such that the overall travel time $c(\mathcal{L}', \mathcal{R})$ is minimized.

(uLPwR) was defined in [SS10b, SS12a]. In [Sch05, SS06], Scholl and Schöbel investigate (scLPwR) with the additional restriction that the number of trains on every edge e of the corresponding PTN is restricted by a maximal edge frequency $f_e^{\max}$.

Individually Optimal Routings

Even ignoring timetabling issues, it is quite unrealistic to assume that in capacitated line planning, the *feasibility* of a routing will be enough to avoid capacity violations in trains. This is demonstrated in the following example.

We consider a linear PTN with three stations s_1, s_2, and s_3 which are connected by edges of length 1. There are three lines: lines l_1 and l_2 go from s_1 to s_3. l_1 has driving time function $\alpha_{l_1}(x) := 2x$ for $x \in \mathbb{R}_0^+$, l_2 has driving time function $\alpha_{l_2}(x) := x$ for $x \in \mathbb{R}_0^+$, and l_3 has a driving time function of $\alpha_{l_3}(x) := 2.5x$ for $x \in \mathbb{R}_0^+$ and goes from s_2 to s_3. The line costs are given by $b_{l_1} = 2$, $b_{l_2} = 3$, and $b_{l_3} = 1$ and we have a budget $B = 4$. 50 passengers want to travel from s_1 to s_3 and 100 passengers travel from s_2 to s_3. We assume a train capacity of Cap $= 100$. See Fig. 2.1 for a picture of the PTN.

2.1 Introduction to Line Planning

In order to minimize the overall travel time, we choose lines l_2 and l_3 with frequency $f_{l_2} = f_{l_3} = 1$ as a solution to scLPwR. Then half of the passengers starting the journey in B can board the fast line l_2, while the others have to take the slower line. As long as we do not assume different prices for traveling on the lines, this solution does not seem to be realizable in practice, as all passengers would prefer to take the faster line.

The optimum calculated with conventional frequency models is a *system optimum* which may force some passengers to take large detours, while the more desirable goal would be to find a collection of *individually optimal routing decisions*. To overcome this difficulty, Borndörfer et al. suggest in [BGP08] to investigate a model that allows only paths with travel time in a predefined range (depending on shortest path distances in the PTN).

In contrast to this approach in the following we present a model from [Sch12] which allows all passengers to travel on shortest paths in the routing network.

For a given line concept $\mathcal{L}'$, we call a line-routing $\mathcal{R}$ *shortest-path line-routing for $\mathcal{L}'$* if for every OD-pair $(u, v) \in \mathcal{OD}$ the travel time on every line-route $P \in \mathcal{R}_{uv}$ equals the minimal possible travel time from u to v on a line-route using lines from $\mathcal{L}'$.

The third line planning problem we consider in this book is defined as follows.

Definition 2.1.6.
Capacitated line planning with routing
An instance $(N, \mathcal{L}, \mathcal{OD}, B, \text{Cap})$ of *capacitated line planning with routing (cLPwR)* consists of a public transportation network $\mathcal{PTN}$, a line pool $\mathcal{L}$, a set of OD-pairs $\mathcal{OD}$, a budget B, and a restriction on the capacity of the trains Cap.

The task is to choose a feasible line concept $\mathcal{L}'$ together with frequencies $f_l \in \mathbb{N}$ for all $l \in \mathcal{L}'$ and a feasible shortest-path routing $\mathcal{R}$ in $N(\mathcal{L}')$, such that the overall travel time $c(\mathcal{L}', \mathcal{R})$ is minimized.

A solution $(\mathcal{L}', \mathcal{R})$ consisting of a feasible line concept and a feasible shortest-path routing in $N(\mathcal{L}')$ does not necessarily exist, even if a solution $(\tilde{\mathcal{L}}', \tilde{\mathcal{R}})$, providing a system optimum, exists, as we see in the following example.

We are given a PTN consisting of four stations, $s_1, s_2, s_3,$ and s_4 connected as shown in Fig. 2.2. There are two lines $l_1 = (s_1, s_2, s_4)$ and $l_2 = (s_1, s_3, s_4)$ with $b_{l_1} = b_{l_2} = 1$ and $\alpha_{l_1}(x) := x$ and $\alpha_{l_2}(x) := x$ for $x \in \mathbb{R}_0^+$. The budget is $B = 2$. The capacities of the trains are Cap $= 10$. The set of OD-pairs is $\mathcal{OD} := \{(s_1, s_2), (s_1, s_3), (s_1, s_4)\}$ with passenger numbers $w_{s_1 s_2} = w_{s_1 s_3} := 5$ and $w_{(s_1, s_4)} := 10$. The only feasible line concept that allows all passenger to reach their destinations is to establish both lines with frequency $f_{l_1} = f_{l_2} := 1$. However, in this solution, five passengers of OD-pair (s_1, s_4) have to take line l_2 and have a travel time of 4 which is not the shortest path distance between s_1^{org} and s_4^{dest} in $\mathcal{N}(\mathcal{L})$.

However, a solution that *in theory* forces some passengers to take long detours will probably not be accepted *in practice*. Most likely, the passengers try to take the shorter paths anyway and hence violate the present capacity restrictions. Thus, whenever there exists a solution to cLPwR which allows all passengers to travel on shortest paths, it improves the quality of the transportation system compared to the system optimum to choose this solution.

Fig. 2.2 Example for nonexistence of a solution that allows passengers to travel on shortest paths. Demand for traveling is indicated by the *dotted arrows* in the PTN

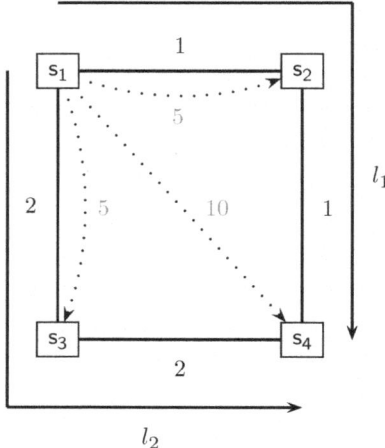

2.1.4 Outline of the Line Planning Chapter

In Sect. 2.2 we introduce the CGN and the line network and show how line planning problems can be modeled using these networks. Section 2.3 explains how line planning can be solved if part of the solution, i.e., the line concept or the routing, is fixed. From the strong NP-completeness result for the line connectivity problem in [BNP09], strong NP-hardness of uLPwR can easily be deduced. Therefore, in Sect. 2.4, we define subclasses of instances of uLPwR and investigate the computational complexity of these subclasses in Sect. 2.5. In particular we perform an extensive complexity analysis of uLPwR in linear PTNs in Sect. 2.5.2.

For capacitated line planning, i.e., scLPwR and cLPwR, we are able to prove strong NP-completeness of the corresponding decision problem even in a very restricted case in Sect. 2.6.1. This motivates Sect. 2.6.2 where we develop an integer program for cLPwR based on an integer programming formulation for scLPwR from [Sch05, SS06].

2.2 Modeling Line Planning with Routing

In this section we introduce the concept of the change-and-go network (*CGN*) which combines the PTN, the line pool, and the OD-pairs into one network model. It is often used for line planning purposes (see, e.g., [Sch05, SS06, SS10b, SS12a, Sch12]) since all information specified in a line-route can be conveniently represented as a path in the CGN.

Under some quite strong restrictions, i.e., if the considered PTN is linear, the line driving time is equal for all lines, and the transfer penalties do not depend on the

2.2 Modeling Line Planning with Routing

stations where the transfers take place, we can condense the information contained in the PTN to a much smaller network, the *line network*. (see, e.g., [Sch05, SS10b, SS12a]). Under the mentioned restrictions, the line network allows faster solution algorithms than the CGN, since it omits parts of the structure which are dispensable in this case.

2.2.1 The Change-and-Go Network

In order to depict the various travel possibilities from the origins to the destinations, the concept of a *CGN* was introduced in [Sch05, SS06]. We follow the slightly different modeling approach of [SS10b, SS12a] to cope with differing requirements like directed lines and different line driving times.

Given a public transportation network $\mathcal{PTN}$, a line pool $\mathcal{L}$, and a set of OD-pairs $\mathcal{OD}$, we construct a *CGN* in the following way:

We first define the node set of the CGN. There are three different types of nodes. The *travel nodes* represent the stopping of lines.

Definition 2.2.1. Let S be the node set of $\mathcal{PTN}$. The set of *travel nodes* is defined as $V_{\text{travel}} := \{[s, l] : s \in S, l \in \mathcal{L}(s)\}$.

Furthermore, origins and destinations have to be integrated in the network. To this end, we define two mappings org and dest that map an OD-pair (u, v) to an *origin node* $u^{\text{org}} := \text{org}(u, v)$ and to a *destination node* $v^{\text{dest}} := \text{dest}(u, v)$ which are added to the CGN.

Definition 2.2.2.

- The set of *origin nodes* is given as $V_{\text{org}} := \text{org}(\mathcal{OD}) = \{u^{\text{org}} : (u, v) \in \mathcal{OD}\}$.
- Analogously, the set of *destination nodes* is $V_{\text{dest}} := \text{dest}(\mathcal{OD}) = \{v^{\text{dest}} : (u, v) \in \mathcal{OD}\}$.

To model passenger flows in the network, we need to represent the lines, provide possibilities for transfers, and include the passengers' demand. All of this is done by means of arcs.

Definition 2.2.3. Let E be the edge set of $\mathcal{PTN}$.

- The set of *driving arcs* $A_{\text{drive}} \subset V_{\text{travel}} \times V_{\text{travel}}$ is defined as

$$A_{\text{drive}} := \{([s, l], [s', l]) : l \in \mathcal{L}, \{s, s'\} \in E(l), s \leq_l s'\}.$$

The length of a driving arc $([s, l], [s', l])$ is given as $c_{([s,l],[s',l])} := L^l_{ss'}$.
- The set of *transfer arcs* $A_{\text{trans}} \subset V_{\text{travel}} \times V_{\text{travel}}$ is defined as

$$A_{\text{trans}} := \{([s, l], [s, l']) : s \in S, l \neq l' \in \mathcal{L}(s)\}$$

and the arc length of a transfer arc $([s, l], [s, l'])$ is given as $c_{([s,l],[s,l'])} := p_s^{ll'}$.

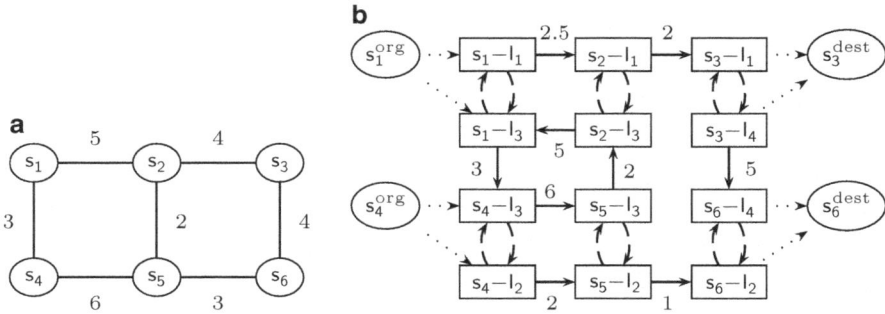

Fig. 2.3 Construction of the change-and-go network N (**a**) PTN (**b**) constructed CGN

- The set of *origin arcs* $A_{\text{org}} \subset V_{\text{org}} \times V_{\text{travel}}$ is defined as

$$A_{\text{org}} := \{(u^{\text{org}}, [u,l]) : u^{\text{org}} \in V_{\text{org}}, [u,l] \in V_{\text{travel}}\}.$$

  The length of every origin arc $(u^{\text{org}}, [u,l])$ is set to $c_{(u^{\text{org}}, [u,l])} := 0$.
- Analogously the set of *destination arcs* $A_{\text{dest}} \subset V_{\text{travel}} \times V_{\text{dest}}$ is defined as

$$A_{\text{dest}} := \{([v,l], v^{\text{dest}}) : v^{\text{dest}} \in V_{\text{dest}}, [v,l] \in V_{\text{travel}}\}$$

and $c_{([v,l], v^{\text{dest}})} := 0$ for every $([v,l], v^{\text{dest}}) \in A_{\text{dest}}$.

For $a \in A_{\text{drive}}$, let $l(a)$ be the (uniquely defined) line belonging to arc a. The *CGN* is now defined as $N = (V, A)$ with

$$V := V_{\text{travel}} \cup V_{\text{org}} \cup V_{\text{dest}} \quad \text{and} \quad A := A_{\text{drive}} \cup A_{\text{trans}} \cup A_{\text{org}} \cup A_{\text{dest}}.$$

We denote $n_N := |V|$ and $m_N := |A|$.

In Fig. 2.3 an example for the construction of a CGN is given: a PTN is depicted in Fig. 2.3a. We have a line pool $\mathcal{L} := \{l_1 = (s_1, s_2, s_3), l_2 = (s_4, s_5, s_6), l_3 = (s_1, s_4, s_5, s_2, s_1), l_4 = (s_3, s_6)\}$ with line driving times $\alpha_{l_1}(x) := \frac{1}{2}x$, $\alpha_{l_2}(x) := \frac{1}{3}x$, $\alpha_{l_3}(x) := x$ and $\alpha_{l_4}(x) := \frac{5}{4}x$ and OD-pairs (s_1, s_6) and (s_4, s_3). In Fig. 2.3b the CGN is shown. The dotted lines represent the origins and destinations of the passengers and the dashed lines stand for the transfer possibilities between two lines.

It can be seen easily that given a public transportation network $\mathcal{PTN} = (S, E)$, a line pool $\mathcal{L}$, and a set of OD-pairs $\mathcal{OD}$, the corresponding CGN is of size $O(|S| \cdot |\mathcal{L}|^2)$ and that the time needed for its construction is of the same order: Since every line contains at most $|S|$ stations and edges, for every line $l \in \mathcal{L}$ it takes time $O(|S|)$ to follow l and draw the nodes V_{travel} and the arcs A_{drive} corresponding to l. This gives $O(|S| \cdot |\mathcal{L}|)$ over all lines. Given the travel nodes, at every station we draw

2.2 Modeling Line Planning with Routing

at most $|\mathcal{L}|^2$ transfer arcs which takes time $O(|S| \cdot |\mathcal{L}|^2)$ for all stations. Finally, we add at most one origin node and $|\mathcal{L}|$ origin arcs at every station, the same holds for the destination nodes and the destination arcs.

2.2.2 Line Planning in the Change-and-Go Network

Let a PTN, a line pool $\mathcal{L}$, and a set of OD-pairs $\mathcal{OD}$ be given and let $N = (V, E)$ be the corresponding CGN. For an OD-pair (u, v) every path P_{uv} from origin node u^{org} to destination node v^{dest} in N represents a line-route for OD-pair (u, v) and vice versa.

Hence we can translate the definitions of driving time, transfer time, travel time, and costs along a line-route to paths in N: Let $\mathcal{L}'$ be a line concept and P a path in the routing network $N(\mathcal{L}')$. We define

- the *driving time along P* as

$$d(\mathcal{L}', P) = \sum_{a \in P \cap A_{\mathrm{drive}}} c_a,$$

- the *transfer time along P* as

$$t(\mathcal{L}', P) = \sum_{a \in P \cap A_{\mathrm{trans}}} c_a,$$

- the *travel time along P* as

$$c(\mathcal{L}', P) = d(\mathcal{L}', P) + t(\mathcal{L}', P) = \sum_{a \in P} c_a,$$

- and the *cost along P* as

$$b(\mathcal{L}', P) = \sum_{l: \exists s \in S \text{ s.t. } [s,l] \in P} b_l.$$

Using these definitions, for a line concept $\mathcal{L}'$, a feasible line-route P'_{uv}, and the corresponding path P_{uv} in N it holds that

$$d(\mathcal{L}', P'_{uv}) = d(\mathcal{L}', P_{uv}), \ t(\mathcal{L}', P'_{uv}) = t(\mathcal{L}', P_{uv}), \ c(\mathcal{L}', P'_{uv}) = c(\mathcal{L}', P_{uv}),$$

and $b(\mathcal{L}', P'_{uv}) = b(\mathcal{L}', P_{uv})$.

Due to this observation, we identify line-routes and the corresponding paths in N.

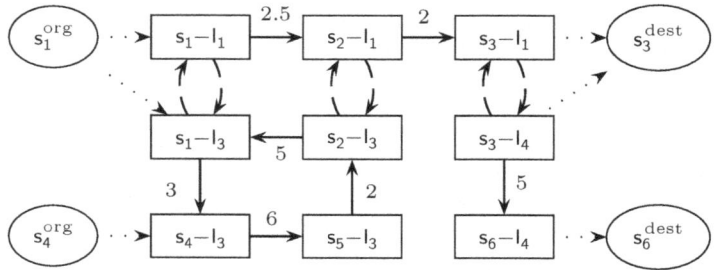

Fig. 2.4 Routing network $N(\mathcal{L}')$ for $\mathcal{L}' = \{l_1, l_3, l_4\}$

We define a *routing* $\mathcal{R}_{uv}$ *for OD-pair* $(u,v) \in \mathcal{OD}$ to be a collection of paths P from u^{org} to v^{dest} in the CGN. w_{uv}^P now denotes the *number of passengers of OD-pair* (u,v) *using path* P. A *routing* $\mathcal{R}$ is then defined as $\mathcal{R} := \bigcup_{(u,v)\in\mathcal{OD}} \mathcal{R}_{uv}$.

Given a line concept $\mathcal{L}'$, the *routing network* $N(\mathcal{L}')$ is obtained by removing all arcs belonging to other lines from the CGN. That is, $N(\mathcal{L}') := (V(\mathcal{L}'), A(\mathcal{L}'))$ with

$$V(\mathcal{L}') := \{[s,l] \in V_{\text{travel}} : l \in \mathcal{L}'\} \cup V_{\text{org}} \cup V_{\text{dest}}$$

and

$$A(\mathcal{L}') := \{(i,j) \in A : i,j \in V(\mathcal{L}')\}.$$

In Fig. 2.4 the routing network $N(\mathcal{L}')$ for the example given in Fig. 2.3 and line concept $\mathcal{L}' = \{l_1, l_3, l_4\}$ is shown. All arcs belonging to line l_2 are removed.

The definition of the routing network allows us to transfer the concept of individually optimal routings to the CGN: For a given line concept $\mathcal{L}'$, a routing $\mathcal{R}$ in the routing network $N(\mathcal{L}')$ is a *shortest-path routing for* $\mathcal{L}'$, if for every OD-pair $(u,v) \in \mathcal{OD}$ the travel time on every path $P \in \mathcal{R}_{uv}$ equals the travel time on a shortest path from u^{org} to v^{dest} in $N(\mathcal{L}')$.

Also the notion of feasibility can be transferred easily from line-routings to routings using the routing network notation:

Definition 2.2.4. A routing $\mathcal{R}$ in N is *feasible for a line concept* $\mathcal{L}'$ (or, vice versa, $\mathcal{L}'$ is *feasible for* $\mathcal{R}$) if

- $\mathcal{R}$ is a routing in $N(\mathcal{L}')$, i.e., $P_{uv} \subset N(\mathcal{L}')$ for all $P_{uv} \in \mathcal{R}$,
- and (in case of capacitated line planning) if

$$\sum_{(u,v)\in\mathcal{OD}} \sum_{P\in\mathcal{R}_{uv}: P\ni a} w_{uv}^P \leq f_{l(a)}\text{Cap} \quad \forall a \in A_{\text{drive}}.$$

That is, given a line concept $\mathcal{L}'$, a routing is feasible for $\mathcal{L}'$ if and only if the corresponding line-routing is feasible for $\mathcal{L}'$.

2.2 Modeling Line Planning with Routing

Let $\mathcal{L}'$ be a line concept and $\mathcal{R} := \bigcup_{(u,v)\in\mathcal{OD}} \mathcal{R}_{uv}$ be a feasible routing with passenger number w_{uv}^P for all $(u,v) \in \mathcal{OD}$, $P \in \mathcal{R}_{uv}$. The overall *travel time* can be calculated as

$$c(\mathcal{L}', \mathcal{R}) = \sum_{(u,v)\in\mathcal{OD}} \sum_{P\in\mathcal{R}_{uv}} w_{uv}^P c(\mathcal{L}', P).$$

We can now restate the line planning problems from Definitions 2.1.5 and 2.1.6.

Definition 2.2.5.

- *Uncapacitated line planning with routing*
  An instance $(N, \mathcal{L}, \mathcal{OD}, B)$ of *uncapacitated line planning with routing (uLPwR)* consists of a CGN N, a line pool $\mathcal{L}$, a set of OD-pairs $\mathcal{OD}$, and a budget B.
  The task is to choose a feasible (in the uncapacitated sense) line concept $\mathcal{L}'$ and a routing $\mathcal{R}$ in $N(\mathcal{L}')$, such that the overall travel time $c(\mathcal{L}', \mathcal{R})$ of the passengers is minimized.
- *Simple capacitated line planning with routing*
  An instance $(N, \mathcal{L}, \mathcal{OD}, B, \text{Cap})$ of *simple capacitated line planning with routing (scLPwR)* consists of a CGN N, a line pool $\mathcal{L}$, a set of OD-pairs $\mathcal{OD}$, a budget B, and a restriction on the capacity of the trains Cap.
  The task is to choose a feasible (in the capacitated sense) line concept $\mathcal{L}'$ together with frequencies $f_l \in \mathbb{N}$ for all $l \in \mathcal{L}'$ and a feasible routing $\mathcal{R}$, such that the overall travel $c(\mathcal{L}', \mathcal{R})$ is minimized.
- *Capacitated line planning with routing (cLPwR)*
  An instance $(N, \mathcal{L}, \mathcal{OD}, B, \text{Cap})$ of *capacitated line planning with routing (cLPwR)* consists of a CGN N, a line pool $\mathcal{L}$, a set of OD-pairs $\mathcal{OD}$, a budget B, and a restriction on the capacity of the trains Cap.
  The task is to choose a feasible line concept $\mathcal{L}'$ together with frequencies $f_l \in \mathbb{N}$ for all $l \in \mathcal{L}'$ and a feasible (in the capacitated sense) shortest-path routing $\mathcal{R}$ in $N(\mathcal{L}')$, such that the overall travel time $c(\mathcal{L}', \mathcal{R})$ is minimized.

Since they describe the same problems, we use Definition 2.2.5 as well as Definitions 2.1.5 and 2.1.6 for uLPwR, scLPwR, and cLPwR equivalently.

2.2.3 The Line Network

In Sect. 2.5 we partly restrict our analysis of uLPwR to *linear* PTNs as defined in Sect. 1.3. Under some additional restrictions, this allows us to condense the information contained in the CGN to a smaller network in the way explained in this section. Similarly, scLPwR and cLPwR can be regarded on line networks. However, since we investigate only uLPwR under these restrictions, we state our definitions in this section in terms of uLPwR.

As we will see in Lemma 2.2.6, in linear PTNs with all lines having the same driving time function, the travel time of an OD-pair depends only on the transfers on its path P_{uv} from origin to destination (remember that in uLPwR all passengers of an OD-pair travel on the same path P_{uv}). We can hence minimize the overall *transfer time*

$$t(\mathcal{L}', \mathcal{R}) := \sum_{P_{uv} \in \mathcal{R}} w_{uv} t(\mathcal{L}', P_{uv})$$

of a solution $(\mathcal{L}', \mathcal{R})$ to an instance $I = (N, \mathcal{L}, \mathcal{OD}, B)$ of uLPwR instead of the overall travel time.

Lemma 2.2.6. *Let $I = (\mathcal{PTN}, \mathcal{L}, \mathcal{OD}, B)$ be an instance of uLPwR with $\mathcal{PTN}$ linear and line driving time functions $\alpha_l := \alpha$ with a fixed $\alpha : \mathbb{R}_0^+ \to \mathbb{R}^+$ for all $l \in \mathcal{L}'$. Let $(\mathcal{L}', \mathcal{R})$ be a solution to I. Then it holds that*

$(\mathcal{L}', \mathcal{R})$ *is an optimal solution to* I

$\Leftrightarrow$

$t(\mathcal{L}', \mathcal{R}) \leq t(\hat{\mathcal{L}}, \hat{\mathcal{R}})$ *for all feasible solutions* $(\hat{\mathcal{L}}, \hat{\mathcal{R}})$ *to* I.

Proof. Let $(u_k, v_k) \in \mathcal{OD}$ be an OD-pair. Note that for any path P_k for $(u_k, v_k) \in \mathcal{N}$, the corresponding path P_k^{PTN} in $\mathcal{PTN}$ is the same, since $\mathcal{PTN}$ is linear. Since $\alpha_l = \alpha$ for all $l \in \mathcal{L}'$

$$d(\mathcal{OD}) := \sum_{(u_k, v_k) \in \mathcal{OD}} w_k \sum_{e \in P_k^{\mathcal{PTN}}} \alpha(L_e)$$

is well-defined and does not depend on the chosen routing. We obtain

$$c(\mathcal{L}', \mathcal{R}) = \sum_{P_k \in \mathcal{R}} w_k c(\mathcal{L}', P_k)$$

$$= \sum_{P_k \in \mathcal{R}} w_k \sum_{a \in P_k} c_a$$

$$= \sum_{P_k \in \mathcal{R}} w_k \sum_{a \in P_k \cap A_{\text{drive}}} c_a + \sum_{P_k \in \mathcal{R}} w_k \sum_{a \in P_k \cap A_{\text{trans}}} c_a$$

$$= \sum_{P_k \in \mathcal{R}} w_k \sum_{e \in P_k^{\mathcal{PTN}}} \alpha(L_e) + \sum_{P_k \in \mathcal{R}} w_k t(\mathcal{L}', P_k)$$

$$= \sum_{(u_k, v_k) \in \mathcal{OD}} w_k \sum_{e \in P_k^{\mathcal{PTN}}} \alpha(L_e) + t(\mathcal{L}', \mathcal{R})$$

$$= d(\mathcal{OD}) + t(\mathcal{L}', \mathcal{R}).$$

2.2 Modeling Line Planning with Routing

Since $d(\mathcal{OD})$ does not depend on the chosen line concept or the chosen routing, the statement of the lemma follows. □

Let $I = (\mathcal{PTN}, \mathcal{L}, \mathcal{OD}, B)$ be an instance of uLPwR. We say that the transfer penalties in I are *station-independent* if for every pair of lines $l, l' \in \mathcal{L}$, there is a $p^{ll'} \in \mathbb{N}_0$ with $p_s^{ll'} := p^{ll'}$ for all $s \in S(l) \cap S(l')$.

If for instance $I = (\mathcal{PTN}, \mathcal{L}, \mathcal{OD}, B)$ of uLPwR the PTN is linear, line driving time functions are equal, and transfer penalties are station-independent; the transfer time of an OD-pair on path P_k only depends on the line sequence which is used on P_k, i.e.,

$$t(\mathcal{L}', P_k) = \sum_{a \in P_k \cap A_{\text{trans}}} c_a = \sum_{([s,l],[s,l']) \in P_k \cap A_{\text{trans}}} p_s^{ll'} = \sum_{(l,l'): \exists s \in S \, s.t. ([s,l],[s,l']) \in P_k \cap A_{\text{trans}}} p^{ll'}.$$

Let $I = (\mathcal{PTN}, \mathcal{L}, \mathcal{OD}, B)$ be an instance of uLPwR. In the remainder of this section we assume that

1. the corresponding public transportation network $\mathcal{PTN}$ is linear,
2. the driving time functions are equal for all lines.
3. the transfer penalties are station-independent.

In this case, we can condense the information contained in the CGN to a smaller network, the *line network*. The nodes in the line network model the lines the passengers can take as well as origins and destinations.

Definition 2.2.7. The node set of the line network is defined as $V_L := \mathcal{L} \cup V_{\text{org}} \cup V_{\text{dest}}$.

Again the arcs are introduced to model travel possibilities.

Definition 2.2.8.

- The set of *transfer arcs* of the line network is defined as

$$(A_L)_{\text{trans}} := \{(l, l') : S(l) \cap S(l') \neq \emptyset\}.$$

The length of a transfer arc $(l, l') \in (A_L)_{\text{trans}}$ is $c_{ll'} := p^{ll'}$.
- The set of *origin arcs* of the line network is defined as

$$(A_L)_{\text{org}} := \{(u^{\text{org}}, l) : u^{\text{org}} \in V_{\text{org}}, l \in \mathcal{L}(u)\}.$$

The arc lengths are $c_{(u^{\text{org}}, l)} := 0$ for all $(u^{\text{org}}, l) \in (A_L)_{\text{org}}$.
- The set of *destination arcs* of the line network is defined as

$$(A_L)_{\text{dest}} := \{(l, v^{\text{dest}}) : v^{\text{dest}} \in V_{\text{dest}}, l \in \mathcal{L}(v)\}.$$

The arc lengths are $c_{(l, v^{\text{dest}})} := 0$ for all $(l, v^{\text{dest}}) \in (A_L)_{\text{dest}}$.

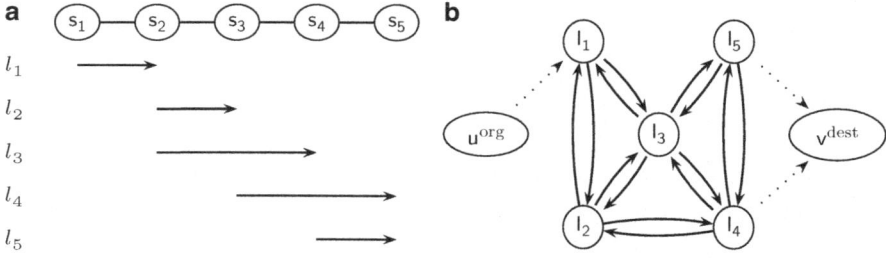

Fig. 2.5 Construction of the LN (**a**) linear PTN (**b**) constructed LN

We now define the *line network (LN)* as $N_L = (V_L, A_L)$ with

$$A_L := (A_L)_{\text{trans}} \cup (A_L)_{\text{org}} \cup (A_L)_{\text{dest}}.$$

We denote $n_{N_L} := |V_L|$ and $m_{N_L} := |A_L|$.

See Fig. 2.5 for an example. In Fig. 2.5a a PTN and five lines l_1, l_2, l_3, l_4, and l_5 are given. Assume that $\mathcal{OD} = \{(s_1, s_5)\}$. Figure 2.5b shows the corresponding LN for the so-defined instance of uLPwR.

Given an instance $(\mathcal{PTN}, \mathcal{L}, \mathcal{OD}, B)$ of uLPwR fulfilling (1)–(3) with $\mathcal{PTN} = (S, E)$, the corresponding LN is of size $O((|\mathcal{L}| + |\mathcal{OD}|) \cdot |\mathcal{L}|)$. The time needed for its construction is in $O((|S| \cdot |\mathcal{L}| + |\mathcal{OD}|) \cdot |\mathcal{L}|)$. This can be seen as follows: Given the nodes $\mathcal{L}$ of the LN, for every station $s \in S$ we draw a pair of directed transfer arcs between all pairs of nodes in $\mathcal{L}(s)$. This can be done in $O(|S| \cdot |\mathcal{L}|^2)$. Then for each of the origin and destination nodes (there are at most $O(|\mathcal{OD}|)$), we add at most $|\mathcal{L}|$ origin or destination arcs. The resulting networks have $O(|\mathcal{L}| + |\mathcal{OD}|)$ nodes and $O(|\mathcal{L}|^2 + |\mathcal{OD}| \cdot |\mathcal{L}|)$ arcs.

2.2.4 Uncapacitated Line Planning in the Line Network

Let $I = (N, \mathcal{L}, \mathcal{OD}, B)$ be an instance of uLPwR with properties (1)–(3) from Sect. 2.2.3 and N_L the corresponding LN. It is easy to check that for every OD-pair $(u, v) \in \mathcal{OD}$, every path in N_L from u^{org} to v^{dest} represents a line-route or a path in N from u^{org} to v^{dest} and vice versa.

Now, for a line concept $\mathcal{L}'$ and a line-route P'_{uv} which is feasible for $\mathcal{L}'$, we can regard the corresponding path P_{uv} in N_L instead and calculate the *transfer time* along P'_{uv} as

$$t(\mathcal{L}', P'_{uv}) = \sum_{a \in P_{uv}} c_a.$$

2.2 Modeling Line Planning with Routing

If P_{uv} is a simple path in N_L, the *cost along* P'_{uv} is

$$b(\mathcal{L}', P'_{uv}) = \sum_{l \in P'_{uv}} b_l.$$

Note that any path corresponding to a "reasonable" line-route is simple. Hence for instances with properties (1)–(3), we identify line-routes, paths in the CGN, and paths in the LN.

Again, we can define a routing on our simplified network structure, which for instances of uLPwR with properties (1)–(3) corresponds to a routing in the CGN or a line-routing. A *routing* in an LN N_L is a collection $\mathcal{R} := \{P_{uv} : (u,v) \in \mathcal{OD}\}$ with P_{uv} being a path from u^{org} to v^{dest} in N_L.

Similar to the routing network in Sect. 2.2.1, we can define a *line routing network* as $N_L(\mathcal{L}') := (V_L(\mathcal{L}'), A_L(\mathcal{L}'))$ with

$$V_L(\mathcal{L}') := \mathcal{L}' \cup V_{\text{org}} \cup V_{\text{dest}}$$

and

$$A_L(\mathcal{L}') := \{(i,j) \in A_L : i, j \in V_L(\mathcal{L}')\}.$$

We can easily transfer the notion of feasibility to routings in the LN: A routing $\mathcal{R}$ in N_L is *feasible for a line concept* $\mathcal{L}'$ (or, vice versa, $\mathcal{L}'$ is feasible for $\mathcal{R}$) if $\mathcal{R}$ is a routing in $N_L(\mathcal{L}')$, i.e., $P_{uv} \subset N_L(\mathcal{L}')$ for all $P_{uv} \in \mathcal{R}$. That is, given a line concept $\mathcal{L}'$, a routing in N_L is feasible for $\mathcal{L}'$ if and only if the corresponding line-routing is feasible for $\mathcal{L}'$.

We can now restate uLPwR from Definitions 2.1.5 and 2.2.5 for instances fulfilling properties (1)–(3), where, due to Lemma 2.2.6, we can take the overall transfer time instead of the overall travel time as an objective.

Definition 2.2.9. An instance $(N_L, \mathcal{L}, \mathcal{OD}, B)$ of *uncapacitated line planning with routing (uLPwR)* fulfilling conditions (1)–(3) consists of a line pool $\mathcal{L}$, a LN N_L, a set of OD-pairs $\mathcal{OD}$, and a budget B.

The task is to choose a feasible line concept $\mathcal{L}'$ and a feasible routing $\mathcal{R}$ in $N_L(\mathcal{L}')$, such that the overall transfer time $t(\mathcal{L}', \mathcal{R})$ of the passengers is minimized.

In the following, we assume that for every instance $I = (\mathcal{PTN}, \mathcal{L}, \mathcal{OD}, B)$ of uLPwR, N and N_L are given, since they can be determined in polynomial time from $(\mathcal{PTN}, \mathcal{L}, \mathcal{OD}, B)$. To solve the problems or to prove complexity results, we use Definitions 2.1.5, 2.2.5, and (in case that conditions (1)–(3) hold) 2.2.9 for uLPwR equivalently. The running time of the algorithms is always stated with respect to the problem formulation used and the time needed to transfer a PTN to a CGN or to an LN is not added.

Also for scLPwR and cLPwR we use both Definitions 2.1.5 and 2.2.5 equivalently and do not add the time needed to transfer instances to the algorithms.

2.3 Solving Line Planning When Part of the Solution Is Fixed

In this section we show that the choice of a line concept or a routing is enough to define a solution to line planning problems.

2.3.1 Solving Uncapacitated Line Planning When Part of the Solution Is Fixed

Definition 2.3.1. Let $I = (N, \mathcal{L}, \mathcal{OD}, B)$ be an instance of uLPwR and let $\mathcal{R}$ be a routing in N:

- We define the *line concept corresponding to* $\mathcal{R}$ as

$$\mathcal{L}(\mathcal{R}) := \bigcup_{P_k \in \mathcal{R}} \{l : \exists a \in P_k \text{ with } l(a) = l\}.$$

- If $\mathcal{L}(\mathcal{R})$ is feasible, we call $(\mathcal{L}(\mathcal{R}), \mathcal{R})$ the *solution corresponding to* $\mathcal{R}$.

Analogously to Definition 2.3.1 we write $\mathcal{L}(P) := \{l : \exists a \in P \text{ with } l(a) = l\}$ for any path P in a CGN $N = (V, A)$.

Since $\mathcal{L}(\mathcal{R})$ is the minimal line concept that contains $\mathcal{R}$, we have the following lemma.

Lemma 2.3.2. Let $I = (N, \mathcal{L}, \mathcal{OD}, B)$ be an instance of uLPwR and let $\mathcal{R}$ be a routing in N. Then the following equivalence holds:

$$\exists \mathcal{L}' \text{ feasible for } \mathcal{R} \Leftrightarrow \mathcal{L}(\mathcal{R}) \text{ is feasible for } \mathcal{R}.$$

Analogously to Definition 2.3.1 we can also specify solutions corresponding to a given line concept.

Definition 2.3.3. Let $I = (N, \mathcal{L}, \mathcal{OD}, B)$ be an instance of uLPwR and let $\mathcal{L}'$ be a feasible line concept. If there exists a feasible routing for $\mathcal{L}'$, for every $(u_k, v_k) \in \mathcal{OD}$ we denote by $\mathcal{P}_k(\mathcal{L}')$ (or $\mathcal{P}_{u_k v_k}(\mathcal{L}')$) the set of shortest paths from u_k^{org} to v_k^{dest} in $N(\mathcal{L}')$.

- We denote by

$$\text{OPT}(\mathcal{L}') := \{\mathcal{R}(\mathcal{L}') : \{P_1(\mathcal{L}'), P_2(\mathcal{L}'), \ldots, P_{|\mathcal{OD}|}(\mathcal{L}') : P_k(\mathcal{L}') \in \mathcal{P}_k(\mathcal{L}')\}$$

the set of *routings corresponding to* $\mathcal{L}'$.
- We call $\{(\mathcal{L}', \mathcal{R}(\mathcal{L}')) : \mathcal{R}(\mathcal{L}') \in \text{OPT}(\mathcal{L}')\}$ the set of *solutions corresponding to* $\mathcal{L}'$.

Since $\mathcal{R}(\mathcal{L}')$ is defined as a shortest-path routing, Lemma 2.3.4 follows directly.

2.3 Solving Line Planning When Part of the Solution Is Fixed

Lemma 2.3.4. *Let* $I = (N, \mathcal{L}, \mathcal{OD}, B)$ *be an instance of uLPwR and let* $\mathcal{L}'$ *be a feasible line concept. Then for every* $\mathcal{R}(\mathcal{L}') \in \mathrm{OPT}(\mathcal{L}')$ *it holds that*

$$c(\mathcal{L}', \mathcal{R}(\mathcal{L}')) \leq c(\mathcal{L}', \mathcal{R}')$$

for all routings $\mathcal{R}'$ *in* $N(\mathcal{L}')$.

Notation 2.3.5. *Let* $I = (N, \mathcal{L}, \mathcal{OD}, B)$ *be an instance of uLPwR. In the following we sometimes describe a solution to I stating only*

- *a line concept* $\mathcal{L}'$, *referring to a corresponding full solution* $(\mathcal{L}', \mathcal{R}(\mathcal{L}'))$ *for any* $\mathcal{R}(\mathcal{L}') \in \mathrm{OPT}(\mathcal{L}')$, *or*
- *a routing* $\mathcal{R}$ *referring to the corresponding full solution* $(\mathcal{L}(\mathcal{R}), \mathcal{R})$.

2.3.2 Solving Capacitated Line Planning When Part of the Solution Is Fixed

It has been shown in [Sch12] that also in scLPwR and cLPwR it is possible to construct a line concept from a given routing and vice versa.

For the first, we proceed analogously to Definition 2.3.1 and additionally set the frequencies just high enough such that every passenger can travel on his or her chosen path.

Definition 2.3.6. *Let* $I = (N, \mathcal{L}, \mathcal{OD}, B, \mathrm{Cap})$ *be an instance of scLPwR or cLPwR and let* $\mathcal{R}$ *be a routing in* N:.

- We define the *line concept corresponding to* $\mathcal{R}$ as

$$\mathcal{L}(\mathcal{R}) := \bigcup_{P_k \in \mathcal{R}} \{l : \exists a \in P_k \text{ with } l(a) = l\}.$$

with frequencies

$$f_l := \max_{a: l(a) = l} \left\lceil \frac{\sum_{(P, w_{uv}^P) \in R} w_{uv}^P}{\mathrm{Cap}} \right\rceil.$$

- If $\mathcal{L}(\mathcal{R})$ is feasible, we call $(\mathcal{L}(\mathcal{R}), \mathcal{R})$ the *solution corresponding to* $\mathcal{R}$.

Since $\mathcal{L}(\mathcal{R})$ is the minimal line concept that contains $\mathcal{R}$, we have the following lemma.

Lemma 2.3.7. *Let* $I = (N, \mathcal{L}, \mathcal{OD}, B, \mathrm{Cap})$ *be an instance of scLPwR or cLPwR and let* $\mathcal{R}$ *be a routing in* N. *Then we have the following equivalence:*

$$\exists \mathcal{L}' \text{ feasible for } \mathcal{R} \Leftrightarrow \mathcal{L}(\mathcal{R}) \text{ is feasible for } \mathcal{R}.$$

Given a line concept $\mathcal{L}'$, the frequencies of $\mathcal{L}'$ determine capacities

$$k_a := \begin{cases} 0 & \text{if } a \in A_{\text{drive}} \text{ and } l(a) \notin \mathcal{L}', \\ f_l(a)\text{Cap} & \text{if } a \in A_{\text{drive}} \text{ and } l(a) \in \mathcal{L}', \\ \infty & \text{otherwise.} \end{cases}$$

on the arcs of the CGN. Finding a feasible routing for scLPwR or cLPwR for a given line concept hence resembles a *multi-commodity flow problem* (see e.g. [GJ79]) instead of a shortest path problem, as in the uncapacitated case. In the following we will see that in fact, finding a feasible line concept for a given routing in uLPwR is not easier than a multi-commodity flow problem, or, in particular, the problem *Directed Two-Commodity Integral Flow*.

An instance $(G, r_1, r_2, t_1, t_2, R_1, R_2)$ of *Directed Two-Commodity Integral Flow* consists of a directed graph $G = (V_G, A_G)$ with capacities k_a for all $a \in A_G$. The question to decide is whether there are two flow functions $f_1, f_2 : A_G \to \mathbb{Z}_0^+$ such that

1. for every $a \in A_G$ $f_1(a) + f_2(a) \leq k_a$,
2. for every $v \in V_G \setminus \{r_1, r_2, t_1, t_2\}$

$$\sum_{a \in \delta^+(v)} f_i(a) - \sum_{a \in \delta^-(v)} f_i(a) = 0 \quad \text{for } i = 1, 2,$$

3. and

$$\sum_{a \in \delta^+(t_i)} f_i(a) \geq R_i \quad \text{for } i = 1, 2.$$

Directed Two-Commodity Integral Flow is (strongly) NP-complete [GJ79].

Theorem 2.3.8. *Let $I = (N, \mathcal{L}, \mathcal{OD}, B, \text{Cap})$ be an instance of scLPwR or cLPwR and let $\mathcal{L}'$ be a line concept. It is strongly NP-complete to decide whether there exists a feasible routing for $\mathcal{L}'$ or not.*

Proof. Let $(G, r_1, r_2, t_1, t_2, R_1, R_2)$ be an instance of Directed Two-Commodity Integral Flow. We now construct an instance $I = (N, \mathcal{L}, \mathcal{OD}, B, \text{Cap})$ of scLPwR or cLPwR and a line concept $\mathcal{L}'$.

- We define the PTN as the undirected graph corresponding to G, i.e., $S := \{s_i : i \in V_G\}$, $E := \{\{s_i, s_j\} : (i, j) \in A_G\}$ and we set $L_e := 1$ for all $e \in E$.
- For every $a = (i, j) \in A_G$ we introduce a line $l_a := (s_i, \{s_i, s_j\}, s_j)$. We set $\alpha_l := 0$ and $b_l := 1$ for every $l \in \mathcal{L}$.
- We set $p_s^{ll'} := 0$ for all occurring transfers in this network.
- We define $\mathcal{OD} := \{(s_{r_1}, s_{t_1}), (s_{r_2}, s_{t_2})\}$ and $w_{s_{r_1} s_{t_1}} := R_1$, $w_{s_{r_2} s_{t_2}} := R_2$.
- We set $B := \sum_{a \in A_G} k_a$ and $\text{Cap} := 1$.

- Let $\mathcal{L}' := \mathcal{L}$ with $f_{l_a} := k_a$ for all $l_a \in \mathcal{L}$ be the given line concept. We see that due to the definitions of B, $\mathcal{L}'$ is feasible.

We show now that there exists a solution to $(G, r_1, r_2, t_1, t_2, R_1, R_2)$ if and only if there exists a solution to I.

- Suppose that $\mathcal{R} := \mathcal{R}_{s_{r_1} s_{t_1}} \cup \mathcal{R}_{s_{r_2} s_{t_2}}$ is a feasible routing for $\mathcal{L}'$. We define

$$f_i(a) := \sum_{P \in \mathcal{R}_{s_{r_i} s_{t_i}} : a \in P} w^P_{s_{r_i} s_{t_i}}.$$

Then we can easily verify (1)–(3). Hence, (f_1, f_2) is a solution to $(G, r_1, r_2, t_1, t_2, R_1, R_2)$.

- Note that if there is a solution (f_1, f_2) to $(G, r_1, r_2, t_1, t_2, R_1, R_2)$, there is also a solution (f_1^*, f_2^*) to $(G, r_1, r_2, t_1, t_2, R_1, R_2)$ with

$$\sum_{a \in \delta^+(t_i)} f_i^*(a) = R_i \quad \text{for } i = 1, 2. \tag{2.1}$$

- Let $f_1^*, f_2^* : A_G \to \mathbb{Z}_0^+$ be a solution to $(G, r_1, r_2, t_1, t_2, R_1, R_2)$ with

$$\sum_{a \in \delta^+(t_i)} f_i^*(a) = R_i \quad \text{for } i = 1, 2.$$

Then due to (2), and (2.1), f_i^* defines a routing $\mathcal{R}_{s_{r_i} s_{t_i}}$ for (s_{r_i}, s_{t_i}). $\mathcal{R} := \mathcal{R}_{s_{r_1} s_{t_1}} \cup \mathcal{R}_{s_{r_2} s_{t_2}}$ is feasible due to (1). Hence, $\mathcal{R}$ is a solution to the instance $(N, \mathcal{L}, \mathcal{OD}, B, \text{Cap})$ of scLPwR.

- Since line driving time functions and transfer penalties are 0, $\mathcal{R}$ is a shortest-path routing in $N(\mathcal{L}')$. Thus, $\mathcal{R}$ is a solution to the instance $(N, \mathcal{L}, \mathcal{OD}, B, \text{Cap})$ of cLPwR.

□

2.4 Classification of Line Planning Problems with Routing

In Theorem 2.5.1 we will see that even the problem of finding a feasible solution to uLPwR is strongly NP-complete in general. Thus, unless P = NP we cannot hope to find a polynomial-time optimization algorithm or even a polynomial-time approximation algorithm for the general line planning problem with routing.

Due to this result, we impose several restrictions on the problem structure of line planning with routing. In the following sections, we give polynomial- or pseudopolynomial-time algorithms for the obtained subcases or show that they are already NP-hard. The considered restrictions are the following:

- *Capacity/frequency restrictions:* We start discussing *uncapacitated line planning with routing (uLPwR)* as defined in Definitions 2.1.5, 2.2.5, and 2.2.9 in

Sect. 2.5. In Sect. 2.6 we investigate the models scLPwR and cLPwR defined in Definitions 2.1.5, 2.1.6, and 2.2.9 for *line planning under capacity restrictions*.
- *Number/structure of OD-pairs:* With respect to the OD-pairs we distinguish between problems where there is only one OD-pair, problems where the OD-pairs have a common origin, and problems with arbitrary sets of OD-pairs.
- *No-line-twice property:* In Sect. 2.5 we see that considering only one OD-pair, uLPwR becomes much easier if the line pool has a certain structure. This structure is called *no-line-twice (NLT)* property which signifies that there must exist an optimal solution in which no passenger enters a line more than once.
- *Network structure:* Since most problems turn out to be hard on general networks, we mostly concentrate on linear PTNs.
- *Line driving time functions:* In some cases we assume equal line driving time functions.
- *Transfer penalties:* In the general case the transfer penalties depend on the lines between which the transfer takes place and on the station where it takes place. We investigate the following special cases: (a) station-independent transfer penalties where the transfer penalties only depend on the lines between which the transfers take place, (b) equal transfer penalties, and (c) no transfer penalties (i.e., all transfer penalties are zero).

While the restrictions on the structure of the OD-pairs are only interesting for theoretical reasons, many of the other assumptions are not as far from reality as they may seem on first glance. Equal line driving functions model the situation that all used trains are of the same type. This may not be completely true in practical instances; nevertheless, we can assume that the number of different train types is small and so is the number of different driving time functions.

Transfers from one train to another are inconvenient for the passengers due to the waiting time at the transfer station and the eventuality of missing connecting trains in case of delays. This is reflected in the transfer penalties. Without too much simplification we can assume that this inconvenience only depends on the number of transfers and not on the stations where the transfers take place nor on the involved trains. Hence, equal transfer penalties seem to be a reasonable choice. Thinking about reasonable passenger routes in public transportation settings, the restriction to instances with *NLT* appears natural. The situation that underlying PTNs (or at least parts of them) are linear frequently occurs in countries with less dense railway systems like New Zealand or India.

Although vehicle capacity is an important issue in public transportation planning, uncapacitated line planning accentuates the question *which lines should be established* for the passengers' convenience and leaves the adjustment of capacity requirements to a subsequent step of frequencies setting.

2.5 Uncapacitated Line Planning with Routing

In this section we analyze the computational complexity of uLPwR under different restrictions based on the results from [SS10b, SS12a]. We start restricting the problem class to instances of uLPwR with only one OD-pair in Sect. 2.5.1. Seeing that this problem is already hard, we discover a property inherent to most practical instances that allows us to develop a pseudopolynomial-time solution algorithm for uLPwR with one OD-pair. If we additionally restrict the problem class to linear networks, in Sect. 2.5.2.1 we see that in some cases a significant speed-up is possible. Furthermore, when we relax our assumption of only one OD-pair in Sects. 2.5.2.2 and 2.5.2.3, we discover that in some cases we can transfer our results for one OD-pair to instances of uLPwR in linear networks with several OD-pairs.

2.5.1 Uncapacitated Line Planning in General Networks

Before we study the complexity of uLPwR, we shortly discuss complexity results regarding the related *line connectivity* problem defined by Borndörfer et al. in [BNP09]. *Line Connectivity* is defined as follows: Given an undirected graph $G = (V, E)$, a set of terminal nodes $T \subseteq V$, and a set of lines $\mathcal{L}$ with nonnegative line costs which cover all edges of G, is there a subset of lines $\mathcal{L}' \subseteq \mathcal{L}$ of minimal cost such that for each pair of distinct terminal nodes $t_1, t_2 \in T$ there exists a path from t_1 to t_2 which is completely covered by lines of $\mathcal{L}'$?

NP-hardness of this problem can be shown by reduction from the well-known *Steiner Tree* problem (see Definition on p. 61) by defining for every edge e in the PTN, a line l_e with unit line cost which serves only edge e.

In case of only two terminals, the Steiner tree problem corresponds to the problem of finding a shortest path between two nodes. Also the line connectivity problem can be solved in this case by a shortest path algorithm in an auxiliary graph. If all nodes are terminal nodes, the Steiner tree problem becomes the *minimum spanning tree* problem which can be solved to optimality by greedy algorithms (see, e.g., [Gib85]). However, it can be shown that line connectivity is strongly NP-hard even if all nodes of the graph are terminals and all line costs are equal using a reduction from *Set Cover* [BNP09].

The latter result implies NP-hardness of our line planning problem uLPwR. More precisely, setting $\mathcal{OD} = \{(t_1, t_2) : t_1, t_2 \in T\}$ we obtain that it is even NP-complete to decide whether a feasible solution to uLPwR exists, if all line costs are equal. Since line driving time functions and transfer penalties do not influence the feasibility of a solution, we can assume that for every line l the line driving time function α_l is defined by $\alpha_l(x) := x$ for $x \in \mathbb{R}_0^+$ and that $p_s^{ll'} := 0$ for all lines l, l' and all stations s to strengthen the theorem.

Theorem 2.5.1. *Finding a feasible solution to uLPwR is strongly NP-complete, even if*

- *all line costs are equal,*
- *the line driving time functions are equal for all lines, and*
- *the transfer penalties are equal for all transfers.*

Hence, unless P = NP, there is no polynomial-time solution algorithm to find a feasible solution to uLPwR, much less one which finds a solution with guaranteed quality or even an optimal one. Theorem 2.5.1 hence justifies our approach of analyzing subclasses of instances of uLPwR according to the classification made in Sect. 2.4.

We now start our analysis of the optimization problem uLPwR by showing that without further limitations on the structure of the line pool, uLPwR is strongly NP-hard, even if we have only one OD-pair.

Theorem 2.5.2. *uLPwR is strongly NP-hard, even if*

- *there is only one OD-pair,*
- *the line driving time functions of all lines are equal,*
- *there are no transfer penalties, and*
- *the line costs are equal for all lines.*

We prove this by reduction from the strongly NP-complete problem *Hitting Set* [GJ79]. An instance (P, Q, K) of *Hitting Set* consists of a set $P = \{p_1, p_2, \ldots, p_m\}$, a set of subsets $Q = \{q_1, q_2, \ldots, q_n\}$ of P, and a natural number $K < |P|$. The question to decide is whether there is a subset $P' \subset P$ with $|P'| \leq K$ such that every $q_j \in Q$ contains at least one $p_i \in P'$.

Proof. Let (P, Q, K) be an instance of Hitting Set with $P = \{p_1, p_2, \ldots, p_m\}$ and $Q = \{q_1, q_2, \ldots, q_n\}$. We construct an instance I of uLPwR with one OD-pair in the following way.

- We set the set of stations to be

$$S := \{s_1, s_{4n+2}\} \cup \{s_{4(j-1)+2}, s_{4(j-1)+3}, s_{4(j-1)+4} : j = 1, \ldots, n\}$$
$$\cup \{s^i_{4(j-1)+2}, s^i_{4(j-1)+4} : j = 1, \ldots, n, i = 1, \ldots, m\}$$
$$\cup \{s^i_{4(j-1)+5} : j = 1, \ldots, n-1, i = 1, \ldots, m\}$$

and define the set of edges of the PTN as

$$E = \{\{s_1, s_2\}\}$$
$$\cup \{\{s_{4(j-1)+2}, s_{4(j-1)+3}\}, \{s_{4(j-1)+3}, s_{4(j-1)+4}\}, \{s_{4(j-1)+4}, s_{4(j-1)+6}\}$$
$$: j = 1, \ldots, n\}$$
$$\cup \{\{s^i_{4(j-1)+2}, s^i_{4(j-1)+4}\} : j = 1, \ldots, n, i = 1, \ldots, m\}$$

2.5 Uncapacitated Line Planning with Routing

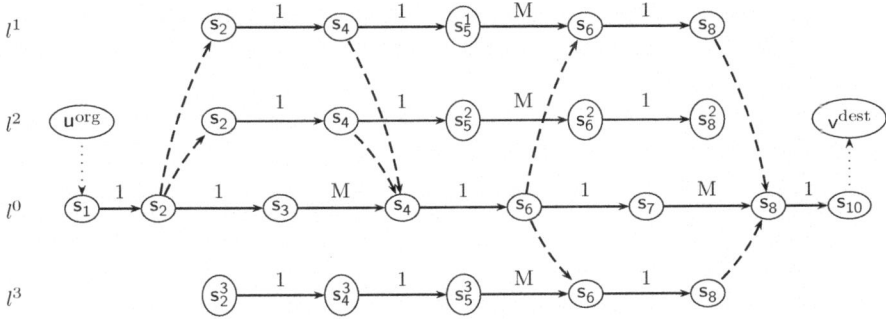

Fig. 2.6 Example for the construction of the CGN in the proof of Theorem 2.6

$$\cup \{\{s^i_{4(j-1)+4}, s^i_{4(j-1)+5}\}, \{s^i_{4(j-1)+5}, s^i_{4(j-1)+6}\} : j = 1, \ldots, n-1,$$
$$i = 1, \ldots, m\}$$

We set $M := 2n + 2$,

$$L_{\{s_{4(j-1)+3}, s_{4(j-1)+4}\}} := M \text{ for } j = 1, \ldots, n,$$

and $L_{\{s^i_{4(j-1)+5}, s^i_{4(j-1)+6}\}} := M \text{ for } i = 1, \ldots, m, \ j = 1, \ldots, n-1.$

All other edge lengths are set to $L_e := 1$.

Now whenever $p_i \in q_j$, we identify $s^i_{4(j-1)+2} := s_{4(j-1)+2}$ and $s^i_{4(j-1)-4} := s_{4(j-1)+4}$.

- We have a line l^0 that visits stations s_k in increasing order of k. Furthermore, we have m more lines $l^1, l^2, \ldots, l^m$. Line l^i visits stations s^i_k in increasing order of k. All lines l have line driving time function α_l with $\alpha_l(x) := x$ for $x \in \mathbb{R}^-_C$ and cost $b_l := 1$. We set $B := K + 1$.
- We set all transfer penalties to 0.
- Our OD-pair is (u, v) with $u := s_1$ and $v := s_{4n+2}$.

An example for the CGN constructed for the set $P = \{1, 2, 3\}$ and the set of subsets $Q = \{\{1, 2\}, \{1, 3\}\}$ where we omit all transfer arcs that cannot be used in a solution $(\mathcal{L}', \mathcal{R})$ with $c(\mathcal{L}', \mathcal{R}') \leq 5$ is shown in Fig. 2.6. For the sake of a compact representation, the lines are omitted in the node labels.

Now we claim that if and only if there is a solution to the constructed instance of uLPwR with objective value better or equal to $2n + 1$, there is a solution to the given instance (P, Q, K) of Hitting Set.

1. Let $(\mathcal{L}', \mathcal{R})$ be a solution to I with $c(\mathcal{L}', \mathcal{R}) \leq 2n + 1$. Let P_{uv} be the path contained in $\mathcal{R}$. No arc of length M can be contained in P_{uv}, because otherwise

$$(\mathcal{L}', P_{uv}) \geq M > 2n + 1.$$

Thus, for every $j = 1, \ldots, n$, $\{[s_{4(j-1)+4}, l^0], [s_{4(j-1)+6}, l^0]\} = \{[s_{4(j-1)+4}, l^0], [s_{4j+2}, l^0]\} \in P_{uv}$ and there exists an i such that $\{[s^i_{4(j-1)+2}, l^i], [s^i_{4(j-1)+4}, l^i]\} \in P_{uv}$. Furthermore $\{[s_1, l^0], [s_2, l^0]\} \in P_{uv}$.

Hence, for every $j = 1, \ldots, n$ there is a line $l^i \in \mathcal{L}'$ such that a transfer from line l^0 to line l^i is possible at station $s_{4(j-1)+2}$. In other words, for every $j = 1, \ldots, n$ there is an index i with $l^i \in \mathcal{L}'$ such that $s^i_{4(j-1)+2} = s_{4(j-1)+2}$.

Setting $P(\mathcal{L}') := \{p_i : l^i \in \mathcal{L}'\}$ we obtain that for every $j = 1, \ldots, n$ there is an index i with $p_i \in q_j$. Furthermore,

$$|P(\mathcal{L}')| = |\mathcal{L}'| - 1 = \sum_{l \in \mathcal{L}'} b_l - 1 \leq B - 1 = K.$$

2. Let P' be a solution to (P, Q, K). We define a line-route $P_{uv}(P')$ for (u, v) as follows. $P_{uv}(P')$ starts in s_1 taking line l^0. For every $j = 1, \ldots, n$ at station $s_{4(j-1)+2}$, $P_{uv}(P')$ transfers to a line l^i for whose index i $p_i \in q_j$ holds and goes back to l^0 at station $s_{4(j-1)+4}$.

Then by construction $c(\mathcal{L}(P_{uv}(P')), P_{uv}(P')) = 2n + 1$, and

$$\sum_{l \in \mathcal{L}(P_{uv}(P'))} b_l \leq \sum_{l \in \{l^0\} \cup \{l^i : p_i \in q_i\}} b_l \leq |P'| + 1 = K + 1.$$

$\square$

Note that the instances constructed in the reduction from Hitting Set in the proof of Theorem 2.5.2 look somehow artificial: although there might be situations where a line l has a long detour and we can take a shortcut using another line and entering an earlier train of line l', this is not very likely to happen in practice.

We hence assume the *NLT* property in most of the following results:

Definition 2.5.3. An instance $I = (N, \mathcal{L}, \mathcal{OD}, B)$ of uLPwR has the *NLT property* if there exists an optimal solution $(\mathcal{L}', \mathcal{R})$ to I such that for every $P \in \mathcal{R}$ the following holds:

If $i, k \in A_{\text{drive}} \cap P$ and $l(i) = l(k)$, then $l(j) = l(i)$ for all $j \in A_{\text{drive}} \cap P$ for which $i \leq_P j \leq_P k$.

The NLT property is not an unrealistic requirement on the line pool. If the line driving time functions are equal for all lines and the line pool only allows lines that run along shortest paths in the PTN, the NLT property is fulfilled, as we see in the following lemma. Such a line pool is not uncommon in the literature (see, e.g., [BKZ97]).

Lemma 2.5.4. *Consider a public transportation network $\mathcal{PTN}$ and a line pool $\mathcal{L}$. If*

- *every line $(s_1, e_1, s_2, e_2, \ldots, e_{k-1}, s_k)$ is a shortest path in $\mathcal{PTN}$ from its start node s_1 to its end node s_k, and*
- *the line driving time functions are equal for all lines in $\mathcal{L}$,*

for every choice of $\mathcal{OD} \subset S \times S$ and $B \in \mathbb{N}$, the instance $I = (\mathcal{PTN}, \mathcal{L}, \mathcal{OD}, B)$ has the NLT property.

2.5 Uncapacitated Line Planning with Routing

Proof. Let $\mathcal{PTN} = (V, E)$. Consider an optimal solution $(\mathcal{L}', \mathcal{R})$ to I. Let $(u, v) \in \mathcal{OD}$ be an OD-pair and let P_{uv} be the path chosen for (u, v) in the corresponding routing network $N(\mathcal{L}') = (V(\mathcal{L}'), A(\mathcal{L}'))$. Suppose that there is a line l which is entered more than once on P_{uv}. Let i_1 be the node where l is left the first time and i_k be the node where it is entered the second time. Let P' be the subpath of P_{uv} from i_1 to i_k and let P be the path from i_1 to i_k contained in l. For the travel time on $P_{uv}^* := (P_{uv} \setminus P') \cup P$ we obtain

$$c(\mathcal{L}', P_{uv}^*) = \sum_{a \in P_{uv}^*} c_a$$

$$= \sum_{a \in P} c_a + \sum_{a \in P_{uv}^* \setminus P} c_a$$

$$= \alpha \left(\sum_{\{i_j, i_{j+1}\} \in E: ([i_j, l], [i_{j+1}, l]) \in P} L_{\{i_j, i_{j+1}\}} \right) + \sum_{a \in P_{uv}^* \setminus P} c_a$$

$$\leq \alpha \left(\sum_{\{i'_j, i'_{j+1}\} \in E: \exists l' \text{ s.t. } ([i_j, l'], [i_{j+1}, l']) \in P'} L_{\{i'_j, i'_{j+1}\}} \right) + \sum_{a \in P_{uv}^* \setminus P} c_a$$

$$\leq \sum_{a \in P'} c_a + \sum_{a \in P_{uv}^* \setminus P} c_a$$

$$= \sum_{a \in P'} c_a + \sum_{a \in P_{uv} \setminus P'} c_a$$

$$= c(\mathcal{L}', P_{uv}).$$

Thus, we can replace P' by P, i.e., consider $P_{uv}^* := (P_{uv} \setminus P') \cup P$ instead of P_{uv}. We iterate this replacement procedure until the resulting path has the required property. □

As we will see in Theorem 2.5.6, the NLT property allows us to find solutions to instances of uLPwR with one OD-pair in pseudopolynomial time. We achieve this by solving uLPwR with one OD-pair as a *Resource-Constrained Shortest Path (RCSP)* problem in the CGN.

An instance $(G, (s, t), R)$ of the optimization problem *RCSP* (see, e.g., [BC89]) consists of a (directed or undirected) network $G = (V, A)$ with two nonnegative labels: the length L_a and the resource consumption r_a, on each edge/arc $a \in A$. An origin node s and a destination node t are specified, and a bound $R \in \mathbb{N}$ on the resource consumption is given. The task is to find a path P of minimal length from s to t, under the constraint that $\sum_{a \in P} r_a \leq R$.

Let $I = (N, \mathcal{L}, \mathcal{OD}, B)$ be an instance of uLPwR with $|\mathcal{OD}| = 1$, say $\mathcal{OD} = \{(u, v)\}$. In order to transform our problem to a RCSP problem, we assign to every transfer arc $a = ([s,l],[s,l'])$ and every origin arc $a = (u^{\text{org}}, [u,l'])$ a second label b_a that represents the line cost of the line l' which is entered, i.e., $b_a := b_{l'}$. For the driving arcs and the destination arcs, this cost label is set to $b_a := 0$. Let P_{uv} be a path from u^{org} to v^{dest} in N.

We denote
$$\tilde{b}(\mathcal{L}(P_{uv}), P_{uv}) := \sum_{a \in P_{uv} \cap (A_{\text{org}} \cup A_{\text{trans}})} b_a.$$

Then
$$\begin{aligned} b(\mathcal{L}(P_{uv}), P_{uv}) &= \sum_{l:E(l) \cap P_{uv} \neq \emptyset} b_l \\ &= \sum_{l:([u^{\text{org}},[u,l]) \in P_{uv}} b_l + \sum_{l:\exists s,l' \text{ s.t. }([s,l'],[s,l]) \in A_{\text{trans}} \cap P_{uv}} b_l \\ &\leq \sum_{a \in P_{uv} \cap (A_{\text{org}} \cup A_{\text{trans}})} b_a \\ &= \tilde{b}(\mathcal{L}(P_{uv}), P_{uv}). \end{aligned} \qquad (2.2)$$

This gives us the following lemma.

Lemma 2.5.5. *Let* $I = (N, \{(u,v)\}, \mathcal{L}, B)$ *be an instance of uLPwR and let* P *be a feasible solution with objective value* c^* *to the instance* $(N, (u^{\text{org}}, v^{\text{dest}}), B)$ *of RCSP with the labels* $L_a := c_a$ *and* $r_a := b_a$ *on the arcs* a *of* N, *origin node* u^{org}, *destination node* v^{dest}, *and the bound* B *on the sum over the* b_a. *Then* $(\mathcal{L}(P), \{P\})$ *is a feasible solution to* I *with* $c(\mathcal{L}(P), \{P\}) = c^*$.

Note that if the NLT property holds, in (2.2) it follows that $b(\mathcal{L}(P_{uv}), P_{uv}) = \tilde{b}(\mathcal{L}(P_{uv}), P_{uv})$. In this case, the optimal solution path P_{uv} to I is also optimal for RCSP. Hence, in this case we obtain the stronger result of Theorem 2.5.6.

Theorem 2.5.6. *Given an instance* $I = (N, \{(u,v)\}, \mathcal{L}, B)$ *of uLPwR which has the NLT property, any optimal solution to the instance* $(N, (u^{\text{org}}, v^{\text{dest}}), B)$ *of RCSP with the labels* $L_a := c_a$ *and* $r_a := b_a$ *on the arcs of* N, *origin node* u^{org}, *destination node* v^{dest}, *and the bound* B *on the sum over the* b_a, *is an optimal solution to* I.

Although RCSP is NP-hard [WC96], it can be solved in pseudopolynomial time. The algorithm given in [Phi93] solves the RCSP problem in time $O(|A|R + |V|R\log(|V|R))$.. For an instance $I' = (G, (s,t), R)$ of RCSP, it constructs a *search graph* $S = (V_S, A_S)$, that is, a graph whose node set V_S consists of R copies of the node set of the initial graph $G = (V, A)$ for the constraint on the resources R. Nodes $[i,r]$ in this graph represent the nodes i from the initial graph as well as the actual level of resource consumption. Consequently, an arc $([i,r],[j,r'])$ is introduced if $(i,j) \in A$ and $r_{ij} = r' - r$. Applying Dijkstra's algorithm on S for

2.5 Uncapacitated Line Planning with Routing

start node, $[s, 0]$ provides us with a set of paths $\{P_r : r = 1, \ldots, R\}$ where P_r is a shortest path from $[s, 0]$ to $[t, r]$ in S if such a path exists. Choosing a path of minimal length from $\{P_r : r = 1, \ldots, R\}$, we obtain an optimal solution to I'.

Note that the structure of the search graph makes it possible to reverse the role of "resource" and "length": that is, given a bound L on the path lengths, e.g., $L := \sum_{a \in A} L_a$ for the set of edges or arcs A, we can construct a search graph $S' = (V'_S, A'_S)$ whose node sets consist of L copies of V. In this case, nodes $[i, l]$ in S' represent the nodes i from the initial graph as well as the traveled distance to the origin, i.e., $([i, l], [j, l']) \in A'_S$, if $(i, j) \in A$ and $L_{ij} = l' - l$. Applying Dijkstra's algorithm on S' for start node, $[s, 0]$ provides us with a set of paths $\{P_l : l = 1, \ldots, L\}$ where P_l is a shortest path from $[s, 0]$ to $[t, l]$ if such a path exists. The path P_l with minimal l among all found paths is an optimal solution to I'.

Using the algorithm from [Phi93] and observing that the travel time in N is bounded by $C := \max\{1, \sum_{a \in A} c_a\}$ and the costs are bounded by $\hat{B} := \min\{B, \sum_{a \in A} b_a\}$, we obtain the following results.

Theorem 2.5.7. *Let $I = (N, \mathcal{L}, \mathcal{OD}, B)$ be an instance of uLPwR with $|\mathcal{OD}| = 1$.*

- *The RCSP algorithm can be applied to I in pseudopolynomial time*
  1. $O(m_N \hat{B} + n_N \hat{B} \log(n_N \hat{B}))$ with $\hat{B} := \min\{B, \sum_{a \in A} b_a\}$, or
  2. $O(m_N C + n_N C \log(n_N C))$ with $C := \max\{1, \sum_{a \in A} c_a\}$.
- *If the RCSP algorithm returns a solution, it is feasible for I and its objective value is an upper bound on the optimal objective value.*
- *If I has the NLT property and there exists a feasible solution to I, the algorithm returns an optimal solution to I.*

Hence, an upper bound on the objective value of an instance I of uLPwR with one OD-pair is provided by the RCSP algorithm. On the other hand, a lower bound for the objective value of any line planning problem with routing is easily obtained by performing a shortest-path routing in the CGN (neglecting line costs).

By modifying a proof for NP-completeness of the decision version of the RCSP problem (see [WC96]), in Theorem 2.5.8 we find that we cannot do better than solving uLPwR with one OD-pair in pseudopolynomial time, even if we require that our instances have the NLT property.

Theorem 2.5.8. *uLPwR is NP-hard, even if*

- *there is only one OD-pair,*
- *the considered instances have the NLT property,*
- *the line driving time functions of all lines are equal, and*
- *there are no transfer penalties.*

We prove Theorem 2.5.8 by reduction from *Partition*. An instance of *Partition* consists of a set $\mathcal{M}$ of n natural numbers that sum up to a number M. The question is whether there is a subset $\mathcal{M}'$ of $\mathcal{M}$ such that the sum of all elements in $\mathcal{M}'$ is $\frac{M}{2}$. Partition is NP-complete [GJ79].

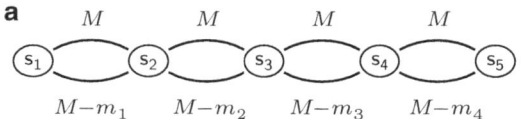
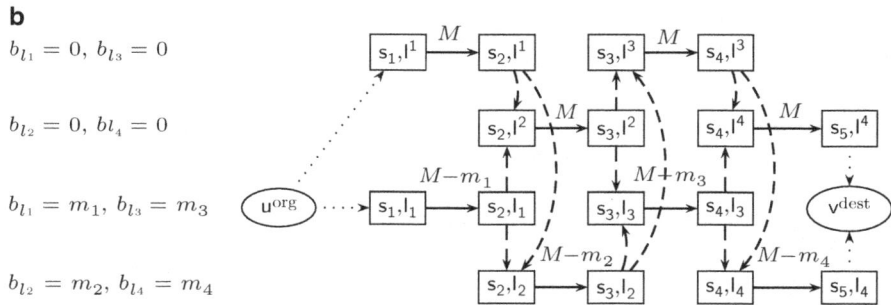

Fig. 2.7 Example for the construction of an instance of uLPwR in the proof of Theorem 2.5.8 (**a**) The constructed PTN (**b**) The constructed CGN

Proof. Let $\mathcal{M} = \{m_1, \ldots, m_n\}$ be an instance of Partition. We construct the following instance I of uLPwR:

- The PTN consists of $n+1$ stations $s_1, s_2, \ldots, s_{n+1}$. For $j = 1, \ldots n$, s_j is connected to s_{j+1} by two edges, e^j and e_j. The length of e^j is set to $L_{e^j} := M$, the length of e_j to $L_{e_j} := M - m_j$.
- The line pool $\mathcal{L}$ consists of $2n$ lines, $l^j = (s_j, e^j, s_{j+1})$ with line cost $b_{l^j} := 0$ and $l_j = (s_j, e_j, s_{j+1})$ with line cost $b_{l_j} := m_j$ for $j = 1, \ldots, n$. We define α_l as $\alpha_l(x) := x$ for all $l \in \mathcal{L}$ and $B := \frac{M}{2}$.
- We set the transfer penalties to 0.
- Let $\mathcal{OD} = \{u, v\}$ with $u := s_1$ and $v := s_n$.

Figure 2.7 shows an example with $\mathcal{M} = \{m_1, m_2, m_3, m_4\}$. We observe that by construction the NLT property holds in all so-constructed instances.

Note that for every line-route P_{uv} from $u = s_1$ to $v = s_{n+1}$, the sum of its line costs $b(\mathcal{L}(P_{uv}), P_{uv})$ and its traveling time $c(\mathcal{L}(P_{uv}), P_{uv})$ is

$$b(\mathcal{L}(P_{uv}), P_{uv}) + c(\mathcal{L}(P_{uv}), P_{uv}) = nM.$$

We show that if and only if there is a line-route P_{uv} from s_1 to s_{n+1} with cost $b(\mathcal{L}(P_{uv}), P_{uv}) \leq B = \frac{M}{2}$ and travel time $c(\mathcal{L}(P_{uv}), P_{uv}) \leq nM - \frac{M}{2}$, there is a solution to the given instance of Partition.

2.5 Uncapacitated Line Planning with Routing

1. Let P_{uv} be such a line-route and let $E_{P_{uv}}$ be the edge set of its corresponding path in $\mathcal{PTN}$. From

$$b(\mathcal{L}(P_{uv}), P_{uv}) \leq \frac{M}{2}, \quad c(\mathcal{L}(P_{uv}), P_{uv}) \leq nM - \frac{M}{2}$$

and

$$b(\mathcal{L}(P_{uv}), P_{uv}) + c(\mathcal{L}(P_{uv}), P_{uv}) = nM,$$

we can conclude that $b(\mathcal{L}(P_{uv}), P_{uv}) = \frac{M}{2}$ holds. We define $\mathcal{M}(P_{uv}) := \{m_j : e_j \in E_{P_{uv}}\}$ and obtain that

$$\sum_{m_j \in \mathcal{M}(P_{uv})} m_j = \sum_{j: e_j \in E_{P_{uv}}} m_j = \sum_{j: l_j \in \mathcal{L}(P_{uv})} m_j = \sum_{j: l_j \in \mathcal{L}(P_{uv})} b_{l_j} = b(\mathcal{L}(P_{uv}), P_{uv}) = \frac{M}{2},$$

hence $\mathcal{M}(P_{uv})$ is a solution to the given instance of Partition.

2. Vice versa for a solution $\mathcal{M}'$ to Partition, we define

$$E_{P_{uv}}(\mathcal{M}') := \{e_j : m_j \in \mathcal{M}'\} \cup \{e^j : m_j \notin \mathcal{M}'\}.$$

Then $E_{P_{uv}}(\mathcal{M}')$ induces a path $P_{uv}(\mathcal{M}')$ in $\mathcal{PTN}$ for which

$$b(\mathcal{L}(P_{uv}(\mathcal{M}')), P_{uv}(\mathcal{M}')) = \sum_{l: E_{P_{uv}(\mathcal{M}')} \cap E(l) \neq \emptyset} b_l$$

$$= \sum_{j: E_{P_{uv}(\mathcal{M}')} \cap E(l^j) \neq \emptyset} b_{l^j} + \sum_{j: E_{P_{uv}(\mathcal{M}')} \cap E(l_j) \neq \emptyset} b_{l_j}$$

$$= 0 + \sum_{m_j \in \mathcal{M}'} m_j$$

$$= \frac{M}{2},$$

and

$$c(\mathcal{L}(P_{uv}(\mathcal{M}')), P_{uv}(\mathcal{M}')) = nM - b(\mathcal{L}(P_{uv}(\mathcal{M}')), P_{uv}(\mathcal{M}')) = nM - \frac{M}{2}.$$

□

Later, in Corollary 2.5.17, we see that the problem from Theorem 2.5.8 can be solved in polynomial time if the driving time function of all lines is equal and the PTN is linear.

Although it allows to reduce the solution time to pseudopolynomial time in the case of only one OD-pair, the NLT property does not help in the general case of uLPwR, where we have more OD-pairs. In fact, even if we assume that every line covers only one edge of the PTN (a case in which the NLT property trivially holds), the problem of finding an optimal line concept and an optimal routing can be reduced to the *Steiner tree* problem (see [Sch05]).

2.5.2 Uncapacitated Line Planning in Linear Networks

In this section we restrict our analysis to linear PTNs as defined in Definition 1.3.4. Before we expose our results regarding the optimization problem uLPwR in the next three sections, we shortly discuss the solvability of the associated feasibility problem. In contrast to the result of Theorem 2.5.1, the restriction to linear PTNs allows polynomial-time solutions, by regarding the problem as a Set Cover problem with consecutive-ones property:

Theorem 2.5.9. *An instance* $I = (\mathcal{PTN}, \mathcal{L}, \mathcal{OD}, B)$ *of the feasibility problem of uLPwR where* $\mathcal{PTN}$ *is a linear network can be solved in polynomial time.*

Proof. Let $I = (\mathcal{PTN}, \mathcal{L}, \mathcal{OD}, B)$ be an instance of uLPwR with $\mathcal{PTN} = (V, E)$ linear. Let E' denote the set of edges that are used by at least one OD-pair. A line concept $\mathcal{L}'$ is a feasible solution to I if and only if $\sum_{l \in \mathcal{L}'} b_l \leq B$ and for all $e \in E'$ there is at least one line $l \in \mathcal{L}'$ with $e \in E(l)$. Thus, we can decide the feasibility problem by solving the following (weighted) Set Cover problem with variables $x_l := \begin{cases} 1 \text{ if } l \in \mathcal{L}' \\ 0 \text{ otherwise} \end{cases}$ for all $l \in \mathcal{L}$:

$$\min \sum_{l=1}^{|\mathcal{L}'|} b_l x_l$$

$$s.t. \ Ax \geq 1,$$

$$x_l \in \{0, 1\},$$

where $A := (a_{el})_{e=1,\ldots,|E'|;\ l=1,\ldots,|\mathcal{L}|}$ with $a_{el} := \begin{cases} 1 \text{ if } e \in l \\ 0 \text{ otherwise.} \end{cases}$

This integer program can be solved in polynomial time: Let the elements e'_i in E' be indexed consecutively in the same order as in E. Then, for a line l, if $e'_i \in l$ and $e'_k \in l$, also $e'_j \in l$ for all $j \in \{i+1, i+2, \ldots, k-1\}$. Hence, A has the consecutive-ones property and is totally unimodular. Thus, the integer program can be solved by linear programming. □

In Sect. 2.5.2.1 we consider the case of uLPwR for only one OD-pair in a linear PTN. In Sects. 2.5.2.2 and 2.5.2.3 we extend our results to situations with more OD-pairs.

2.5.2.1 Uncapacitated Line Planning in Linear Networks for One OD-Pair

The first theorem shows that even in the case of linear PTNs, the NLT property is indispensable for solving instances of uLPwR with one OD-pair efficiently.

Theorem 2.5.10. *uLPwR is strongly NP-hard, even if*

- *the PTN is linear, and*
- *there is only one OD-pair.*

Proof. Similarly to Theorem 2.5.2 we prove this by reduction from Hitting Set. Let (P, Q, K) be an instance of Hitting Set with $P = \{p_1, p_2, \ldots, p_m\}$ and $Q = \{q_1, q_2, \ldots, q_n\}$. We construct an instance I of uLPwR in the following way:

- We have $2n + 2$ stations $s_1, \ldots, s_{2n+2}$ which are connected in this order by edges. The edges $\{s_i, s_{i+1}\}$ for i odd have length 1 and for i even have length 2.
- There are m lines $l_1, \ldots, l_m$ running from station s_1 to station s_{2n+2} with line driving time functions $\alpha_{l_i}(x) := x$ and line costs $b_{k_i} := 1$ for $i = 1, \ldots, m$. Additionally there are n more lines $l^1, \ldots, l^n$ with l^j running from station s_{2j} to s_{2j+1} with line driving time functions $\alpha_{l^j}(x) := \frac{1}{2}x$ and line costs $b_{l^j} := 0$ for $j = 1, \ldots, n$. We set $B := K$.
- The OD-pair is (u, v) with $u := s_1$ and $v := s_{2n+2}$.
- The transfer penalty to transfer from l_i to l^j is 0 for all i, j in all stations, the transfer penalty from l_i to $l_{i'}$ is 1 for every $i \neq i'$ and every station. Now we set

$$p_{s_{2j+1}}^{l^j l_i} := \begin{cases} 0 \text{ if } p_i \in q_j \\ 1 \text{ otherwise.} \end{cases}$$

See Fig. 2.8 for an example of this construction with element set $P = \{p_1, p_2, p_3, p_4\}$ and set of subsets $Q = \{\{p_2, p_4\}, \{p_1, p_3\}, \{p_2, p_3, p_4\}\}$. The PTN is depicted in Fig. 2.8a. The CGN is shown in Fig. 2.8b where we only show transfer arcs with zero penalty.

We now claim that there is a solution $(\mathcal{L}', \mathcal{R})$ to the described instance of uLPwR with objective value $c(\mathcal{L}', \mathcal{R}) \leq 2n + 1$ if and only if there is a solution to the instance (P, Q, K) of Hitting Set.

1. Let $(\mathcal{L}', \mathcal{R})$ be a solution to I with $c(\mathcal{L}', \mathcal{R}) \leq 2n + 1$. Let P_{uv} denote the path from u^{org} to v^{dest} contained in $\mathcal{R}$. Note that P_{uv} can contain only driving arcs of length 1 and transfer arcs of length 0. Thus, P_{uv} uses all lines l^j. Furthermore, in every station s_{2j+1} there has to be a transfer arc of length 0 from line l^j to a line $l_i \in \mathcal{L}'$. In other words, for every j there is an index i such that we can transfer from l^j to l_i at cost 0.

   Hence, for the set $P(\mathcal{L}') := \{p_i : l_i \in \mathcal{L}'\}$ it holds that for every q_j there is a $p_i \in P(\mathcal{L}')$ such that $p_i \in q_j$. Furthermore,

$$|P(\mathcal{L}')| = |\{p_i : l_i \in \mathcal{L}'\}| = \sum_{l \in \mathcal{L}'} b_l \leq K.$$

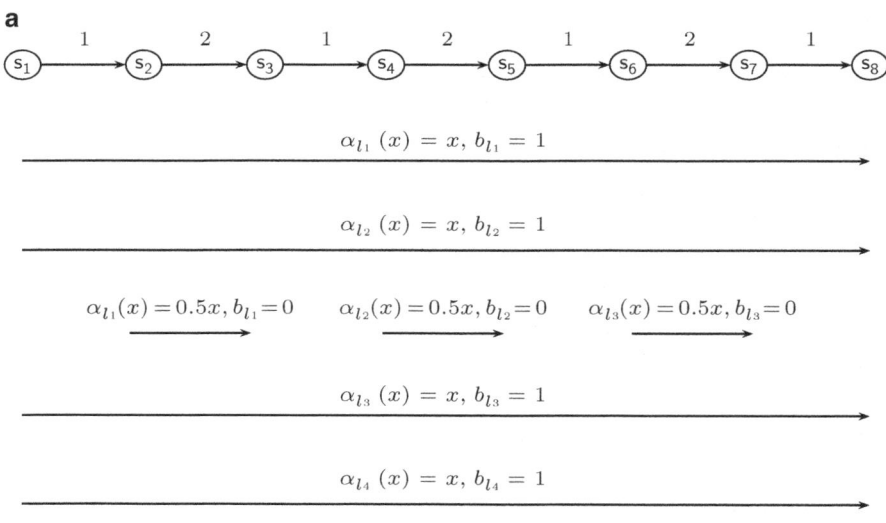

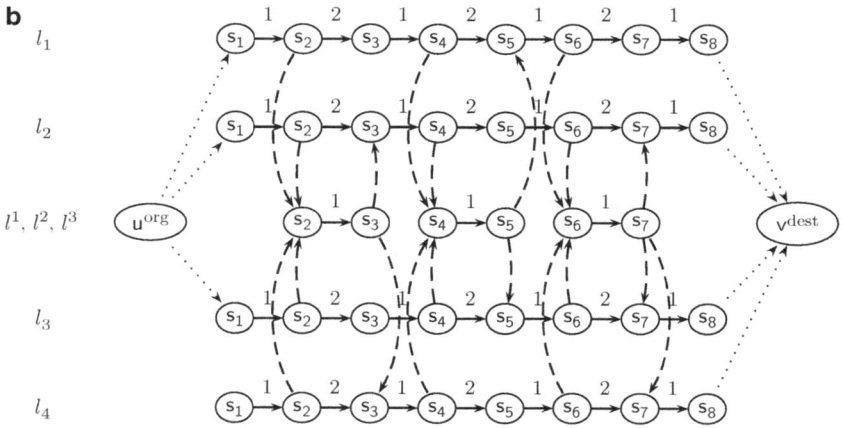

Fig. 2.8 Example for the construction of an instance of uLPwR in Theorem 2.5.10 (**a**) constructed PTN (**b**) constructed CGN

2. Let P' be a solution to the considered instance of Hitting Set. We construct a line-route $P_{uv}(P')$ as follows: on edge $\{s_1, s_2\}$, P_{uv} uses line l_1. For every j, on edge $\{s_{2j}, s_{2j+1}\}$ $P_{uv}(P')$ uses line l^j. On edge $\{s_{2j}, s_{2j+1}\}$ $P_{uv}(P')$ uses a line l^i with $p_i \in q_j$. Then by construction the transfer penalties along the line-route are 0 and $c(\mathcal{L}(P_{uv}(P')), \{P_{uv}(P')\}) = 2n + 1$. Furthermore,

2.5 Uncapacitated Line Planning with Routing

$$\sum_{l \in \mathcal{L}(P_{uv})(P')} b_l = |\{l_i \in \mathcal{L}(P_{uv}(P'))\}| \leq |P'| = K.$$

□

Due to this hardness result we assume in the following that the NLT property holds. Applying the result of Lemma 2.5.4 to linear PTNs, we obtain the following corollary.

Corollary 2.5.11. *Let $I = (\mathcal{PTN}, \mathcal{L}, \mathcal{OD}, B)$ be an instance of uLPwR with*

- *linear PTN, and*
- *all lines having the same driving time function.*

Then I has the NLT property.

Theorem 2.5.12. *uLPwR is NP-hard, even if*

- *the PTN is linear,*
- *there is only one OD-pair,*
- *the NLT property holds,*

and one of the following conditions is fulfilled:

1. *either there are no transfer penalties,*
2. *or the line driving time functions of all lines are equal (which implies the NLT property) and the transfer penalties are station-independent.*

This can be shown by reduction from Partition analogously to the proof of Theorem 2.5.8 (see [SS12a]). Thus, under the assumption that $P \neq NP$, we cannot get substantially better than the pseudopolynomial running time of the RCSP algorithm proposed in Lemma 2.5.7.

Certainly, if the NLT property is fulfilled, for uLPwR with one OD-pair, we can apply the RCSP algorithm also on linear networks. However, compared to the running time of the algorithm presented in Theorem 2.5.7 we can obtain a speed-up if the line driving time functions of all lines are equal, due to the property shown in Lemma 2.2.6. There, we saw that in case of a linear PTN and equal line driving time functions for all lines, only the transfer penalties need to be considered.

Consequently, without loss of generality, when searching for an optimal solution $(\mathcal{L}', \mathcal{R})$ we can assume all driving arc lengths to be 0 in the situation of Lemma 2.2.6. Thus, we obtain the following corollary to Theorem 2.5.7.

Corollary 2.5.13. *An optimal solution to an instance of uLPwR with*

- *linear PTN, and*
- *all lines having the same line driving time function*

can be found in time

1. $O(m_N Q + n_N Q \log(n_N Q))$ *with* $Q := \max\{1, \sum_{l \in \mathcal{L}} \max_{l' \in \mathcal{L}} \max_{s \in S(l)} p_s^{l'l}\}$, *or*
2. $O(m_N \hat{B} + n_N \hat{B} \log(n_N \hat{B}))$ *with* $\hat{B} := \min\{B, \sum_{a \in A} b_a\}$.

Proof. Due to Lemma 2.2.6 we can assume that all driving arcs in the CGN have length 0. Since the NLT property holds (Lemma 2.5.11), every line is entered at most once. Thus, the length of our path (counting only transfer arcs) in the CGN is at most $Q := \max\{1, \sum_{l \in \mathcal{L}} \max_{l' \in \mathcal{L}} \max_{s \in S(l)} p_s^{l'l}\}$. Applying the RCSP algorithm yields the result of the corollary. □

Now we consider station-independent transfer penalties and equal line driving time functions. We are hence in the situation of Sect. 2.2.3 and can apply the RCSP algorithm to the problem in the LN instead of the CGN which leads to further speed-up.

Theorem 2.5.14. *Given an instance $I = (N_L, (u, v), \mathcal{L}, B)$ of uLPwR with*

- *linear PTN and corresponding line network $N_L = (V_L, A_L)$,*
- *one OD-pair (u, v),*
- *equal line driving time functions, and*
- *station-independent transfer penalties,*

any optimal solution to the instance $(N_L, (u^{\mathrm{org}}, v^{\mathrm{dest}}), B)$ of RCSP with the labels c_a for all $a \in A_L$ and $b_{ll'} := b_l$ for all $(l, l') \in A_L$ is an optimal solution to $(N_L, (u, v), \mathcal{L}, B)$.

Proof. Consider an instance $I = (N_L, (u, v), \mathcal{L}, B)$ of uLPwR as described in the lemma. Again, we assign to every arc $a = (u^{\mathrm{org}}, l')$ or $a = (l, l')$ the second label $b_{ll'} := b_{l'}$ and set $b_{l, v^{\mathrm{dest}}} := 0$ for every $l, l' \in \mathcal{L}'$ for which the corresponding arcs are defined in the LN.

Since $c_a, l_a \geq 0$ for every $a \in A_L$, without loss of generality we can assume that any optimal path P_{uv} for OD-pair $(u, v) \in \mathcal{OD}$ is simple. Hence,

$$b(\mathcal{L}(P_{uv}), P_{uv}) = \sum_{l \in P_{uv}} b_a = \sum_{a \in P_{uv}} b_a =: \tilde{b}(\mathcal{L}(P_{uv}), P_{uv}). \tag{2.3}$$

Therefore, any optimal solution to the instance $(N_L, (u^{\mathrm{org}}, v^{\mathrm{dest}}), B)$ of RCSP is an optimal solution to the instance $(N_L, (u, v), \mathcal{L}, B)$ of uLPwR. □

Applying the RCSP algorithm from [Phi93] described on p. 40 we obtain the following corollary.

Corollary 2.5.15. *An instance $I = (N_L, \mathcal{L}, \mathcal{OD}, B)$ of uLPwR with*

- *linear PTN,*
- *one OD-pair (u, v),*
- *equal line driving time functions, and*
- *station-independent transfer penalties*

can be solved in pseudopolynomial time $O(m_{N_L} \hat{Q} + n_{N_L} \hat{Q} \log(n_{N_L} \hat{Q}))$ with

2.5 Uncapacitated Line Planning with Routing

$$\hat{Q} := \max\left\{1, \sum_{l \in \mathcal{L}} \max_{l' \in \mathcal{L}} p^{l'l}\right\}.$$

If additionally the transfer penalties are equal, we can assume all transfer penalties to be 1:

Lemma 2.5.16. *Let* $I^p = (N_L^p, \mathcal{L}, \mathcal{OD}, B)$ *be an instance of uLPwR with*

- *linear PTN,*
- *equal line driving time functions, and*
- *equal transfer penalties p.*

Let $(\mathcal{L}^p, \mathcal{R}^p)$ *be an optimal solution to* I^p. *Then* $t(\mathcal{L}^p, \mathcal{R}^p) = p \cdot t(\mathcal{L}^1, \mathcal{R}^1)$ *for an optimal solution* $(\mathcal{L}^1, \mathcal{R}^1)$ *to the instance* $I^1 := (N_L^1, \mathcal{L}, \mathcal{OD}, B)$ *where* N_L^1 *is the LN* N_L^p *with transfer penalties* 1 *instead of p.*

Proof. For any routing $\mathcal{R}^p$

$$t(\mathcal{L}(\mathcal{R}^p), \mathcal{R}^p) = p \cdot \sum_{P_i \in \mathcal{R}^p} |A_{\text{trans}}^p \cap P_i| = p \cdot t(\mathcal{L}(\mathcal{R}^1), \mathcal{R}^1)$$

which implies the statement of the lemma. □

Due to Lemma 2.2.6, this leads to the following corollary to Theorem 2.5.15.

Corollary 2.5.17. *Let* $I = (N_L, \mathcal{L}, \mathcal{OD}, B)$ *be an instance of uLPwR with*

- *linear PTN,*
- *one OD-pair, and*
- *equal line driving time functions*

I can be solved in time

1. $O(m_{N_L} + n_{N_L} \log(n_{N_L}))$ *if there are no transfer penalties.*
2. $O(m_{N_L}|\mathcal{L}| + n_{N_L}|\mathcal{L}| \log(n_{N_L}))$ *if the transfer penalties are all equal.*

2.5.2.2 Uncapacitated Line Planning with OD-Pairs Having the Same Origin in Linear Networks

In the next two sections we investigate whether the results of the previous section can be generalized. We stick to the restriction that the underlying PTN is linear but relax the strong assumption of only one OD-pair by allowing a set of OD-pairs $\mathcal{OD} = \{(u, v_k) : k = 1\ldots, m\}$ which all have the same origin u and start traveling in the same direction. Thereby, for the sake of a simpler notation, we assume that the destinations of the OD-pairs as well as the nodes in the PTN are labeled in the direction the OD-pairs are traveling.

Given a public transportation network $\mathcal{PTN}$ with a specified direction of travel, we write $s \leq_{\mathcal{PTN}} s'$, or $s' \geq_{\mathcal{PTN}} s$ to indicate that station s precedes station s' in the direction the OD-pairs travel in $\mathcal{PTN}$, and $s <_{\mathcal{PTN}} s'$, or, $s' >_{\mathcal{PTN}} s$, respectively, to emphasize that additionally $s \neq s'$ holds.

In case of several OD-pairs with the same origin, the NLT property alone is not enough to obtain pseudopolynomial running time anymore. By reduction from Hitting Set, similar to the proof of Theorem 2.5.2, uLPwR can be shown to be strongly NP-hard in general, even for instances which have the NLT property. (See [SS12a] for a proof.)

Theorem 2.5.18. *uLPwR is strongly NP-hard, even if*

- *the PTN is linear,*
- *all OD-pairs have the same origin and go to the same direction, and*
- *the NLT property holds.*

Similarly to the reduction in the proof of Theorem 2.5.8 we can show that when we have several OD-pairs having the same origin, uLPwR is NP-hard, even if all lines have the same line driving time function and if all transfer penalties are equal. Details of the proof are provided in [SS12a].

Theorem 2.5.19. *uLPwR is NP-hard, even if*

- *the PTN is linear,*
- *all OD-pairs have the same origin and go to the same direction*
- *the line driving time functions of all lines are equal, and*
- *all transfer penalties are equal.*

This contrasts the result of Corollary 2.5.17, where we could solve the case of only one OD-pair in polynomial time.

However, in the situation of Theorem 2.5.19, there is an optimal solution such that the routing is *nested*, as we will see in Lemma 2.5.20. We say that a collection of paths in a directed network N is *nested* if there is a simple path P^* in N such that

$$P^* = \bigcup_{P \in \mathcal{R}} \left(P \cap (A_{\text{drive}} \cup A_{\text{trans}}) \right).$$

Lemma 2.5.20. *Consider an instance $I = (N, \mathcal{L}, \mathcal{OD}, B)$ of uLPwR with*

- *linear PTN,*
- *set of OD-pairs, $\mathcal{OD} = \{(u, v_i) : i = 1, \ldots, m\}$ where all OD-pairs travel in the same direction,*
- *equal line driving time functions and*
- *equal transfer penalties.*

For every solution $(\mathcal{L}', \mathcal{R})$ there exists a routing $\mathcal{R}^* := \{P_i^* : i = 1, \ldots, m\} \in \text{OPT}(\mathcal{L}')$ which is nested.

2.5 Uncapacitated Line Planning with Routing

Proof. Let a solution $(\mathcal{L}', \mathcal{R})$ be given. For every $l \in \mathcal{L}'$ denote by s_l the last station on l. For every $s \in \bigcup_{l \in \mathcal{L}'} S(l)$ choose a line $l_s \in (\mathcal{L}' \cap \mathcal{L}(s))$ such that $s_{l_s} \geq_{\mathcal{PTN}} s_{l'}$ for all $l' \in \mathcal{L}' \cap \mathcal{L}(s)$. Then we construct a line-route P^* as follows:

1. Set $s := u$.
2. Take the chosen line l_s and follow l_s until either v_m or $s_{l'_s}$ is reached.
3. If $v_m \in l'_s$ stop. Otherwise, set $s := s_{l'_s}$ and go to (2).

We now show that P^* contains optimal line-routes from u to v_i for $i = 1, \ldots, m$ with regard to the line concept $\mathcal{L}'$. Denote by P_i^* the line-route for OD-pair (u, v_i) contained in P^* and by P_i the corresponding path in the CGN. Assume that there is an i such that there exists a path P_i' from u^{org} to v_i^{dest} with $c(\mathcal{L}', P_i') < c(\mathcal{L}', P_i)$, or, equivalently,

$$|A_{\text{trans}} \cap P_i'| < |A_{\text{trans}} \cap P_i|. \tag{2.4}$$

Let $S^{P_i} := (s^1, \ldots, s^{k_i})$ denote the sequence of stations where passengers on P_i enter or leave a line, i.e., $s^1 = u$, $s^{k_i} = v_i$ and for $j = 2, \ldots, k_i - 1$, $s^j \in S^{P_i}$ if and only if $P_i \cap \{([s^j, l], [s^j, l']) : l, l' \in \mathcal{L}'\} \neq \emptyset$. Let $S^{P_i'} := ((s')^1, \ldots, (s')^{k_i'})$ denote the analogous sequence for P_i'.

Then $s^1 = u = (s')^1$, $s^{k_i} = v_i = (s')^{k_i'}$ and $k_i' < k_i$ due to (2.4). Thus, there is a $j \in \{1, 2, \ldots, k_i'\}$ such that $(s')^j \leq_{\mathcal{PTN}} s^j$ and $(s')^{j+1} >_{\mathcal{PTN}} s^{j+1}$. Let l_j' denote the line taken by P_i' between $(s')^j$ and $(s')^{j+1}$. Then $s^j \in S(l_j')$ and $s_{l_j'} \geq_{\mathcal{PTN}} (s')^{j+1} >_{\mathcal{PTN}} s^{j+1} = s_{l_s}$ which is a contradiction to the definition of l_s. □

Note that Lemma 2.5.20 implies that for every instance of uLPwR restricted as described there, there is an optimal solution $(\mathcal{L}', \mathcal{R}^*)$ such that $P^* := \bigcup_{P_i^* \in \mathcal{R}} P_i^*$ is a path in $N(\mathcal{L}')$. This enables us to transfer the instances of uLPwR with OD-pairs with the same origins to instances of uLPwR with only one OD-pair.

Lemma 2.5.21. *Consider an instance* $I^1 = (N^1, \mathcal{OD}, \mathcal{L}, B)$ *of uLPwR with*

- *linear public transportation network* $\mathcal{PTN}$,
- *set of OD-pairs* $\mathcal{OD} = \{(u, v_i) : i = 1, \ldots, m\}$,
- *equal line driving time functions for all lines, and*
- *equal transfer penalties p.*

Denote by N^2 *the CGN built from* $\mathcal{PTN}$ *for the only OD-pair* $(u, v_m) \in \mathcal{OD}^1$ *and transfer penalties given as*

$$p_s^{ll'} := p_s := \sum_{(u, v_j) \in \mathcal{OD}: s <_{\mathcal{PTN}} v_j} w_{uv_j} \, p. \tag{2.5}$$

Then there is a bijection between

1. *nested routings* $\mathcal{R}^1$ *for* I^1, *and*
2. *routings* $\mathcal{R}^2$ *for* $I^2 := (N^2, \mathcal{L}, (u, v_m), B)$.

fulfilling the property that

- $\mathcal{L}(\mathcal{R}^1) = \mathcal{L}(\mathcal{R}^2)$, and
- $c(\mathcal{L}(\mathcal{R}^1), \mathcal{R}^1) = c(\mathcal{L}(\mathcal{R}^2), \mathcal{R}^2)$.

Proof. Without loss of generality we can assume that $c_a = 0$ for all $a \in A_{\text{drive}}$ due to Lemma 2.2.6.

1. Let $(\mathcal{L}^2, \mathcal{R}^2)$ be a solution to I^2. Let P^2 denote the path contained in $\mathcal{R}^2$. For $(u, v_i) \in \mathcal{OD}$ we define the path $P_i^1(P^2)$ to consist of the path P^2 seen as path in N^1 ending at the first node in the CGN which belongs to station v_i and add a destination arc to v_i^{org}. We set $\mathcal{R}^1 := \{P_i^1(P^2) : i = 1, \ldots, m\}$. Then, by definition, $\mathcal{L}(\mathcal{R}^1) = \mathcal{L}(\mathcal{R}^2)$ and $P^1(P^2) := \bigcup_{P_i \in \mathcal{R}^1} \left[P_i^1(P^2) \cap (A_{\text{drive}} \cup A_{\text{trans}}) \right]$ is a simple path in N^1. Hence, $\mathcal{R}^1$ is nested.

   Furthermore setting $P_0^1(P^2) = \emptyset$ we obtain

$$c(\mathcal{L}(\mathcal{R}^2), \mathcal{R}^2) = t(\mathcal{L}(\mathcal{R}^2), P^2)$$

$$= \sum_{([s,l],[s,l']) \in P^2} p_s^{ll'}$$

$$= \sum_{([s,l],[s,l']) \in P^2} \sum_{j=1,\ldots,m : s <_{\mathcal{PTN}} v_j} w_{uv_j} p$$

$$= \sum_{r=1}^{m} \sum_{([s,l],[s,l']) \in (P_r^1(P^2) \setminus P_{r-1}^1(P^2))} \sum_{j=1,\ldots,m : s <_{\mathcal{PTN}} v_j} w_{uv_j} p$$

$$= \sum_{r=1}^{m} \sum_{([s,l],[s,l']) \in (P_r^1(P^2) \setminus P_{r-1}^1(P^2))} \sum_{j=r}^{m} w_{uv_j} p$$

$$= \sum_{r=1}^{m} |[P_r^1(P^2) \setminus P_{r-1}^1(P^2)] \cap A_{\text{trans}}| \sum_{j=r}^{m} w_{uv_j} p$$

$$= \sum_{r=1}^{m} t(\mathcal{L}(\mathcal{R}^1), P_r^1(P^2) \setminus P_{r-1}^1(P^2)) \sum_{j=r}^{m} w_{uv_j}$$

$$= \sum_{r=1}^{m} \sum_{j=r}^{m} t(\mathcal{L}(\mathcal{R}^1), P_r^1(P^2) \setminus P_{r-1}^1(P^2)) w_{uv_j}$$

$$= \sum_{j=1}^{m} \sum_{r=1}^{j} t(\mathcal{L}(\mathcal{R}^1), P_r^1(P^2) \setminus P_{r-1}^1(P^2)) w_{uv_j}$$

$$= \sum_{j=1}^{m} w_{uv_j} \sum_{r=1}^{j} t(\mathcal{L}(\mathcal{R}^1), P_r^1(P^2) \setminus P_{r-1}^1(P^2))$$

2.5 Uncapacitated Line Planning with Routing

$$= \sum_{j=1}^{m} w_{uv_j} t(\mathcal{L}(\mathcal{R}^1), P_j^1(P^2))$$

$$= c(\mathcal{L}(\mathcal{R}^1), \mathcal{R}^1).$$

2. Vice versa, let $\mathcal{R}^1 := \{P_i^1 : i = 1, \ldots, m\}$ be a nested routing in N^1 and denote

$$P^1 := \bigcup_{P_i^1 \in \mathcal{R}^1} \left[P_i^1 \cap (A_{\text{drive}} \cup A_{\text{trans}}) \right].$$

We define $P^2(P^1)$ as the path $P_i^1 \cap (A_{\text{drive}} \cup A_{\text{trans}})$ regarded in N^2 adding an origin and a destination arc and set $\mathcal{R}^2 := \{P^2(P^1)\}$. Then like above, we have $\mathcal{L}(\mathcal{R}^1) = \mathcal{L}(\mathcal{R}^2)$ and $c(\mathcal{L}(\mathcal{R}^2), \mathcal{R}^2) = c(\mathcal{L}(\mathcal{R}^2), P^2(P^1)) = \sum_{j=1}^{m} w_{uv_j} t(\mathcal{L}(\mathcal{R}^1), P_j^1) = c(\mathcal{L}(\mathcal{R}^1), \mathcal{R}^1)$.

□

Applying Lemmas 2.5.20 and 2.5.21 we obtain the following theorem.

Theorem 2.5.22. *Consider an instance* $I = (N, \mathcal{L}, \mathcal{OD}, B)$ *of uLPwR with*

- *linear PTN,*
- *all OD-pairs having the same origin and going to the same direction,*
- *equal line driving time functions for all lines, and*
- *equal transfer penalties p.*

The instance is solvable in pseudopolynomial time

1. $O(m_N \hat{B} + n_N \hat{B} \log(n_N \hat{B}))$ with $\hat{B} := \min\{B, \sum_{a \in A} b_a\}$, or
2. $O(m_N W + n_N W \log(n_N W))$ with

$$W := \sum_{s \in S} \sum_{\{l_i, l_j\}, l_i, l_j \in \mathcal{L}(s)} \sum_{(u, v_j) \in \mathcal{OD}, s <_{\mathcal{PTN}} v_j} w_{uv_j}.$$

Proof. Without loss of generality, we can assume that $c_a = 0$ for all $a \in A_{\text{drive}}$ because of Lemma 2.2.6. Due to Lemma 2.5.20, there exists an optimal solution $(\mathcal{L}, \mathcal{R})$ to I with $\mathcal{R}$ being nested. Thus, we can apply Lemma 2.5.21 and solve the instance I^2, constructed as described there, instead. This can be done in time

1. $O(m_N \hat{B} + n_N \hat{B} \log(n_N \hat{B}))$ with $\hat{B} := \min\{B, \sum_{a \in A} b_a\}$, or
2. $O(m_N W + n_N W \log(n_N W))$ with

$$W := \max\left\{1, \sum_{a \in A} c_a\right\} = \max\left\{1, \sum_{a \in A_{\text{trans}}} c_a\right\}$$

$$= \max\left\{1, \sum_{\{l_i, l_j\}, l_i, l_j \in \mathcal{L}(s)} c_a\right\}$$

$$= \max \left\{ 1, \sum_{s \in S} \sum_{\{l_i, l_j\}, l_i, l_j \in \mathcal{L}(s)} \sum_{(u, v_j) \in \mathcal{OD}, s <_{\mathcal{PTN}} v_j} w_{uv_j} p \right\}. \tag{2.6}$$

due to Theorem 2.5.7. Since we can assume that $p = 1$ due to Lemma 2.5.16, we obtain the statement of the lemma. □

The property that for every instance there is an optimal solution $(\mathcal{L}', \mathcal{R})$ such that the paths in $\mathcal{R}$ are nested can also be proven for instances without transfer penalties for which the NLT property holds. However, since this is possible even in the case of general sets of OD-pairs, this special case will be analyzed in the next section.

2.5.2.3 Uncapacitated Line Planning with General Sets of OD-Pairs in Linear Networks

We finally investigate the case in which we allow a general structure of the OD-pairs. We make the following assumptions on the instances $I = (\mathcal{PTN}, \mathcal{L}, \mathcal{OD}, B)$ considered in this section.

- We assume that the lines after going to one direction turn and go back (i.e., that all lines are bi-directional) and that the transfer penalties are symmetric, i.e., $p_s^{ll'} = p_s^{l'l}$ for all $s \in S, l, l' \in \mathcal{L}(s)$. Then due to the fact that we do not have any capacity restrictions, we can add up w_{uv} and w_{vu} for the OD-pairs (u, v) and (v, u) and hence without loss of generality assume that all OD-pairs travel in the same direction.
- Without loss of generality we assume that every $\{s_i, s_{i+1}\}$ is used on at least one line-route, i.e, for every $\{s_i, s_{i+1}\}$ there exists $(u_j, v_j) \in \mathcal{OD}$ with $u_j \leq_{\mathcal{PTN}} s_i$ and $v_j \geq_{\mathcal{PTN}} s_{i+1}$. Note that if this is not the case for an edge $\{s_i, s_{i+1}\}$, we can just shrink s_i, $\{s_i, s_{i+1}\}$, and s_{i+1} to one node, setting the transfer penalties at the new node to an arbitrary nonnegative value (since no passenger will transfer there anyway) and thus successively shrink the network until the property is fulfilled.

In Lemma 2.5.20 we saw that for some instances of uLPwR in a linear PTN, we can assume that the solution paths are nested. Based on this observation, we developed an algorithm with pseudopolynomial running time in Theorem 2.5.22.

In the following we see that if we do not have transfer penalties, the property that there is an optimal solution with nested routing also holds for arbitrary sets of OD-pairs and lines with arbitrary line driving time functions.

Lemma 2.5.23. *Consider an instance $I = (N, \mathcal{L}, \mathcal{OD}, B)$ of uLPwR*

- *with linear PTN, and*
- *without transfer penalties.*

For every solution $(\mathcal{L}', \mathcal{R})$ to I, there exists a nested routing $\mathcal{R}^* := \{P_k^* : k = 1, \ldots, m\} \in \mathrm{OPT}(\mathcal{L}')$.

2.5 Uncapacitated Line Planning with Routing

Proof. Let $(\mathcal{L}', \mathcal{R})$ be a solution to I. Let s_n be the last node in $\mathcal{PTN}$, i.e., $s \leq s_n \forall s \in S$. We consider a path P^* from s_1^{org} to s_n^{dest} in $N(\mathcal{L}')$ defined by the following line-route: for every edge e we choose a line $l_e \in \arg\min_{l \in \mathcal{L}(e) \cap \mathcal{L}'} \alpha_l(L_e)$.

Let $N(u_k, v_k)$ denote the subnetwork induced by

$$\{[s, l] : s \in S \text{ s.t. } s \geq_{\mathcal{PTN}} u_k \text{ and } s \leq_{\mathcal{PTN}} v_k; \, l \in \mathcal{L}' \cap \mathcal{L}(s)\}.$$

For every $(u_k, v_k) \in \mathcal{OD}$ we define P_k^* to consist of the subpath $(P^* \cap N(u_k, v_k)) \subset P^*$ and the corresponding origin and destination arcs.

We show now that indeed $c(\mathcal{L}', P_k^*) \leq c(\mathcal{L}', P_k)$ for any path P_k for $(u_k^{\text{org}}, v_k^{\text{dest}})$: since the transfer penalties are 0 we have

$$c(\mathcal{L}', P_k) = \sum_{([s,l],[s',l]) \in P_k \cap A_{\text{drive}}} c_a$$

$$= \sum_{([s,l],[s',l]) \in P_k \cap A_{\text{drive}}} \alpha_l(L_{ss'})$$

$$\geq \sum_{([s,l_{\{s,s'\}}],[s',l_{\{s,s'\}}]) \in P_k^* \cap A_{\text{drive}}} \alpha_{l_{\{s,s'\}}}(L_{ss'})$$

$$= c(\mathcal{L}', P_k^*).$$

Thus, $c(\mathcal{L}', \mathcal{R}^*) \leq c(\mathcal{L}', \mathcal{R})$. □

In particular, this implies that there is an optimal solution with nested paths. Like in Lemma 2.5.21 we use this property to establish a correspondence between the described problem and a class of instances for which a solution approach is known.

Lemma 2.5.24. *Consider an instance $I^1 := (N^1, \mathcal{L}, \mathcal{OD}, B)$ of uLPwR*

- *for which the corresponding public transportation network $\mathcal{PTN}$ is linear, and*
- *without transfer penalties.*

Denote by N^2 the CGN N^1 with arc lengths

$$\tilde{c}_{([s_i,l],[s_{i+1},l])} := \left(\sum_{k=1,\dots,m : (u_k, v_k) \in OD : u_k \leq_{\mathcal{PTN}} s_i+1 <_{\mathcal{PTN}} v_k} w_{u_k v_k} \right) \cdot c_{([s_i,l],[s_{i+1},l])}.$$

We define $I^2 := (N^2, \mathcal{L}, (s_1, s_n), B)$.
Then there is a bijection between

1. *nested routings $\mathcal{R}^1$ for I^1, and*
2. *routings $\mathcal{R}^2$ for I^2.*

with the properties that

- $\mathcal{L}(\mathcal{R}^1) = \mathcal{L}(\mathcal{R}^2)$,
- *and $c(\mathcal{L}(\mathcal{R}^2), \mathcal{R}^2) = c(\mathcal{L}(\mathcal{R}^1), \mathcal{R}^1)$.*

Proof. We use the following notation: For an CGN $N = (V, A)$ with set of OD-pairs $\mathcal{OD}$ we define

$$A(u_j, v_j) := \{([s_i, l], [s_{i+1}, l]) \in A : \{s_i, s_{i+1}\} \in \mathcal{PTN},$$
$$u_j \leq_{\mathcal{PTN}} s_i, s_{i+1} <_{\mathcal{PTN}} v_j, l \in \mathcal{L}(s_i) \cap \mathcal{L}(s_{i+1})\}$$

to be the subset of driving arcs that can be used by OD-pair (u_j, v_j).

1. Let a routing $\mathcal{R}^2$ for instance I^2 be given. Consider the path P^2 defined by $\mathcal{R}^2$ in N^2. Let $P^1(P^2)$ denote the path in N^1 that is obtained by transferring P^2 to N^1. We denote by $P_k^1(P^2)$ the subpath $(P^1(P^2) \cap A(u_k, v_k)) \subset P^1(P^2)$ with additional origin and destination arcs from u_k^{org} and to v_k^{dest} and define $R^1 := \{P_k^1(P^2) : (u_k, v_k) \in \mathcal{OD}\}$.

    Then $\mathcal{L}(\mathcal{R}^2) = \mathcal{L}(\mathcal{R}^1)$. For the objective values given as the (cumulated) travel time of the OD-pairs we obtain

$$c(\mathcal{L}(\mathcal{R}^1), \mathcal{R}^1) = \sum_{j=1}^{m} w_{u_j v_j} c(\mathcal{L}(\mathcal{R}^1), P_j^1(P^2))$$

$$= \sum_{j=1}^{m} w_{u_j v_j} \sum_{([s_i,l],[s_{i+1},l]) \in P_j^1(P^2)} c_{([s_i,l],[s_{i+1},l])}$$

$$= \sum_{([s_i,l],[s_{i+1},l]) \in P^1(P^2)} \left(\sum_{u_j \leq_{\mathcal{PTN}} s_i <_{\mathcal{PTN}} v_j} w_{u_j v_j} \right) c_{([s_i,l],[s_{i+1},l])}$$

$$= \sum_{([s_i,l],[s_{i+1},l]) \in P^1(P^2)} \tilde{c}_{([s_i,l],[s_{i+1},l])}$$

$$= \sum_{([s_i,l],[s_{i+1},l]) \in P^2} \tilde{c}_{([s_i,l],[s_{i+1},l])}$$

$$= c(\mathcal{L}(\mathcal{R}^2), \mathcal{R}^2).$$

2. Vice versa, let $\mathcal{R}^1 = \{P_k^1 : (u_k, v_k) \in \mathcal{OD}\}$ be a nested routing in I^1. Let $P_k^2(P_k^1)$ be the path P_k^1 transferred to N^2 for $k = 1, \ldots, m$ without origin and destination arc. Then we define $\tilde{P}^2 := (\bigcup_{k=1,\ldots,m} P_k^2(P_k^1))$ and add transfer arcs and an origin and a destination arc to $\tilde{P}^2$ to obtain a path P^2 from u_k^{org} to v_k^{dest}. We set $\mathcal{R}^2 := \{P^2\}$. Then $\mathcal{L}(\mathcal{R}^2) = \mathcal{L}(\mathcal{R}^1)$. Analogously to the calculations before we see that $c(\mathcal{L}(\mathcal{R}^1), \mathcal{R}^1) = c(\mathcal{L}(\mathcal{R}^2), \mathcal{R}^2)$. □

We want to use the result of Lemma 2.5.24 to solve instances I^1 of uLPwR in linear PTNs without transfer penalties. If the NLT property holds for the instance I^2 with one OD-pair constructed from I^1 as described in Lemma 2.5.24, we can apply the RCSP algorithm to solve I^2 (and thus also I^1) in pseudopolynomial time. This is the case if I^1 has the *OD-NLT property*:

2.5 Uncapacitated Line Planning with Routing

Definition 2.5.25. Let $I = (\mathcal{PTN}, \mathcal{L}, \mathcal{OD}, B)$ be an instance of uLPwR with $\mathcal{PTN} = (S, E)$ linear with $S = \{s_1, \ldots, s_n\}$. Consider the instance $I' = (\mathcal{PTN}, \mathcal{L}, \{(s_1, s_n)\}, B)$ which is equal to I, except that we exchanged $\mathcal{OD}$ with one OD-pair (s_1, s_n). If the so-constructed instance I' has the NLT property, we say that I has the *OD-NLT property*.

Similarly to Theorem 2.5.22 we can use this property to solve uLPwR.

Theorem 2.5.26. *Let $I = (N, \mathcal{L}, \mathcal{OD}, B)$ be an instance of uLPwR with*

- *linear PTN,*
- *OD-NLT property, and*
- *without transfer penalties.*

Then an optimal solution to I can be found in pseudopolynomial time

1. $O(m_N \hat{B} + n_N \hat{B} \log(n_N \hat{B}))$ with $\hat{B} := \min\{B, \sum_{a \in A} b_a\}$, or
2. $O(m_N \hat{C} + n_N \hat{C} \log(n_N \hat{C}))$ with

$$\hat{C} := \max \left\{ 1, \sum_{l \in \mathcal{L}} \sum_{i=1}^{n-1} \left(\sum_{j=1,\ldots,m:(u_j,v_j) \in \mathrm{OD}: u_j \leq_{\mathcal{PTN}} s_{i+1} <_{\mathcal{PTN}} v_j} w_{u_j v_j} \right) \cdot c_{([s_i,l],[s_{i+1},l])} \right\}.$$

Proof. Due to Lemma 2.5.23 there is an optimal solution $(\mathcal{L}', \mathcal{R})$ to I such that $\mathcal{R}$ is nested. Due to Lemma 2.5.24 this solution corresponds to a solution of an instance I^2 of uLPwR with one OD-pair in an instance $I^2 = (N^2, \mathcal{L}, (s_1, s_n), B)$ as defined there, where the arc lengths in N^2 are given as

$$\tilde{c}_{s_i s_{i+1}} := \left(\sum_{k=1,\ldots,m:(u_k,v_k) \in \mathrm{OD}^1: u_k \leq_{\mathcal{PTN}} s_{i+1} <_{\mathcal{PTN}} v_k} w_{u_k v_k} \right) \cdot c_{s_i s_{i+1}}.$$

with c_a denoting the arc lengths in N. Since I has the OD-NLT property, I^2 has the NLT property. According to Theorem 2.5.7, an optimal solution can be found in time

1. $O(m_N \hat{B} + n_N \hat{B} \log(n_N \hat{B}))$ with $\hat{B} := \min\{B, \sum_{a \in A} b_a\}$, or
2. $O(m_N \hat{C} + n_N \hat{C} \log(n_N \hat{C}))$ with

$$\hat{C} := \max \left\{ 1, \sum_{l \in \mathcal{L}} \sum_{i=1}^{n-1} \left(\sum_{j=1,\ldots,m:(u,v_j) \in \mathrm{OD}: u_j \leq_{\mathcal{PTN}} s_{i+1} <_{\mathcal{PTN}} v_j} w_{u_j v_j} \right) \cdot c_{([s_i,l],[s_{i+1},l])} \right\}.$$

$\square$

Adding the assumption of equal line driving time functions, we can even obtain polynomial running time in the LN.

Table 2.1 Complexity of uLPwR in linear PTNs with NLT or OD-NLT property

Restrictions line driving time function	Restrictions transfer penalties	Complexity one OD-pair	Complexity same origin	Complexity general OD-pairs
Equal	No penalties	$O(m_{N_L} + n_{N_L}\log(n_{N_L}))$ (2.5.17.1)	$O(m_{N_L} + n_{N_L}\log(n_{N_L}))$ (2.5.27)	$O(m_{N_L} + n_{N_L}\log(n_{N_L}))$ (2.5.27)
Equal	Equal penalties	$O(m_{N_L}|\mathcal{L}| + n_{N_L}|\mathcal{L}|\log(n_{N_L}))$ (2.5.17.2)	$O(m_{N_L}\hat{B} + n_N\hat{B}\log(n_N\hat{B}))$, $O(m_NW + n_NW\log(n_NW))$ (2.5.22), NP-hard (2.5.19)	NP-hard (2.5.19)
Equal	Station-independent penalties	$O(m_{N_L}\hat{B} + n_{N_L}\hat{B}\log(n_{N_L}\hat{B}))$ (2.5.13), $O(m_{N_L}\hat{Q} + n_{N_L}\hat{Q}\log(n_{N_L}\hat{Q}))$ (2.5.15), NP-hard (2.5.12.2)	NP-hard (2.5.12.2, 2.5.19)	NP-hard (2.5.12.2,2.5.19)
Equal	No restrictions	$O(m_N\hat{B} + n_N\hat{B}\log(n_N\hat{B}))$, $O(m_NQ + n_NQ\log(n_NQ))$, (2.5.13) NP-hard (2.5.12.2)	NP-hard (2.5.12.2,2.5.19)	NP-hard (2.5.12.2,2.5.19)

2.5 Uncapacitated Line Planning with Routing

Restrictions	Penalties			
No restrictions	No penalties	$O(m_N \hat{B} + n_N \hat{B} \log(n_N \hat{B}))$, $O(m_N C + n_N C \log(n_N C))$ (2.5.7), NP-hard (2.5.12.1)	$O(m_N \hat{B} + n_N \hat{B} \log(n_N \hat{B}))$, $O(m_N \hat{C} + n_N \hat{C} \log(n_N \hat{C}))$ (2.5.26), NP-hard (2.5.12.1), NP-hard (2.5.12.1, 2.5.19)	$O(m_N \hat{B} + n_N \hat{B} \log(n_N \hat{B}))$, $O(m_N \hat{C} + n_N \hat{C} \log(n_N \hat{C}))$ (2.5.26), NP-hard (2.5.12.1), NP-hard (2.5.12.1, 2.5.19)
No restrictions	Equal penalties	$O(m_N \hat{B} + n_N \hat{B} \log(n_N \hat{B}))$, $O(m_N C + n_N C \log(n_N C))$ (2.5.7), NP-hard (2.5.12.1)		
No restrictions	Station-independent penalties	$O(m_N \hat{B} + n_N \hat{B} \log(n_N \hat{B}))$, $O(m_N C + n_N C \log(n_N C))$ (2.5.7), NP-hard (2.5.12.1, 2.5.12.2)	NP-hard (2.5.12.1, 2.5.12.2, 2.5.19)	NP-hard (2.5.12.1, 2.5.12.2, 2.5.19)
No restrictions	No restrictions	$O(m_N \hat{B} + n_N \hat{B} \log(n_N \hat{B}))$, $O(m_N C + n_N C \log(n_N C))$ (2.5.7), NP-hard (2.5.12.1, 2.5.12.2)	Strongly NP-hard (2.5.18)	Strongly NP-hard (2.5.18)

Corollary 2.5.27. *Let $I = (N_L, \mathcal{L}, \mathcal{OD}, B)$ be an instance of uLPwR*
- *with linear PTN,*
- *with equal line driving time functions, and*
- *without transfer penalties.*

I can be solved in time $O(m_{N_L} + n_{N_L} \log(n_{N_L}))$.

Proof. Without loss of generality, we can assume that $c_a = 0$ for all $a \in A_{\text{drive}}$ because of Lemma 2.2.6. Due to Lemma 2.5.23, there exists an optimal solution $(\mathcal{L}, \mathcal{R})$ to I with $\mathcal{R}$ being nested. Thus, we can apply Lemma 2.5.24 and solve the instance I^2, constructed as described there, instead. Since the line driving time functions are equal for I^2 and there are no transfer penalties, I^2 can be solved in the line network. Hence, we can directly regard I as an instance in terms of the line network N_L and transfer it to an instance of I^2 in the corresponding line network N_L^2. According to Corollary 2.5.17, I^2 can be solved in time $O(m_{N_L^2} + n_{N_L^2} \log(n_{N_L^2}))$; hence, the statement of the lemma follows. □

The results of Sect. 2.5 for uLPwR in linear PTNs with NLT or OD-NLT property are summarized in Table 2.1 where

- $C := \max\{1, \sum_{a \in A} c_a\}$,
- $\hat{C} := \max\left\{1, \sum_{l \in \mathcal{L}} \sum_{i=1}^{n-1} (\sum_{j=1,\ldots,m:(u,v_j)\in\mathcal{OD}: u_j \leq_{\mathcal{PTN}} qs_{i+1} <_{\mathcal{PTN}} v_j} w_{u_j v_j}) \cdot c_{([s_i,l],[s_{i+1},l])}\right\}$,
- $\hat{B} := \min\{B, \sum_{a \in A} b_a\}$,
- $Q := \max\{1, \sum_{l \in \mathcal{L}} \max_{l' \in \mathcal{L}} \max_{s \in S(l)} p_s^{l'l}\}$,
- $\hat{Q} := \max\{1, \sum_{l \in \mathcal{L}} \max_{l' \in \mathcal{L}} p^{l'l}\}$, and
- $W := \sum_{s \in S} \sum_{\{l_i,l_j\}, l_i,l_j \in \mathcal{L}(s)} \sum_{(u,v_j) \in \mathcal{OD}, s<_{\mathcal{PTN}} v_j} w_{uv_j}$.

Note that the request of the NLT or OD-NLT property is only an additional restriction for the cases with non-equal line driving time functions as shown in Lemma 2.5.11 and that for instances where all OD-pairs have the same origin the NLT property implies the OD-NLT property.

2.6 Capacitated Line Planning with Routing

In this section we analyze the complexity of capacitated line planning with integrated routing and present integer programming formulations for scLPwR and cLPwR.

2.6.1 Complexity of Capacitated Line Planning with Routing

uLPwR can be regarded as a special case of scLPwR and cLPwR where the capacity is set to $\text{Cap} := \sum_{(u,v)\in\mathcal{OD}} w_{uv}$. Thus, the hardness results from Sect. 2.5 generalize to scLPwR and cLPwR.

2.6 Capacitated Line Planning with Routing

However, the capacity restrictions add extra complexity to the line planning problem: a while uLPwR with only one OD-pair can be modeled as a RCSP problem (Theorem 2.5.6) and solved in pseudopolynomial time (Theorem 2.5.7) if the NLT property holds, scLPwR and cLPwR are strongly NP-hard even under these restrictions.

Analogously to Definition 2.5.3, we define the NLT property for instances of scLPwR or cLPwR.

Definition 2.6.1. An instance $(N, \mathcal{L}, \mathcal{OD}, B, \text{Cap})$ of scLPwR or cLPwR has the *NLT property* if there exists an optimal solution $(\mathcal{L}', \mathcal{R})$ to I such that for every $P \in \mathcal{R}$ the following holds:

If $i, k \in A_{\text{drive}} \cap P$ and $l(i) = l(k)$, then $l(j) = l(i)$ for all $j \in A_{\text{drive}} \cap P$ for which $i \leq_P j \leq_P k$.

Theorem 2.6.2. *It is strongly NP-complete to decide whether there exists a feasible solution to scLPwR or cLPwR, even if*

- *we consider only instances with the NLT property,*
- *there is only one OD-pair,*
- *the driving time function of all lines is equal, and*
- *there are no transfer penalties.*

This can be proven by reduction from the strongly NP-complete problem *Steiner Tree* [GJ79] as done in [Sch12]. An instance (G, V', K) of Steiner Tree consists of an undirected graph $G = (\tilde{V}, \tilde{E})$, a set of *terminal nodes* $V' \subset \tilde{V}$, and a number K. The question is whether there is a subtree $T \subset G$ with at most K edges which contain all terminals V'. Such a tree is called *Steiner Tree*.

Proof. Given an instance (G, V', K) of Steiner Tree with terminals $V' := \{t_0, t_1, \ldots, t_k\}$, we construct an instance I of cLPwR (or scLPwR, respectively) in the form required by the theorem as follows:

- We have a public transportation network $\mathcal{PTN} = (S, E)$ consisting of k copies of G and some extra stations and edges that are specified later. For a node $x \in \tilde{V}$ we denote by x^j the corresponding node in the jth copy of G, analogously e^j is the jth copy of edge $e \in \tilde{E}$. For every edge $\{x, y\} \in \tilde{E}$ we define an arbitrary order (x, y) of x and y and add extra edges $\{y^1, x^2\}, \{y^2, x^3\}, \ldots, \{y^{k-1}, x^k\}$ to E.

  We choose an arbitrary terminal node $t_0 \in V'$ and connect all of its copies to a new station u. Then we connect the kth copies of all other terminals $t_1^k, \ldots, t_k^k$ by edges, inserting one other station in the middle of each edge, i.e., we add stations $i_1, i_2, \ldots, i_k$ and edges $\{t_1^k, i_1\}, \{i_1, t_2^k\}, \{t_2^k, i_2\}, \{i_2, t_3^k\}, \ldots, \{i_{k-1}, t_k^k\}, \{t_k^k, i_k\}$ to G. We connect t_j^j to t_j^k by an edge for all $j = 1, \ldots, k-1$. We add one more station v and edges (i_j, v) for all $j = 1, \ldots, k$ to G to obtain $\mathcal{PTN}$.

  All edges e have length $L_e := 0$.

- The line pool $\mathcal{L}$ now consists of
  - a line $l_0 := (t_1^k, i_1, t_2^k, i_2, \ldots, t_k^k, i_k)$ with $b_{l_0} := K + k$,
  - lines $l_u^j := (u, t_0^j)$ with line cost $b_{l_u^j} := 0$ for $j = 1, \ldots, k$,
  - lines $l_v^j := (i_j, v)$ with line costs $b_{l_v^j} := 1$ for $j = 1, \ldots, k$,
  - bi-directional lines $l_e := (x^1, y^1, x^2, y^2, \ldots, x^k, y^k, x^k, y^{k-1}, \ldots, x^1)$ for every edge $\{x, y\} \in E$ with order (x, y), i.e., the line runs along all copies of e in both directions, with line cost $b_{l_e} := 1$ for all $e \in E$, and
  - lines l_t^j covering $\{t_j^j, t_j^k\}$ with line costs $b_{l_t^j} =: 0$ for $j = 1, \ldots, k$.

  All lines l are assumed to have equal line driving time functions, e.g., $\alpha_l(x) := x$ for all $x \in \mathbb{R}_0^+$. We set the budget to $B := 2(K + k)$. Note that every edge in the PTN is covered by exactly one line, thus any path in the PTN uniquely defines the used line-route.
- The transfer penalties are 0 for all transfers.
- Our OD-pair is (u, v) with $w_{uv} := k$.
- The capacity of each train is Cap $:= 1$.

An example for this construction is given in Fig. 2.9. The considered instance of Steiner Tree with six nodes and four terminals $t_0 := 6$, $t_1 := 1$, $t_2 := 2$, $t_3 := 4$ that are marked in gray is depicted in Fig. 2.9a. Figure 2.9b shows the constructed instance of cLPwR. For the sake of simplicity, the edges $\{y^i, x^{i+1}\}$ connecting the copies of the graph are omitted. The parts of the bi-directional lines l_{xy} contained in the copies of the graph are marked by thick gray lines. The line l_0 is drawn in light gray. All other lines are depicted as black lines.

Note that in this setting the travel time on every line-route for (u, v) is 0; hence, every line-route is a shortest one. We conclude that in this instance every solution to cLPwR is a solution to scLPwR and vice versa. Therefore, in the following we continue our argumentation considering I as an instance of cLPwR. However, the result follows equivalently for an instance of scLPwR.

We show that there is a solution to the instance (G, V', K) of Steiner Tree if and only if there is a solution to the constructed instance I of cLPwR.

1. Suppose T is a solution to the considered instance of Steiner Tree with edge set E'. We consider the line concept

$$\mathcal{L}(E') := \{l_u^j : j = 1, \ldots, k\} \cup \{l_e : e \in E'\} \cup \{l_0\}$$

$$\cup \{l_t^j : j = 1, \ldots, k-1\} \cup \{l_v^j : j = 1, \ldots, k\}$$

and set the frequency to $f_l := 1$ for every $l \in \mathcal{L}$. This line concept has costs

$$\sum_{l \in \mathcal{L}(E')} b_l = \sum_{j=1}^{k} b_{l_u^j} + \sum_{e \in E'} b_{l_e} + b_{l_0} + \sum_{j=1}^{k} b_{l_t^j} + \sum_{j=1}^{k} b_{l_v^j} = 0 + K + (K+k) + 0 + k$$

$$= 2(K+k)$$

2.6 Capacitated Line Planning with Routing

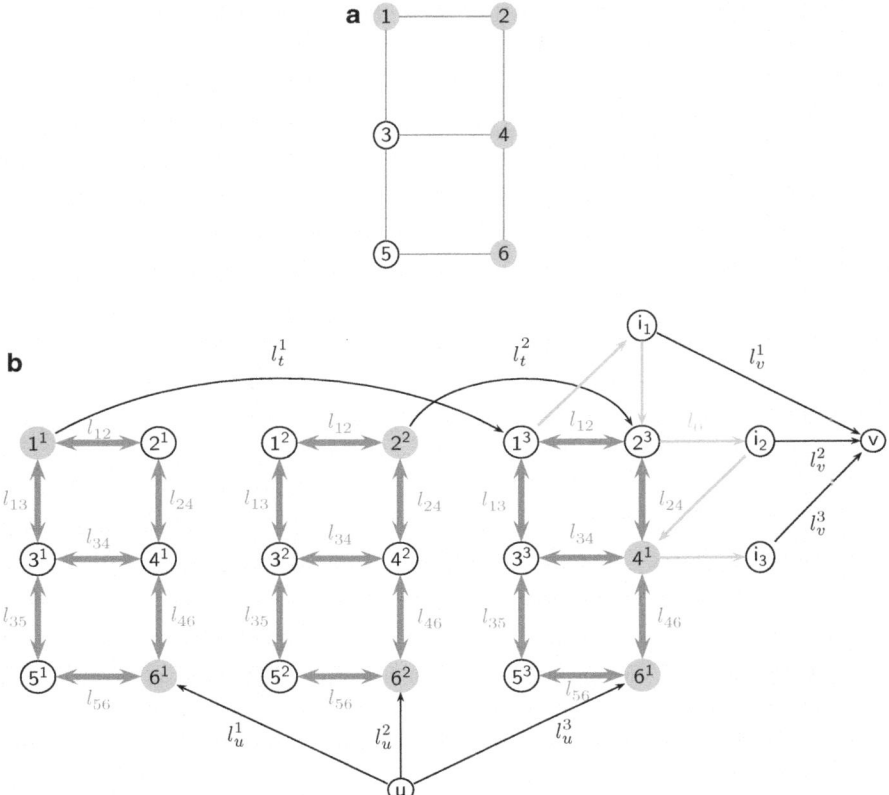

Fig. 2.9 Example for the instance constructed in the proof of Theorem 2.6.2 (**a**) Instance of Steiner Tree (**b**) Constructed PTN

and thus fulfills the budget restriction. Let P_j be the (uniquely determined) path from t_0 to t_j in T and P_j^j its jth copy in $\mathcal{PTN}$. Then for passenger p_j of OD-pair (u, v) we define

$$P_{uv}^j := \begin{cases} (u, t_0^j) \cup P_j^j \cup (t_j^j, t_j^k) \cup (t_j^k, i_j, v) & \text{for } j = 1, \ldots, k-1, \\ (u, t_0^k) \cup P_k^k \cup (t_j^k, i_k, v) & \text{for } j = k, \end{cases}$$

where $\cup$ is meant as composition of node sequences. Then $(\mathcal{L}(E'), \mathcal{R}_{uv})$ with $\mathcal{R}_{uv} := \{P_{uv}^j : j = 1, \ldots, k\}$ and $w_{uv}^{P_{uv}^j} = 1$ for $j = 1, \ldots, k$ is a feasible solution to the constructed instance of cLPwR.

2. Now suppose that $(\mathcal{L}', \mathcal{R})$ is a feasible solution to the considered instance of cLPwR. Line l_0 is contained in $\mathcal{L}'$ because else there would be no path covered by $\mathcal{L}'$ from u to v in $\mathcal{PTN}$. We also note that due to the budget constraint, the frequency of l_0 is $f_{l_0} = 1$. This forces every passenger p_j to enter l_0 at a different terminal (without loss of generality this terminal is t_j^k) and leave it at i_j to go directly to v from there. Thus, $l_v^j \in \mathcal{L}'$ for all $j = 1, \ldots, k$ and without loss of generality we can assume that $f_{l_v^j} = 1$.

We conclude that for every j the line concept $\mathcal{L}'$ allows a path P_j^{uk} from u to t_j^k for every $j = 1, \ldots, k$. Due to the construction of the lines l_e, we can transfer such a path P_j^{uk} to a path P_j from t_0 to t_j in G, transferring the sequence of lines of type l_e used in P_j^{uk} to a sequence of edges e in G. Thus, defining

$$\tilde{E}'(\mathcal{L}') := \{e \in E : l_e \in \mathcal{L}'\}$$

we obtain an edge set connecting t_0 to all other terminals. Because of the budget constraint, the cost for establishing the lines l_e cannot exceed $B - (K+k) - k = K$, and as the line costs are 1 for these lines, we conclude $|\tilde{E}'(\mathcal{L}')| \leq K$. The graph $\tilde{G}'$ with edge set $|\tilde{E}'|$ is connected, contains all terminals, and $|\tilde{E}'| \leq K$. Thus, $\tilde{G}'$ contains a Steiner tree T with $|T| \leq K$.

Note that due to the construction of the Steiner tree it follows that if there exists a feasible solution to the constructed instance of cLPwR, there is also a feasible solution such that every passenger j does not enter the ith copy of G for $i \neq j$ and thus no line is entered more than once by a passenger. Since the travel time for all solutions is equal, this implies the NLT property for the constructed instance. □

2.6.2 Integer Programming Formulations

Due to the high complexity of line planning with routing in general and scLPwR and cLPwR in particular, we develop integer programming formulations for these problems.

When modeling passenger flows in integer linear programming, there are mainly two approaches: the first one is to introduce a variable z_{P_k} for every *path* P_k from u_k^{org} to v_k^{dest} in the CGN and every OD-pair $(u, v) \in \mathcal{OD}$ (see e.g. [PB06, BGP07, BGP08]). As there can be an exponential number of paths, this can lead to an exponential number of variables which is often addressed by column generation.

The second approach is to use an *arc-based* formulation where variables q_a^k indicate the number of passengers of OD-pair (u_k, v_k) on arc $a \in A$. Routings for the OD-pairs are then found using multi-commodity flow conditions. We use the second approach to model the considered capacitated line planning problems.

2.6.2.1 Integer Programming Formulation for Simple Capacitated Line Planning with Routing

In this section we present a slight modification of an integer programming formulation for scLPwR developed in [Sch05, SS06] leaving out the upper bounds on frequencies used there. In addition to the model there, we allow lines going only to one direction and different line driving time functions.

The set of variables consists of integer variables

$$f_l := \text{frequency of line } l \quad \forall l \in \mathcal{L},$$

where $f_l := 0$ for all $l \notin \mathcal{L}'$, and

$x_a^k :=$ number of passengers of OD-pair (u_k, v_k) using arc a $\forall (u_k, v_k) \in \mathcal{OD}, \forall a \in N$.

The IP formulation for solving scLPwR then reads

$$\min \sum_{(u_k,v_k) \in \mathcal{OD}} \sum_{a \in A} c_a x_a^k, \tag{2.7}$$

$$s.t. \sum_{a \in \delta^-(u_k^{\mathrm{org}})} x_a^k = w_k \quad \forall (u_k, v_k) \in \mathcal{OD}, \tag{2.8}$$

$$\sum_{a \in \delta^+(v)} x_a^k - \sum_{a \in \delta^-(v)} x_a^k = 0 \quad \forall (u_k, v_k) \in \mathcal{OD}, v \in V \setminus \{u_k^{\mathrm{org}}, v_k^{\mathrm{dest}}\}, \tag{2.9}$$

$$\sum_{a \in \delta^+(v_k^{\mathrm{dest}})} x_a^k = w_k \quad \forall (u_k, v_k) \in \mathcal{OD}, \tag{2.10}$$

$$\frac{1}{\mathrm{Cap}} \sum_{(u_k,v_k) \in \mathcal{OD}} x_a^k \leq f_{l(a)} \quad \forall a \in A_{\mathrm{drive}}, \tag{2.11}$$

$$\sum_{l \in \mathcal{L}} f_l b_l \leq B, \tag{2.12}$$

$$x_a^k, f_l \in \mathbb{N} \quad \forall (u_k, v_k) \in \mathcal{OD}, a \in A, l \in \mathcal{L}. \tag{2.13}$$

Constraints (2.8)–(2.10) are flow constraints. They ensure that each passenger of an OD-pair (u_k, v_k) travels along a path from u_k^{org} to v_k^{dest} in the CGN. Constraints (2.11) make sure that the constructed routing is feasible. The budget constraint is enforced in (2.12).

For each OD-pair (u_k, v_k) the term $\sum_{a \in A} c_a x_a^k$ denotes the travel time. Thus, the objective (2.7) is the minimization over the overall travel time of the OD-pairs.

Lemma 2.6.3. *The solution $(\mathcal{L}', \mathcal{R})$ defined by an optimal solution (x, f) of the IP (2.7)–(2.13) is an optimal solution to scLPwR.*

Note that since uLPwR is a special case of scLPwR with unlimited capacity it can also be solved using formulation (2.7)–(2.13), setting Cap $\geq \sum_{(u_k,v_k)\in \mathcal{OD}} w_k$. In this case, without loss of generality, the frequency variables f_l can be assumed to take only the values $f_l = 1$ if line l is established and $f_l = 0$ otherwise.

2.6.2.2 Integer Programming Formulation for Capacitated Line Planning with Routing

We modify the approach of Sect. 2.6.2.1 to calculate an optimal solution to cLPwR. To this end, we introduce additional constraints that find shortest paths in the network and ensure that only these paths are taken by the passengers. In this section we assume that $c_{ij} \in \mathbb{N}$ for all arcs (i, j) of the considered CGN.

For a network $N = (V, E)$ and an OD-pair (u_k, v_k), let $V(k)$ be the set of nodes i in V for which there exists a directed path from u_k^{org} to i and from i to v_k^{dest} in N. Let $N(k) = (V(k), A(k))$ denote the network induced by $V(k)$.

The idea of our approach is based on the following optimality conditions for shortest paths (see, e.g., [AMO93]), formulated in terms of a node potential d.

Definition 2.6.4. Let $N = (V, A)$ be a directed network. A *node potential* is a function $d : V \to \mathbb{R}$.

Theorem 2.6.5. *Let $N = (V, A)$ be a directed network and $u \in N$. For $j \geq_N u$ let $d^u(j)$ denote the length of some directed path from u to j. The numbers $d^u(j)$ represent shortest path distances if and only if they satisfy*

$$d^u(j) - d^u(i) \leq c_{ij}.$$

It follows directly that if $(i, j) \in A$ lies on a shortest path from u to j, it holds that

$$d^u(j) - d^u(i) + c_{ij} = 0. \qquad (2.14)$$

In our approach, in addition to the variables

$$f_l := \text{frequency of line } l \quad \forall l \in \mathcal{L},$$

where $f_l := 0$ for all $l \notin \mathcal{L}'$, and

$$x_a^k := \text{number of passengers of OD-pair } (u_k, v_k) \text{ using arc } a \ \forall (u_k, v_k) \in \mathcal{OD}, \forall a \in A(k)$$

also used in (2.7)–(2.13), we introduce variables r_i^k for $(u_k, v_k) \in \mathcal{OD}$ and $i \in V(k)$ which take the role of backwards-directed node potentials: whenever there exists a directed path from i to v_{dest}^k in the routing network, r_i^k indicates the shortest path distance from i to v_{dest}^k in the routing network.

2.6 Capacitated Line Planning with Routing

Furthermore, we introduce variables z_{ij}^k for every $(u_k, v_k) \in \mathcal{OD}$ and $(i, j) \in A(k)$ that are forced to take the value 1 if (i, j) lies on a shortest path from i to v_k^{dest} in the routing network.

Since we are only interested in paths consisting of arcs from the routing network, we introduce additional binary variables

$$y_l := \begin{cases} 1 & \text{if line } l \text{ is established} \\ 0 & \text{otherwise.} \end{cases} \quad \forall l \in \mathcal{L}.$$

Then our problem can be represented using the following integer program:

$$\min \sum_{(u_k, v_k) \in \mathcal{OD}} \sum_{a \in A(k)} c_a x_a^k, \tag{2.15}$$

$$\text{s.t.} \sum_{a \in \delta^-(u_k^{\text{org}}) \cap A(k)} x_a^k = w_k \quad \forall (u_k, v_k) \in \mathcal{OD}, \tag{2.16}$$

$$\sum_{a \in \delta^+(v) \cap A(k)} x_a^k - \sum_{a \in \delta^-(v) \cap A(k)} x_a^k = 0 \quad \forall (u_k, v_k) \in \mathcal{OD}, \ v \in V \setminus \{u_k^{\text{org}}, v_k^{\text{dest}}\}, \tag{2.17}$$

$$\sum_{a \in \delta^+(v_k^{\text{dest}}) \cap A(k)} x_a^k = w_k \quad \forall (u_k, v_k) \in \mathcal{OD}, \tag{2.18}$$

$$\frac{1}{\text{Cap}} \sum_{(u_k, v_k) \in \mathcal{OD}} x_a^k \leq f_{l(a)} \quad \forall a \in A_{\text{drive}} \cap A(k), \tag{2.19}$$

$$\sum_{l \in \mathcal{L}} f_l b_l \leq B, \tag{2.20}$$

$$y_l \leq f_l \quad \forall l \in \mathcal{L}, \tag{2.21}$$

$$M_1 \cdot y_l \geq f_l \quad \forall l \in \mathcal{L}, \tag{2.22}$$

$$z_a^k \leq y_{l(a)} \quad \forall a \in A_{\text{drive}} \cap A(k), (u_k, v_k) \in \mathcal{OD}, \tag{2.23}$$

$$r_j^k + c_{ij} - r_i^k + z_{ij}^k + M_2^{ij}(1 - y_{l(i,j)}) \geq 1 \quad \forall (u_k, v_k) \in \text{OD}, (i, j) \in A(k), \tag{2.24}$$

$$r_j^k + c_{ij} - r_i^k - M_2^{ij} \cdot (1 - z_{ij}^k) \leq 0 \quad \forall (u_k, v_k) \in \text{OD}, (i, j) \in A(k), \tag{2.25}$$

$$r_{v_k^{\text{dest}}}^k = 0 \quad \forall (u_k, v_k) \in \text{OD}, \tag{2.26}$$

$$x_a^k \leq z_a^k \cdot w_k \quad \forall (u_k, v_k) \in \text{OD}, a \in A(k), \tag{2.27}$$

$$y_l, z_a^k \in \{0, 1\} \quad \forall (u_k, v_k) \in \text{OD}, a \in A_{\text{drive}} \cap A(k), l \in \mathcal{L}, \tag{2.28}$$

$$r_i^k \in \mathbb{N} \quad \forall (u_k, v_k) \in \text{OD}, i \in E, \tag{2.29}$$

$$f_l, x_a^k \in \mathbb{N} \quad \forall (u_k, v_k) \in \text{OD}, a \in A(k), l \in \mathcal{L}. \tag{2.30}$$

Here, the objective function (2.15) and the constraints (2.16)–(2.20) and (2.30) remain unchanged compared to the model described in Sect. 2.6.2.1, while constraints (2.21)–(2.29) are introduced to force the solution to contain only lines where enough capacity can be provided for all passengers that would choose this line because of shortest path considerations.

Before we prove the correctness of the formulation we shortly describe the meaning of the additional constraints and of the numbers M appearing in these constraints.

The variable y_l is meant to indicate whether a line is established or not, and thus closely linked to the frequency variable f_l. If we choose M_1 big enough, we ensure in constraints (2.21) and (2.22) that $y_l = 1$ if line f_l is established, and $y_l = 0$ otherwise. Constraints (2.23) force us to choose only arcs for the shortest path calculations that belong to established lines. Constraints (2.24) and (2.25) are introduced to calculate shortest paths in the routing network which is defined implicitly by the chosen y. Note that if r_j^k and r_i^k denote shortest path distances from i to v_k^{dest} or from j to v_k^{dest}, respectively, $r_j^k + c_{ij} - r_i^k$ measures the difference between the shortest direct path from i to v_k^{dest} and the shortest path from i to v_k^{dest} that contains arc (i, j). The addition and subtraction of the constant M_2^{ij} avoids conflicts for arcs which are not contained in the routing network or which are not lying on shortest paths if M_2^{ij} is big enough. Constraints (2.26) set the distance from a destination node to itself to be 0 and hence initialize the implicit distance calculations. Inequalities (2.27) act a linking constraints of the passenger flow assignment and the shortest path calculations allowing only arcs on shortest paths as candidates for the passenger flows.

As the correctness of the IP formulation (2.15)–(2.30) might not be obvious, we prove it formally in Lemmas 2.6.6 and 2.6.7.

Lemma 2.6.6. *Let $I = (N, \mathcal{L}, \mathcal{OD}, B, \text{Cap})$ with $N = (V, A)$ be an instance of cLPwR and $M_1 \in \mathbb{N}_0$ and $M_2^{ij} \in \mathbb{N}_0$ for all $(i, j) \in \bigcup_{(u_k,v_k)\in\mathcal{OD}} A(k)$. Every feasible solution (x, f, y, z, r) to the constraints (2.16)–(2.30) of the corresponding IP defines a feasible solution $(\mathcal{L}', \mathcal{R})$ to I with*

$$c(\mathcal{L}', \mathcal{R}) = \sum_{(u_k,v_k)\in\mathrm{OD}} \sum_{a\in A} c_a x_a^k.$$

Proof. Let (x, f, y, z, r) be a feasible solution to the constraints (2.16)–(2.30). We define $\mathcal{L}' := \{l \in \mathcal{L} : y_l = 1\}$ and denote by $\mathcal{R}$ the routing defined by x.

Due to constraint (2.20), $\mathcal{L}'$ is a feasible line concept. Because of constraints (2.16)–(2.18), $\mathcal{R}$ is a routing and constraints (2.19) ensure that $\mathcal{R}$ is feasible for $\mathcal{L}'$.

It remains to show that $\mathcal{R}$ is a shortest-path routing in $N(\mathcal{L}')$. To this end, we analyze the constraints (2.23)–(2.26) for every $(u_k, v_k) \in \mathcal{OD}$.

- Due to constraints (2.16)–(2.18) for every $(u_k, v_k) \in \mathcal{OD}$, there is at least one path $\tilde{P}^k$ from u_k^{org} to v_k^{dest} with $x_a^k = 1$ for all $a \in \tilde{P}^k$.
- Constraints (2.27) imply that for an arc a $z_a = 1$ whenever $x_a^k = 1$. Therefore, and due to (2.23), every path contained in the routing $\mathcal{R}$ defined by x is entirely contained in $N(\mathcal{L}')$.
- Consider an OD-pair (u_k, v_k) and an arc $(i, j) \in A(k)$. If $z_{ij}^k = 1$ it follows from (2.23) that $y_{l(i,j)} = 1$ and constraints (2.24) and (2.25) read

2.6 Capacitated Line Planning with Routing

$$r_j^k + c_{ij} - r_i^k + 1 \geq 1 \text{ and } r_j^k + c_{ij} - r_i^k \leq 0.$$

Consequently

$$r_j^k + c_{ij} - r_i^k = 0. \tag{2.31}$$

On the other hand, if $y_{l(i,j)} = 1$ and $z_{ij}^k = 0$ constraint (2.24) implies

$$r_j^k + c_{ij} - r_i^k \geq 1. \tag{2.32}$$

- Let i' be a node and $P_{i'}^k$ be a path from i' to v_k^{dest} with $z_{ij}^k = 1$ for all $(i, j) \in P_{i'}^k$. Using (2.31) and $r_{v_k^{\text{dest}}}^k = 0$ we obtain

$$r_{i'}^k = \sum_{(i,j) \in P_{i'}^k} c_{ij}. \tag{2.33}$$

- It follows directly that every other path P from i' to v_k^{dest} in $N(\mathcal{L}')$ with $z_{ij}^k = 1$ for all $(i, j) \in P$ must also satisfy $r_{i'}^k = \sum_{(i,j) \in P} c_{ij}$; hence, it also has length $\sum_{(i,j) \in P_{i'}^k} c_{ij}$.

- Now suppose that $P_{i'}^k \subset N(\mathcal{L}')$ is not a shortest path in $N(\mathcal{L}')$, i.e., there exists a path $P^* \subset N(\mathcal{L}')$ from i' to v_k^{dest} with $\sum_{a \in P^*} c_a < \sum_{a \in P_{i'}^k} c_a$. Then, due to (2.31) and (2.32), we have $r_j^k + c_{ij} - r_i^k \geq 0$ for all $(i, j) \in P^*$. Analogously to (2.33), we obtain the contradiction

$$r_{i'}^k \leq \sum_{(i,j) \in P^*} c_{ij} < \sum_{(i,j) \in P_{i'}^k} c_{ij} = r_{i'}^k.$$

Hence, there can be no path P^* that is shorter than P^k from i' to v_k^{dest} in $N(\mathcal{L}')$. Thus, in particular, every path P from u_k^{org} to v_k^{dest} with $z_a^k = 1$ for all $a \in P$ is the shortest one in $N(\mathcal{L}')$.

Hence, every feasible solution (x, f, y, z, r) to the constraints (2.16)–(2.30) defines a feasible solution $(\mathcal{L}', \mathcal{R})$ to cLPwR. Furthermore,

$$\begin{aligned} c(\mathcal{L}', \mathcal{R}) &= \sum_{(u_k, v_k) \in \mathcal{OD}} \sum_{P \in \mathcal{R}_k} \sum_{a \in P} w_k^P c_a \\ &= \sum_{(u_k, v_k) \in \mathcal{OD}} \sum_{a \in P} c_a \sum_{P \in \mathcal{R}_k : a \in P} w_k^P \\ &= \sum_{(u_k, v_k) \in \text{OD}} \sum_{a \in A(k)} c_a x_a^k. \end{aligned} \tag{2.34}$$

□

Lemma 2.6.7. *Let $I = (N, \mathcal{L}, \mathcal{OD}, B, \text{Cap})$ with $N = (V, A)$ be an instance of cLPwR and let*

$$M_1 \geq \frac{1}{\text{Cap}} \left(\sum_{(u_k, v_k) \in \mathcal{OD}} w_k \right)$$

and for every $(i, j) \in \bigcup_{(u_k, v_k) \in \mathcal{OD}} A(k)$

$$M_2^{ij} \geq \left(\sum_{a \in A} c_a \right) + c_{ij} + 1.$$

Then a solution $(\mathcal{L}', \mathcal{R})$ defined by an optimal solution (x, f, y, r, z) to the IP (2.15)–(2.30) is an optimal solution to I and

$$c(\mathcal{L}', \mathcal{R}) = \sum_{(u_k, v_k) \in \mathrm{OD}} \sum_{a \in A(k)} c_a x_a^k.$$

Proof. We first show that every solution $(\mathcal{L}', \mathcal{R})$ to cLPwR can be transformed to a solution (x, f, y, z, r) that fulfills constraints (2.16)–(2.30). Given $\mathcal{L}'$ and $\mathcal{R}$ we define for all $(u_k, v_k) \in \mathcal{OD}, a \in A(k)$, and $l \in \mathcal{L}$

$x_a^k :=$ number of passengers of OD-pair (u_k, v_k) using arc a in routing $\mathcal{R}$,

$$f_l := \max_{a:l(a)=l} \left\lceil \frac{1}{\text{Cap}} \sum_{(u_k, v_k) \in \mathcal{OD}} x_a^k \right\rceil,$$

$$y_l := \begin{cases} 1 & \text{if } f_l > 0 \\ 0 & \text{if } f_l = 0. \end{cases}$$

For every $(u_k, v_k) \in \mathcal{OD}$ and every i in $V(k)$, we set r_i^k to the shortest path distance of i to v_k^{dest} in $N(\mathcal{L}')$ or to $M := \sum_{a \in A} c_a + 1$ if no such shortest path exists. We furthermore set $z_{ij}^k := 1$ if (i, j) lies on a shortest path from i to v_k^{dest} in $N(\mathcal{L}')$ and $z_{ij}^k := 0$ otherwise.

Since $\mathcal{L}'$ is a feasible line concept, constraints (2.20) are fulfilled. Since $\mathcal{R}$ is a feasible routing for $\mathcal{L}'$ constraints (2.16)–(2.19) and (2.23) are fulfilled. Constraints (2.21) and (2.22) follow directly from the definitions of f, y, and x^k.

Constraints (2.26) are trivially satisfied due to the definition of $r_{v_k^{\text{dest}}}^k$. Also constraints (2.27) are satisfied because x was defined by a routing on shortest paths in $N(\mathcal{L}')$.

It only remains to check constraints (2.24) and (2.25).

2.6 Capacitated Line Planning with Routing

1. First suppose that $(i, j) \in N(\mathcal{L}')$ and that (i, j) is contained in a shortest path from i to v_k^{dest} in $N(\mathcal{L}')$. Hence, $y_{l(i,j)} = 1$, $z_{ij}^k = 1$, and $r_i^k = r_j^k + c_{ij}$. Thus, in this case constraints (2.24) and (2.25) are satisfied.
2. If $(i, j) \in N(\mathcal{L}')$ and (i, j) are contained in a path from i to v_k^{dest} in $N(\mathcal{L}')$, but not in a shortest one, according to the above definition, we have $y_{l(i,j)} = 1$, $z_{ij}^k = 0$ and $r_j^k + c_{ij} - r_i^k \geq 1$ due to the integer arc lengths. Thus

$$r_j^k + c_{ij} - r_i^k + z_{ij}^k + M_2^{ij}(1 - y_{l(i,j)}) = r_j^k + c_{ij} - r_i^k \geq 1$$

and $r_j^k + c_{ij} - r_i^k - (1 - z_{ij}^k)M_2^{ij} \leq r_j^k - r_i^k - \sum_{a \in A} c_a \leq 0$.

3. If $(i, j) \in N(\mathcal{L}')$, but there is no path from j to v_k^{dest} in $N(\mathcal{L}')$, we have $y_{l(i,j)} = 1$, $z_{ij}^k = 0$, and $r_j^k = \sum_{a \in A} c_a + 1$. Hence, since $r_i^k \leq \sum_{a \in A} c_a$,

$$r_j^k + c_{ij} - r_i^k + z_{ij}^k + M_2^{ij}(1 - y_{l(i,j)}) = \sum_{a \in A} c_a + 1 + c_{ij} - r_i^k \geq 1$$

and $r_j^k + c_{ij} - r_i^k - (1 - z_{ij}^k)M_2^{ij} = -r_i^k \leq 0$.

4. If $(i, j) \notin N(\mathcal{L}')$ we have $r_i^k, r_j^k \leq \sum_{a \in A} c_a + 1$ and $y_{l(i,j)} = 0$, and due to constraint (2.23) it follows that $z_{ij}^k = 0$. Hence,

$$r_j^k + c_{ij} - r_i^k + z_{ij}^k + M_2^{ij}(1 - y_{l(i,j)}) \geq \sum_{a \in A} c_a + 1 + c_{ij} - r_i^k \geq 1$$

and $r_j^k + c_{ij} - r_i^k - (1 - z_{ij}^k)M_2^{ij} \leq r_j^k + c_{ij} - \left(\sum_{a \in A} c_a + c_{ij} + 1\right) \leq 0$.

Hence, constraints (2.24) and (2.25) are satisfied.
Furthermore, we see analogously to (2.34) that

$$\sum_{(u_k, v_k) \in \text{OD}} \sum_{a \in A} c_a x_a^k = c(\mathcal{L}', \mathcal{R}).$$

In combination with Lemma 2.6.6, this yields the statement of the lemma. □

Chapter 3
Timetabling

3.1 Introduction to Timetabling

3.1.1 Timetabling Problems

Given a public transportation system and a line concept with frequencies, the next step in public transportation planning is to establish a timetable, i.e., to fix the exact points in time when the trains should arrive and depart at the stations. This decision process is known under the name of *timetabling* or *train scheduling*.

Periodicity

The timetabling literature distinguishes between *periodic* and *aperiodic timetabling*. In *periodic timetabling* it is assumed that the timetable repeats after a (relatively short) time period, e.g., 1 h. For the sake of simplicity in modeling, some periodic timetabling models assume that every train is run once per time period [NV96, CR11, HPB13]. Periodic timetabling models where lines have different frequencies are also possible but require more modeling effort [Odi98]. See also [Lin00, CFT02, Pee03, Lie05, LM07b] for models used in periodic timetabling.

In *aperiodic timetabling* (see, e.g., [WYFL08, FM09, FSZ09, SK09, CCT10, SS10b, SS12b]) in contrast, arrival and departure times are fixed independently for each train of a line. In this book, we consider aperiodic timetabling. For this reason we state definitions and concepts for the timetabling problem with respect to the aperiodic timetabling problem. Many notions can easily be transferred to periodic timetabling, some require more modeling effort.

Timetables

A *timetable* specifies the points in time at which trains arrive and depart at stations. For timetabling purposes, lower bounds on driving times are derived from the track lengths and the train speed as the minimal driving time needed. While some models assume that trains drive at full speed and hence always achieve these driving times (see, e.g., [NV96, Pee03, HPB13]), in most models it is allowed to slow down a train up to an upper bound on the driving time. This upper bound is interpreted as a constraint imposed to secure passengers' acceptance of the public transportation system. Some models (e.g., [LLMS09]), however, do not impose any upper bound on the driving time between stations.

Also for waiting times, lower and upper bounds are given. Again, the latter can be interpreted as an acceptance bound, while the former represents a physical bound on the time needed for passengers to board and alight the train. In the timetabling problem considered in this book, it is furthermore specified between which trains and at which stations it should be possible to transfer from one train to another train, and lower and upper bounds on changing times are given. The time difference between a driving, waiting, or changing time and the corresponding lower bound on its length is called *slack time*.

A timetable is called *feasible*, if it respects capacity restrictions and the upper and lower bounds on driving, waiting, and changing times.

Capacity Restrictions

When planning train schedules, capacity restrictions have to be taken into account: two trains cannot be on the same track or on the same platform at the same time: safety margins between two consecutive trains have to be respected.

Some timetabling models like the ones presented in [NO08, SK09, SS10b, SS12b] do not consider capacity restrictions. Also in this book, we will study the integration of passenger routing in timetabling without capacity constraints.

In other models, precedence relations between trains which use the same track or part of a track are treated as input to the timetabling problem (see, e.g., [Odi98]). In most cases, however, they are considered an important part of the timetabling process.

While some papers consider infrastructural capacity constraints on a macroscopic level [CFT02, CCT10, CR11], there also exist very detailed models of the railway infrastructure; see, e.g., [BLNN98, KP03, Pee03, BK10]. Since precedence decisions tend to make timetabling problems hard, the incorporation of these restrictions is often studied on railway corridors, i.e., a linear PTN with single or dual tracks [BLNN98, CFT02, BK10, HPB13]. Models which consider capacity restrictions on general PTNs can be found, e.g., in [KP03, LM07b].

For more details on reasonable constraints in train timetabling and how to integrate them in the periodic event scheduling framework for periodic timetabling, see [LM07b].

3.1 Introduction to Timetabling

Demand

In the presence of capacity restrictions or periodicity constraints, finding a timetable which respects all constraints is a difficult problem itself. Hence, in some timetabling approaches, passengers are not considered at all [Odi96, Odi98, Lin00, Pee03, LZ05].

To simplify the timetabling problem, almost all models in the literature which take into account the passengers' demand assume that *passenger loads* for each train are given as an input to the problem. That is, it is assumed that the number of passengers who travel in a train between each pair of stations and transfer at each station is known. These numbers can, e.g., be estimated by using a traffic-assignment procedure before starting the construction of the timetable and then summing up passenger loads.

However, in practice, passengers' travel routes depend on the chosen timetable. Hence, similar to the line planning problem, some models assume that demand is given as *OD-pairs* and passenger routing is integrated in the timetabling procedure [Nac98, Kin08, Lüb09, SS10b, SS12b, SG13].

Cordone and Redaelli [CR11] model *variable demand* by a discrete-choice model where the quality of the timetable influences a passenger's choice to travel by train or to use an alternative means of transport. They assume that every passenger is assigned a path in the PTN and a sequence of trains and that he/she can take either this path or an alternative means of transportation; hence in contrast to the above-mentioned models, passenger routing is not incorporated in this model.

Objectives

Since it is already a hard problem to generate a *feasible* periodic timetable, some approaches in periodic timetabling do not formulate an objective at all but concentrate on finding feasible timetables (see, e.g., [Odi96, Odi98, Lin00, LZ05]).

A few cost-oriented models in timetabling aim at the minimization of *costs* related to rolling stock [Lin00, Pee03, LZ05]. Passengers are neglected in these approaches.

Another direction of research is to try to exploit the *infrastructural capacity* as much as possible. To this end, Heydar et al. [HPB13] minimize the length of the period after which the schedule repeats as a primary and the stopping times of the trains at intermediate stations as a secondary objective. Burdett and Kozan [BK10] use a make-span objective to achieve a high utilization of tracks. In [BLNN98, CFT02, FSZ09, CCT10], the distance to an ideal timetable is minimized.

Most timetabling models, however, aim at generating timetables which provide a high quality for the passengers. Objectives are to minimize the passengers' *travel time* [Pee03, Lie06, LPW08, LLMS09, SK09, SS10b, SS12b, SG13], *slack time* [Nac98, NO08], or *waiting time* during the transfers [DV95, WYFL08].

Cordone and Redaelli [CR11] assume that passengers choose an alternative means of transport if the quality of the timetable is not high enough and maximize the total *demand captured*.

Another objective that is receiving increasing attention in the literature is the *robustness* of timetables against disturbances and delays; see, e.g., [ORv06, CDD+07, KMH+08, FSZ09, FM09, SK09, LLMS09, LSS+10, GS10, Sch10].

3.1.2 Literature Overview

Section 3.1.1 gives an overview on different objectives and constraints in timetabling. We complete this overview by a short sketch of different network models, hardness results, and solution techniques used to solve timetabling problems. For more comprehensive surveys on timetabling and train scheduling, see [CTV98, Lie06, LLER11, CT12] and references therein or the survey papers on optimization in public transportation cited in Sect. 1.2.

Timetabling Models

In some timetabling models, the times for departures and arrivals are chosen explicitly. For example, in [DV95], the problem to minimize waiting times in transfer nodes is modeled using one binary variable for every possible departure time in a node. In [CFT02, CCT10], the timetable is given as a set of paths in a graph where the nodes specify arrivals or departures of trains at a given time instant.

However, timetabling problems are mostly modeled using *event-activity networks (EANs)* (also called *constraint graphs* or *disjunctive graphs*), where arrivals and departures are represented as nodes in the network and the timetable is understood as an assignment of times to the nodes (see Sect. 3.2 for the definition of an EAN). Aperiodic versions of these networks can be found in [BLNN98, SK09, SS10b, SS12b].

EANs in periodic timetabling rely on the *periodic event scheduling (PESP)* framework developed by [SU89] and are used for periodic timetabling by, e.g., [Odi96, Nac98, Lin00, Pee03, Lie06, CR11].

When using the EAN as a modeling framework, constraints ensuring feasibility of the timetable (both in the aperiodic and in the periodic case) can be modeled easily by introducing variables for the scheduled times at the events [Lin00, Pee03, Lie06, CDD+07, LPW08, FM09, SK09], for the activity lengths [Lie06, LPW08, Lüb09, SG13], or for the slack times [Nac98, SK09].

When passenger routes are not fixed before the optimization step, but only demand in form of OD-pairs is given, this demand can be included in the EAN by adding additional origin and destination events and activities and routing constraints (see, e.g., [Nac98, Lüb09, SS12b]).

Routing

To simplify the timetabling problem almost all models in the literature which take into account the passengers' demand assume that passengers' routes are fixed before the optimization steps. However, there are some exceptions.

In [Nac98], Nachtigall remarks that passenger routes depend on the choice of the timetable and dedicates one chapter to a periodic timetabling model with integrated passenger routing. In this model, passenger routes are modeled as flows in the EAN. In order to include the passenger flows in the considered periodic timetabling model, flow constraints are added, resulting in a non-convex problem. A decomposition in two linear programs for an aperiodic version with fixed arrival and departure events is given. Furthermore Nachtigall shortly discusses special cases, in particular the case of fixed modulo parameters and fixed departure and arrival events that he proves to be solvable in polynomial time determining passenger flows and slack independently by linear programming.

In [Kin08], a heuristic solution iterating timetabling and routing step is proposed, shown to produce non-increasing objective values, and tested on real-world data. Also in [Lüb09, SG13], the passenger routing is included in the timetabling step by adding variables for passenger flows on arcs which at first leads to a nonlinear formulation, because the total travel time in this case is given as the sum over the product of variable arc length and variable passenger numbers. However, both authors succeed to linearize their formulations by introducing new variables combining both values.

In [Kas10], a heuristic that generates feasible line concepts and timetables simultaneously is used, and the output solution is compared by routing the passengers according to shortest paths and adding up the travel times. In [SS10b, SS12b], the model for aperiodic timetabling with routing (TTwR) considered in this book is investigated.

Computational Complexity

In [Roc84] Rockafellar discusses the problem of finding a feasible aperiodic timetable under the name of *feasible differential problem (FDP)*. He shows that the problem can be solved in polynomial time. Furthermore, it is known that the constraints ensuring feasibility of the timetable for the aperiodic timetabling problem (or the FDP) can be modeled as a polyhedron using a totally unimodular matrix. Thus, for any linear objective function which does not require additional constraints, aperiodic timetabling can be solved by linear programming (see, e.g., [NO08]). Schmidt and Schöbel [SS10b, SS12b] investigate the computational complexity of aperiodic TTwR: the results are presented in Sects. 3.4 and 3.5 of this book.

It is strongly NP-complete to find feasible periodic timetables in the PESP framework as shown in [SU89] by reduction from Hamiltonian circuit. Odijk [Odi98] proved by reduction from graph-coloring that this result holds even for a fixed length of the time period. In [Lie05], Liebchen derives (in-)approximability results for the problem of satisfying as many PESP constraints as possible.

In [CFT02], Caprara et al. show that there is no polynomial-time approximation algorithm for the problem of scheduling trains on a single line such that capacity constraints are fulfilled and the distance to an ideal timetable is minimized, unless $P = NP$.

IP Formulations and Solution Approaches

If no capacity restrictions on tracks or platforms are considered, the problem of finding an aperiodic timetabling can be modeled as an FDP and solved by the *Feasible Differential Algorithm* given in [Roc84].

Furthermore, it is well known (see, e.g., [NO08]) that in case of no capacity constraints and fixed passenger loads on the EAN, a timetable minimizing passengers' travel time can be found in polynomial time by linear programming. The corresponding linear program is given in Sect. 3.3. In contrast to aperiodic timetabling with fixed passenger loads, TTwR is NP-hard and can hence not be solved by linear programming. Two different integer programming formulations are given and compared in [SS12b, Anh12].

Solution approaches for aperiodic timetabling under capacity restrictions include Lagrangian relaxation [BLNN98, CFT02], column generation and separation techniques [CCT10], optimization-based heuristics [CFT02, WYFL08], and metaheuristics [HKF97, BK10]. Different integer programs are listed and compared in [CCT10]. A broader overview about models and solution approaches in aperiodic timetabling is given in [CT12].

Solution techniques for periodic timetabling are in most cases based on the PESP model and the paper [SU89]. To generate feasible periodic timetables, mostly constraint-generation and cutting-plane techniques are used [Odi96, Lin00].

Many integer programming formulations for periodic timetabling [Nac98, Lin00, Pee03, KP03, LZ05] are based on cycle-bases as investigated in Nachtigall's cycle periodicity formulation [Nac98]. A powerful method for solving periodic timetabling is the modulo simplex approach developed by Nachtigall and Opitz [NO08] which was extended in [GS11, GS13]. For the case that passenger loads are not fixed, iterative approaches which solve traffic assignment and timetabling problems, iteratively, are proposed in [Lüb09, CR11, Anh12, SG13]. See [LPW08, CT12, SG13] for comparisons of different solution approaches in periodic timetabling.

3.1.3 The Timetabling Problem Considered in This Book

In this book, we consider an aperiodic timetabling problem with integrated passenger routing which does not take into account capacity restrictions. We now describe the problem in more detail and give some definitions used in the later parts of this chapter.

Timetables

Since in aperiodic timetabling, arrival and departure times are fixed independently for each train of a line, from now on we consider a set of trains TR $:= \{\text{tr}_i^l : l \in \mathcal{L}, i = 1, \ldots, f_l\}$. We denote by $l(\text{tr}_i^l) := l$ the line that a train tr_i^l belongs to.

We denote the arrival of a train tr at a station s with $(\text{tr} - s - \text{Arr})$ and its departure from this station with $(\text{tr} - d - \text{Dep})$.

Definition 3.1.1. Given a public transportation network $\mathcal{PTN} = (S, E)$ and a set of trains TR, an (aperiodic) *timetable* π is a schedule which for every train $\text{tr} \in \text{TR}$ belonging to line $l(\text{tr}) = (s_1, s_2, \ldots, s_j)$ defines a *departure time* $\pi_{(\text{tr}-s_i-\text{Dep})} \in \mathbb{Z}$ for stations s_i with $i = 1, \ldots, j-1$ and an *arrival time* $\pi_{(\text{tr}-s_i-\text{Arr})} \in \mathbb{Z}$ for stations s_i with $i = 2, \ldots, j$.

In this book, we assume that the arrival and departure times are integers which can be interpreted as the time difference to a fixed point in time in minutes. We say that a departure or an arrival is *scheduled* at a certain time t to express that the timetable assigns t to that event.

For a train of line $l = (s_1, s_2, \ldots, s_j) \in L$, the *driving time* from s_i to s_{i+1} is given as $\pi_{(\text{tr}-s_{i+1}-\text{Arr})} - \pi_{(\text{tr}-s_i-\text{Dep})}$ for $i = 1, \ldots, j-1$, and the *waiting time* at station s_i is given as $\pi_{(\text{tr}-s_i-\text{Dep})} - \pi_{(\text{tr}-s_i-\text{Arr})}$ for $i = 2, \ldots, j-1$. To specify between which trains and at which stations it should be possible to transfer from one train to another train, we define a set of *planned connections* $\mathcal{C}$ which consists of triples $(\text{tr}_1, \text{tr}_2, s)$ of trains tr_1, tr_2, and a station s. For every $(\text{tr}_1, \text{tr}_2, s) \in \mathcal{C}$, the *changing time* is defined as $\pi_{(\text{tr}_2-s-\text{Dep})} - \pi_{(\text{tr}_1-s-\text{Arr})}$.

As described in Sect. 3.1.1, we are given upper and lower bounds on driving, waiting, and changing times. Since we do not require periodicity nor consider capacity restrictions, we call a timetable feasible if it respects the upper and lower bounds:

Given a PTN, a set of trains, and a set of planned connections, together with upper and lower bounds on driving, waiting, and changing times, a timetable is *feasible*, if

- For the driving times, it holds that

$$\pi_{(\text{tr}-s'-\text{Arr})} - \pi_{(\text{tr}-s-\text{Dep})} \in [l_{(\text{tr}-s-\text{Dep},\text{tr}-s'-\text{Arr})}, u_{(\text{tr}-s-\text{Dep},\text{tr}-s'-\text{Arr})}]$$

for all $\text{tr} \in \text{TR}$ and $(s, s') \in l(\text{tr})$,

- For the waiting times it holds that

$$\pi_{(\text{tr}-s-\text{Dep})} - \pi_{(\text{tr}-s-\text{Arr})} \in [l_{(\text{tr}-s-\text{Arr},\text{tr}-s-\text{Dep})}, u_{(\text{tr}-s-\text{Arr},\text{tr}-s-\text{Dep})}]$$

for all $\text{tr} \in TR$ and $s \in l(\text{tr})$,
- For the changing times, it holds that

$$\pi_{(\text{tr}'-s-\text{Dep})} - \pi_{(\text{tr}-s-\text{Arr})} \in [l_{(\text{tr}-s-\text{Arr},\text{tr}'-s-\text{Dep})}, u_{(\text{tr}-s-\text{Arr},\text{tr}'-s-\text{Dep})}]$$

for all $(\text{tr}, \text{tr}', s) \in \mathcal{C}$.

Passenger Routes and Travel Time

Like in line planning, our objective in timetabling is to minimize the overall travel time of the passengers. Again, the demand for traveling is given as a set of *OD-pairs* $\mathcal{OD} \subset S \times S$ where for every OD-pair $(u, v) \in \mathcal{OD}$, the *number of passengers* or *passenger weight* w_{uv} is specified.

A passenger's journey can be described in the following way.

Definition 3.1.2. A *train-route* P_{uv} for an OD-pair $(u, v) \in \mathcal{OD}$ specifies

- A path $P = (s_1, s_2, \ldots, s_j)$ with $s_1 := u$ and $s_j := v$ in $\mathcal{PTN}$, and
- The trains $\text{tr}(s_i, \text{Dep})$ which are used to depart at stations s_i for $i = 1, \ldots, j - 1$ and the trains $\text{tr}(s_i, \text{Arr})$ which are used to arrive at stations s_i for $i = 2, \ldots, j$.

Thereby,

- $\text{tr}(s_i, \text{Dep}) = \text{tr}(s_{i+1}, \text{Arr})$ must hold for all $\{s_i, s_{i+1}\} \in P$, and
- At every station s_i, $\text{tr}(s_i, \text{Dep}) = \text{tr}(s_i, \text{Arr})$ or $(\text{tr}(s_i, \text{Arr}), \text{tr}(s_i, \text{Dep}), s_i) \in \mathcal{C}$.

P is called the *path corresponding to P_{uv} in the PTN*.

When the set of OD-pairs is indexed, e.g., $\mathcal{OD} = \{(u_k, v_k) : k \in K\}$ for an index set K, for the sake of simplicity, we also use the notation $w_k := w_{u_k v_k}$ and $P_k := P_{u_k v_k}$.

Let P_{uv} be a train-route and $P = (s_1, \ldots, s_j)$ be the corresponding path in the PTN. For a given timetable π, the *travel time* along P_{uv} is given as the sum over all driving, waiting, and changing times associated to P_{uv}:

$$c(\pi, P_{uv}) := \sum_{\{s_i, s_{i+1}\} \in P} \left(\pi_{(\text{tr}(s_{i+1}, \text{Arr})-s_{i+1}-\text{Arr})} - \pi_{(\text{tr}(s_i, \text{Dep})-s_i-\text{Dep})}\right)$$

$$+ \sum_{s_i \in P} \left(\pi_{(\text{tr}(s_i, \text{Dep})-s_i-\text{Dep})} - \pi_{(\text{tr}(s_i, \text{Arr})-s_i-\text{Arr})}\right). \quad (3.1)$$

In (3.1), the terms for all stations s_i for $i = 2, \ldots, j - 1$ cancel out. As a result, instead of summing up the lengths of all driving, waiting, and changing times of the paths, we can obtain the travel time on a path by just subtracting the departure time at the origin from the arrival time at the destination.

3.1 Introduction to Timetabling

Lemma 3.1.3. *Let P_{uv} be a train-route and $P = (s_1, \ldots, s_j)$ be the path in the PTN associated to P_{uv}. Then, for a given timetable π,*

$$c(\pi, P_{uv}) = \pi_{(\text{tr}(s_j, \text{Arr}) - s_j - \text{Arr})} - \pi_{(\text{tr}(s_1, \text{Dep}) - s_1 - \text{Dep})}.$$

A *train-routing* $\mathcal{R}$ is a set which contains exactly one train-route P_{uv} for every OD-pair $(u, v) \in \mathcal{OD}$, i.e., $\mathcal{R} := \{P_{uv} : (u, v) \in \mathcal{OD}\}$. Given a timetable π and a train-routing $\mathcal{R} = \{P_{uv} : (u, v) \in \mathcal{OD}\}$, the overall *travel time* is defined as

$$c(\pi, \mathcal{R}) := \sum_{(u,v) \in \mathcal{OD}} w_{uv} c(\pi, P_{uv}).$$

Timetabling with Routing

Most traditional approaches in timetabling assume that passenger routes are given in the input, e.g., that they can be predicted before the timetable optimization step by traffic assignment procedures. Then, the goal of timetabling is to calculate a timetable which minimizes the sum of travel times over these given train-routes. This approach has the advantage that in aperiodic timetabling, the determination of optimal (with regard to the predefined passenger routes) timetables can be implemented very efficiently as we see in Sect. 3.3. However, the design of the timetable has a strong influence on travel times on the passenger routes in the transportation system. Hence it is not possible to predict which path is best for a passenger before knowing the timetable. For this reason, also in the timetabling problem, we integrate the routing step and the timetabling step.

Definition 3.1.4. *An instance $(\mathcal{PTN}, \text{TR}, \mathcal{C}, l, u, \mathcal{OD})$ of timetabling with routing (TTwR) consists of a public transportation network $\mathcal{PTN} = (S, E)$; a set of trains TR with specified lines in the PTN; a set of planned connections $\mathcal{C}$; upper and lower bounds u, l on the driving, waiting, and changing times; and a set of OD-pairs $\mathcal{OD}$. The task is to find a feasible timetable π and a train-routing $\mathcal{R}$ which minimize the overall travel time $c(\pi, \mathcal{R})$.*

TTwR was introduced in [SS10b].

In Sect. 3.2 we define the EAN which allows us to visualize timetables and routings much better. Similarly to the proceeding in Chap. 2, we restate the definitions given in this section there in terms of the EAN.

3.1.4 Outline of the Timetabling Chapter

To model TTwR, in Sect. 3.2 we extend the EAN, often used for timetabling problems, by adding *origin* and *destination events* which model the passengers' origins and destinations. In Sect. 3.3 we summarize shortly how TTwR can be solved in polynomial time when either the routing or the timetable is fixed. Contrasting this result, in Sect. 3.4.1 we show strong NP-hardness of TTwR using a reduction from Set Cover which also allows us to transfer inapproximability results from the Set Cover problem to TTwR in Sect. 3.4.2. Furthermore, we show that we can easily find a solution to TTwR whose approximation ratio is bounded in terms of the lower and upper bounds on the activities.

In Sect. 3.5 we define the problem *timetabling with routing between events (TTwRE)*. Although this problem looks quite similar to TTwR, it turns out to be solvable in polynomial time. We shortly discuss how this result can be used to solve TTwR.

Based on a model for aperiodic timetabling with fixed routing, in Sect. 3.7 we propose two integer programming formulations for TTwR. The first formulation is based on multi-commodity flow constraints to model the passenger routes. The second integer programming formulation uses a different approach, exploiting the insights of Sect. 3.5.

3.2 Modeling Timetabling with Routing Using Event-Activity Networks

To determine the arrival and departure times for all vehicles at all stations, the information coded in the change-and-go network, which is used to model line planning problems in Chap. 2, is not enough. Instead we use a modified event-activity network (EAN). EANs are widely used in the literature for timetabling problems; see, e.g., [Nac98,Lie06]. To include passengers' origins and destinations in the standard EAN model for timetabling, *origin* and *destination events* which model the passengers' origins and destinations can be added (see, e.g., [Nac98, Lüb09, SS10b, SS12b]).

Given a public transportation network $\mathcal{PTN}$, a set of trains TR, a set of planned connections $\mathcal{C}$, and a set of OD-pairs $\mathcal{OD}$, we construct an EAN as explained in the following.

The nodes of the network are called *events*. They mainly represent the arrivals and departures of the trains at stations.

Definition 3.2.1. Let S be the node set of $\mathcal{PTN}$. We define

- The set of *departure events* $\mathcal{E}_{\mathrm{dep}} := \{(\mathrm{tr} - s - \mathrm{Dep}) : \mathrm{tr} \in \mathrm{TR}, s \in l(\mathrm{tr})\}$ and
- The set of *arrival events* $\mathcal{E}_{\mathrm{arr}} := \{(\mathrm{tr} - s - \mathrm{Arr}) : \mathrm{tr} \in \mathrm{TR}, s \in l(\mathrm{tr})\}$.

The set of *operational events* is $\mathcal{E}_{\mathrm{op}} := \mathcal{E}_{\mathrm{dep}} \cup \mathcal{E}_{\mathrm{arr}}$.

We add nodes to the network which represent the passengers' origins and destinations.

Definition 3.2.2. We define

- The set of *origin events* as $\mathcal{E}_{\mathrm{org}} := V_{\mathrm{org}} = \{u^{\mathrm{org}} = org(u,v) : (u,v) \in \mathcal{OD}\}$, and
- The set of *destination events* as $\mathcal{E}_{\mathrm{dest}} := V_{\mathrm{dest}} = \{v^{\mathrm{dest}} = dest(u,v) : (u,v) \in \mathcal{OD}\}$.

3.2 Modeling Timetabling with Routing Using Event-Activity Networks

The arcs of the EAN are called *activities* since they represent physical activities of the trains, i.e., driving between stations and waiting at stations to let passengers get on and off or, in case of the changing activities, the movement of passengers from one train to another.

Definition 3.2.3. We define

- The set of *driving activities* $\mathcal{A}_{\text{drive}} \subset \mathcal{E}_{\text{dep}} \times \mathcal{E}_{\text{arr}}$ as

$$\mathcal{A}_{\text{drive}} := \{(\text{tr} - s - \text{Dep}, \text{tr} - s' - \text{Arr}) : \text{tr} \in \text{TR}, (s, s') \in l(\text{tr})\},$$

- The set of *waiting activities* $\mathcal{A}_{\text{wait}} \subset \mathcal{E}_{\text{arr}} \times \mathcal{E}_{\text{dep}}$ as

$$\mathcal{A}_{\text{wait}} := \{(\text{tr} - s - \text{Arr}, \text{tr} - s - \text{Dep}) : \text{tr} \in \text{TR}, s \in l(\text{tr})\},$$

- The set of *changing activities* $\mathcal{A}_{\text{change}} \subset \mathcal{E}_{\text{arr}} \times \mathcal{E}_{\text{dep}}$ as

$$\mathcal{A}_{\text{change}} := \{(\text{tr} - s - \text{Arr}, \text{tr}' - s - \text{Dep}) : (\text{tr}, \text{tr}', s) \in \mathcal{C}\}.$$

We denote by $\mathcal{A}_{\text{op}} := \mathcal{A}_{\text{drive}} \cup \mathcal{A}_{\text{wait}} \cup \mathcal{A}_{\text{change}}$ the set of *operational activities*.

The origin events are connected by means of *origin activities* to all departure events at the corresponding station; analogously, the destination events are connected by *destination activities* to corresponding arrival events.

Definition 3.2.4. We define

- The set of *origin activities* $\mathcal{A}_{\text{org}} \subset \mathcal{E}_{\text{org}} \times \mathcal{E}_{\text{dep}}$ as

$$\mathcal{A}_{\text{org}} := \{(u^{\text{org}}, \text{tr} - u - \text{Dep}) : u^{\text{org}} \in \mathcal{E}_{\text{org}}, (\text{tr} - u - \text{Dep}) \in \mathcal{E}_{\text{dep}}\},$$

- And the set of *destination activities* $\mathcal{A}_{\text{dest}} \subset \mathcal{E}_{\text{arr}} \times \mathcal{E}_{\text{dest}}$ as

$$\mathcal{A}_{\text{dest}} := \{(\text{tr} - v - \text{Arr}, v^{\text{dest}}) : v^{\text{dest}} \in \mathcal{E}_{\text{dest}}, (\text{tr} - v - \text{Arr}) \in \mathcal{E}_{\text{arr}}\}.$$

The EAN is defined as $\mathcal{N} := (\mathcal{E}, \mathcal{A})$ with

$$\mathcal{E} := \mathcal{E}_{\text{op}} \cup \mathcal{E}_{\text{org}} \cup \mathcal{E}_{\text{dest}} \quad \text{and} \quad \mathcal{A} := \mathcal{A}_{\text{op}} \cup \mathcal{A}_{\text{org}} \cup \mathcal{A}_{\text{dest}}.$$

For $e \in \mathcal{E}_{\text{dep}} \cup \mathcal{E}_{\text{arr}}, a \in \mathcal{A}_{\text{drive}} \cup \mathcal{A}_{\text{wait}}$, let tr($e$) or tr($a$) denote the train that e or a, respectively, belongs to.

Since there are no fixed edge lengths, the EAN is not a network in the sense of Definition 1.3.3, but an directed graph. However, we use the term "EAN" or "event-activity network," since this is the common term in the literature.

We call $\mathcal{N}_{\text{op}} := (\mathcal{E}_{\text{op}}, \mathcal{A}_{\text{op}})$ the *operational event-activity network/EAN*. In timetabling problems where a fixed routing or fixed weights on the activities are part of the input, origin and destination events and activities are not needed, and the operational EAN is used and simply referred to as "EAN" (see, e.g.,[Odi96, Nac98, Lin00, Pee03, Lie06]).

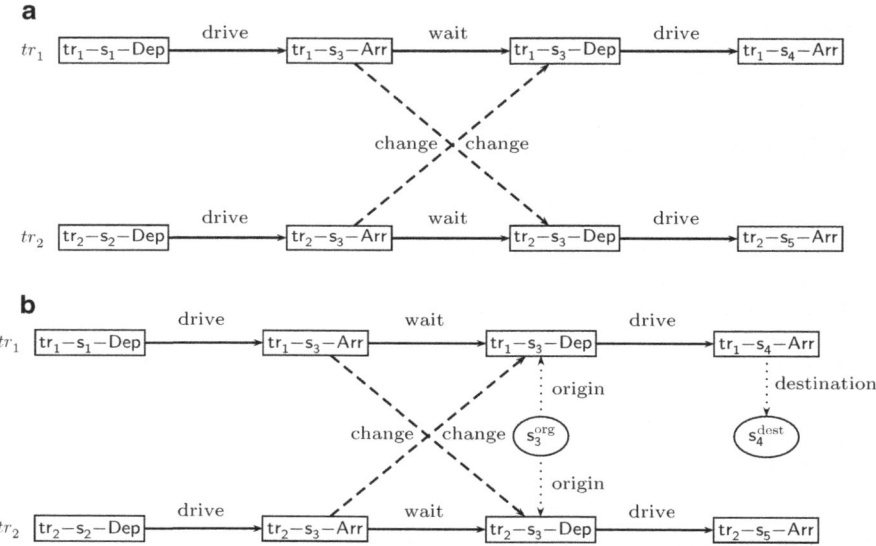

Fig. 3.1 Cutout of an EAN. (**a**) A station with two passing trains and changing possibilities between the trains. (**b**) Additional origin and destination events

Figure 3.1 shows a cutout of an EAN. A station which is visited by two trains is shown. The trains drive into the station, arrive there, wait for passengers to get on and off, and then depart again and drive to the next station. In the station, changing between both trains is possible which is represented by the dashed lines. Note that changing activities always take place between the arrival of a train at a station and the departure of another train at the same station. In Fig. 3.1a, only the operational network is shown. In Fig. 3.1b, origin and destination nodes and arcs for OD-pair (s_3, s_4) are added.

It is easy to see that given an instance $(\mathcal{PTN}, \text{TR}, \mathcal{C}, l, u, \mathcal{OD})$ of TTwR, the corresponding EAN is of size $O(|S| \cdot |\text{TR}| + |\mathcal{C}|)$. The time needed for its construction is of the same order.

We can now understand a timetable as an assignment $\pi : \mathcal{E}_{\text{op}} \to \mathbb{Z}$ of points in time to the nodes in the network. The driving, waiting, and changing times can be understood as the arc lengths $L_{ij} := \pi_j - \pi_i$ for $(i, j) \in \mathcal{A}_{\text{op}}$ resulting from the timetable. We call the network $\mathcal{N}$ with arc lengths

$$L_{ij} := \begin{cases} \pi_j - \pi_i & \text{if } (i, j) \in \mathcal{A}_{\text{op}}, \\ 0 & \text{otherwise.} \end{cases}$$

routing network $\mathcal{N}(\pi)$. The *slack time* for an activity $(i, j) \in \mathcal{A}_{\text{op}}$ can now be calculated as $s_{ij} := L_{ij} - l_{ij}$ where l_{ij} denotes the lower bound on (i, j). By induction, the following lemma can be shown.

Lemma 3.2.5. *Let P be a path in $\mathcal{N}$ from i' to j'. Then*

$$\pi_{j'} = \pi_{i'} + \sum_{(i,j) \in P} (l_{ij} + s_{ij}).$$

Given the time bounds on the driving, waiting, and changing times, expressed in terms of the EAN, a timetable π is *feasible* if

$$L_{ij} = \pi_j - \pi_i \in [l_{ij}, u_{ij}] \quad \forall (i,j) \in \mathcal{A}_{\text{op}}. \tag{3.2}$$

Let an instance $(\mathcal{PTN}, \text{TR}, \mathcal{C}, l, u, \mathcal{OD})$ of TTwR be given, and let $\mathcal{N} = (\mathcal{E}, \mathcal{A})$ be the corresponding EAN. For an OD-pair $(u_k, v_k) \in \mathcal{OD}$, every possible train-route is represented by a path $P_k = (u_k^{\text{org}}, i_1^k, \ldots, i_{n_k}^k, v_k^{\text{dest}})$ from u_k^{org} to v_k^{dest} in $\mathcal{N}$ and vice versa.

Given a timetable π, we can calculate the *travel time* (or *length*) of P_k as the time difference between departure at the origin and arrival at the destination, i.e.,

$$c(\pi, P_k) = \pi_{i_{n_k}^k} - \pi_{i_1^k} \tag{3.3}$$

according to Lemma 3.1.3.

We now transfer the definition of a train-routing to the EAN: A *routing* in the EAN is a collection $\mathcal{R} := \{P_k : (u_k, v_k) \in \mathcal{OD}\}$ with P_k being a path from u_k^{org} to v_k^{dest} in the EAN. Then, as defined in (3.1), given a timetable π and a routing $\mathcal{R}$, the overall *travel time* is calculated as

$$c(\pi, \mathcal{R}) = \sum_{P_k \in \mathcal{R}} w_{uv} c(\pi, P_k). \tag{3.4}$$

We now give an alternative definition of TTwR which is equivalent to Definition 3.1.4. We use this definition to simplify notation in the algorithms, integer programming approaches, and hardness proofs in the following sections.

Definition 3.2.6. *An instance $(\mathcal{N}, l, u, \mathcal{OD})$ of TTwR consists of an EAN $\mathcal{N} = (\mathcal{E}, \mathcal{A})$; upper and lower bounds u, l on the lengths of the operational activities; and a set of OD-pairs $\mathcal{OD}$. The task is to find a feasible timetable π and a routing $\mathcal{R}$ that minimize the overall travel time $c(\pi, \mathcal{R})$.*

3.3 Solving Timetabling When Part of the Solution Is Fixed

In contrast to the hardness result for uLPwR in Theorem 2.5.1, for an EAN with upper and lower bounds, a feasible timetable can be found in polynomial time using, e.g., the *Feasible Differential Algorithm* given in [Roc84].

Furthermore, it is well known (see, e.g., [NO08]) that for a fixed routing $\mathcal{R}$, an optimal timetable $\pi(\mathcal{R})$ can be found in polynomial time. Since we use this result in Sect. 3.5, we shortly sketch it here.

Definition 3.3.1. An instance $(\mathcal{N}, l, u, \mathcal{R})$ of the *simplified timetabling problem (sTT)* consists of an EAN $\mathcal{N} = (\mathcal{E}, \mathcal{A})$, upper and lower bounds u, l on the lengths of the operational activities, and a routing $\mathcal{R}$ (with passenger weights w_k for every $P_k \in \mathcal{R}$).

The task is to find a feasible timetable $\pi := \pi(\mathcal{R})$ such that

$$c(\pi, \mathcal{R}) \leq c(\pi', \mathcal{R})$$

for all feasible timetables π' on $\mathcal{N}$.

Lemma 3.3.2. *Given an EAN $\mathcal{N} = (\mathcal{E}, \mathcal{A})$, a routing $\mathcal{R}$, and a timetable π, we define activity weights $w_a := \sum_{P_k \in \mathcal{R}} \sum_{P_k \ni a} w_{u_k v_k}$. Then the overall travel time can be calculated as*

$$c(\pi, \mathcal{R}) = \sum_{(i,j) \in \mathcal{A}_{op}} w_{ij}(\pi_j - \pi_i).$$

Proof. For every $P_k \in \mathcal{R}$, let i^k denote the first departure event on path P^k and j^k the last arrival event. Then we have

$$c(\pi, \mathcal{R}) = \sum_{P_k \in \mathcal{R}} w_k c(\pi, P_k)$$

$$= \sum_{P_k \in \mathcal{R}} w_k (\pi_{j^k} - \pi_{i^k})$$

$$= \sum_{P_k \in \mathcal{R}} \left(w_k \sum_{(i,j) \in P_k} (\pi_j - \pi_i) \right)$$

$$= \sum_{P_k \in \mathcal{R}} \left(\sum_{(i,j) \in P_k} w_k (\pi_j - \pi_i) \right)$$

$$= \sum_{(i,j) \in \mathcal{A}_{op}} \left(\sum_{P_k \in \mathcal{R}: P_k \ni (i,j)} w_k (\pi_j - \pi_i) \right)$$

$$= \sum_{(i,j) \in \mathcal{A}_{op}} \left((\pi_j - \pi_i) \sum_{P_k \in \mathcal{R}: P_k \ni (i,j)} w_k \right)$$

$$= \sum_{(i,j) \in \mathcal{A}_{op}} w_{ij} (\pi_j - \pi_i).$$

□

3.3 Solving Timetabling When Part of the Solution Is Fixed

Note that in most traditional timetabling models, the travel demand is even directly given in the form of activity weights w_a for $a \in \mathcal{A}_{\text{op}}$ instead of providing a routing (or a set of OD-pairs) (see, e.g., [Nac98, Lie06]).

Due to Lemma 3.3.2, we can now formulate sTT as an integer program, using the variables

$$\pi_i := \text{scheduled time at event } i \quad \forall\, i \in \mathcal{E}_{\text{op}}.$$

The IP reads

$$\min \sum_{(i,j)\in\mathcal{A}_{\text{op}}} w_{ij}(\pi_j - \pi_i), \tag{3.5}$$

$$\text{s.t.} \quad \pi_j - \pi_i \le u_a \qquad \forall (i,j) \in \mathcal{A}_{\text{op}}, \tag{3.6}$$

$$\pi_j - \pi_i \ge l_a \qquad \forall (i,j) \in \mathcal{A}_{\text{op}}, \tag{3.7}$$

$$\pi_i \in \mathbb{Z} \qquad \forall i \in \mathcal{E}_{\text{op}}, \tag{3.8}$$

where (3.5) is the objective function from Lemma 3.3.2 and (3.6) and (3.7) are the feasibility constraints for the timetable.

The constraint matrix of (3.5)–(3.8) is the transposed of a node–arc incidence matrix and thus totally unimodular. Thus, when solving the LP-relaxation of problem (3.5)–(3.8) for integer bounds, the resulting timetable is also integer:

Lemma 3.3.3. *sTT can be solved in polynomial time by linear programming.*

Definition 3.3.4. Let $I = (\mathcal{N}, l, u, \mathcal{OD})$ be an instance of TTwR and let $\mathcal{R}$ be a routing.

- The set of *timetables corresponding to $\mathcal{R}$* is defined as

  $\text{OPT}(\mathcal{R}) := \{\pi(\mathcal{R}) : \text{ is an optimal solution to the instance } I = (\mathcal{N}, l, u, \mathcal{R}) \text{ of sTT}\}.$

- The set of *solutions corresponding to $\mathcal{R}$* is $\{(\pi(\mathcal{R}), \mathcal{R}) : \pi(\mathcal{R}) \in \text{OPT}(\mathcal{R})\}$.

Due to the definition of problem sTT, we have the following corollary.

Corollary 3.3.5. *Let $I = (\mathcal{N}, l, u, \mathcal{OD})$ be an instance of TTwR and let $\mathcal{R}$ be a routing. For every $\pi(\mathcal{R}) \in OPT(\mathcal{R})$, it holds that*

$$c(\pi(\mathcal{R}), \pi) \le c(\pi', \mathcal{R})$$

for all feasible timetables π' in $\mathcal{N}$.

On the other hand, for a *given timetable π*, it is even easier to calculate a corresponding routing $\mathcal{R}$.

Definition 3.3.6. Let $I = (\mathcal{N}, l, u, \mathcal{OD})$ be an instance of TTwR and let π be a timetable in $\mathcal{N}$. For every OD-pair (u_k, v_k), let $\mathcal{P}_k(\pi) = \mathcal{P}_{u_k v_k}(\pi)$ denote the set of shortest paths from u_k^{org} to v_k^{dest} in the network $\mathcal{N}(\pi)$.

- The set of *routings corresponding to* π is defined as

$$\text{OPT}(\pi) := \{\mathcal{R}(\pi) : \mathcal{R}(\pi) = \{P_1(\pi), \ldots, P_{|\mathcal{OD}|}(\pi)\} \text{ with } P_k(\pi) \in \mathcal{P}_k(\pi), k = 1, \ldots, |\mathcal{OD}|\}.$$

- The set of *solutions corresponding to* π is defined as $\{(\pi, \mathcal{R}(\pi)) : \mathcal{R}(\pi) \in \text{OPT}(\pi)\}$.

Since $\mathcal{R}(\pi)$ is defined as a shortest-path routing, Lemma 3.3.7 follows directly.

Lemma 3.3.7. Let $I = (\mathcal{N}, l, u, \mathcal{OD})$ be an instance of TTwR and let π be a timetable in $\mathcal{N}$. For every $\mathcal{R}(\pi) \in \text{OPT}(\pi)$, it holds that

$$c(\pi, \mathcal{R}(\pi)) \leq c(\pi, \mathcal{R}')$$

for all routings R' for $\mathcal{OD}$.

Let $I = (\mathcal{N}, l, u, \mathcal{OD})$ be an instance of TTwR. In the following we sometimes describe a solution to I stating only

- a timetable π, referring to a corresponding full solution $(\pi, \mathcal{R}(\pi))$ for any $\mathcal{R}(\pi) \in \text{OPT}(\pi)$, or
- a routing $\mathcal{R}$, referring to a corresponding full solution $(\pi(\mathcal{R}), \mathcal{R})$ for any $\pi(\mathcal{R}) \in \text{OPT}(\mathcal{R})$.

3.4 Complexity of Timetabling with Routing

3.4.1 NP-Hardness of Timetabling with Routing

In this section we present a proof from [SS12b] which shows that, in contrast to the polynomial solvability of sTT, TTwR is strongly NP-hard.

Theorem 3.4.1. *TTwR is strongly NP-hard, even if*

- *all OD-pairs have the same origin, and*
- *the changing time is the same for all connections.*

Theorem 3.4.1 is shown by reduction from the problem *Set Cover*. An instance (P, Q, K) of the decision problem *Set Cover* (called *Minimum Cover* in [GJ79]) consists of a set $P = \{p_1, \ldots, p_m\}$, a set of subsets $Q = \{q^1, \ldots, q^n\}$ with $q^j \subset P$, and a natural number K. The question to decide is whether there is a subset $Q' \subset Q$ with $|Q'| \leq K$ such that for every $p_i \in P$, there is at least one $q^j \in Q'$ such that p_i is contained in q^j. The decision problem Set Cover is strongly NP-complete [GJ79].

3.4 Complexity of Timetabling with Routing

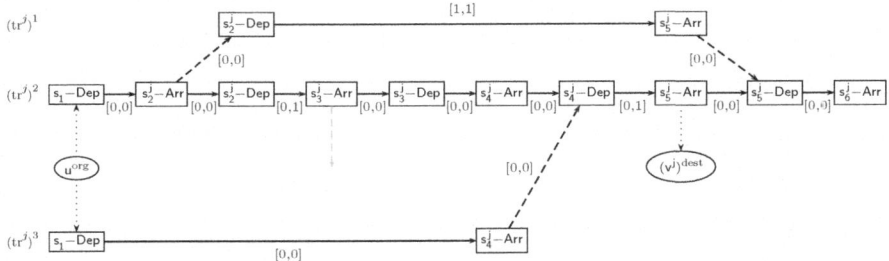

Fig. 3.2 The gadget g^j representing a set q^j in the reduction from Set Cover to TTwR in the proof of Theorem 3.4.1

Proof. Let (P, Q, K) be an instance of Set Cover with $P = \{p_1, \ldots, p_m\}$ and $Q = \{q^1, \ldots, q^n\}$. Without loss of generality we assume that $K \leq n$. Based on P and Q we construct an instance of TTwR in the following way:

We represent the elements $p_i \in P$ by OD-pairs (u, v_i) with passenger numbers $w_{uv_i} := n + 1$ where n denotes the number of sets in Q.

The sets $q^j \in Q$ are represented by a gadget consisting of three trains $(\mathrm{tr}^j)^1$, $(\mathrm{tr}^j)^2$, and $(\mathrm{tr}^j)^3$, six stations $s_1, s_2^j, s_3^j, s_4^j, s_5^j$, and s_6^j (where s_1 is the same station for all OD-pairs); and an OD-pair (u, v^j) with $u := s_1$ and $v^j := s_5^j$ and $w_{uv^j} := 1$, in the way depicted in Fig. 3.2.

The square nodes are the departure and arrival events where the train names are omitted in the node labels for the sake of a compact representation. The origin and destination events are represented by ovals. The dotted lines are the origin and destination activities, the solid lines represent driving and waiting activities, changing activities are represented by dashed lines. The gray line indicates where it is possible to leave this gadget.

For every $p_i \in q^j$, we introduce a train tr_i^j running from s_3^j to a station v_i and connect $((\mathrm{tr}^j)^2 - s_3^j - \mathrm{Arr})$ to a departure event $(\mathrm{tr}_i^j - s_3^j - \mathrm{Dep})$ by a changing activity with upper and lower bound $u_{((\mathrm{tr}^j)^2 - s_3^j - \mathrm{Arr}), (\mathrm{tr}_i^j - s_3^j - \mathrm{Dep})} := 0$ and $l_{((\mathrm{tr}^j)^2 - s_3^j - \mathrm{Arr}), (\mathrm{tr}_i^j - s_3^j - \mathrm{Dep})} := 0$. $(\mathrm{tr}_i^j - s_3^j - \mathrm{Dep})$ is connected to the arrival event of train tr_i^j in v_i by a driving activity a with upper and lower bound $l_a := 0, u_a := 0$.

See Fig. 3.3 (in combination with Fig. 3.2) for an example of the construction for an instance of Set Cover with $P = \{1, 2, 3, 4\}$ and $Q = \{\{2, 3, 4\}, \{1, 4\}, \{2, 3\}\}$. The square nodes are the departure and arrival events. The origin and destination events are represented by ovals. The dotted lines are the origin and destination activities, the solid lines represent driving and waiting activities, changing activities are represented by dashed lines. The nodes g^j represent the gadgets from Fig. 3.2.

Summarizing, our instance I of TTwR is defined by the following input, which defines an EAN N with upper and lower bounds l, u and a set of OD-pairs $\mathcal{OD}$:

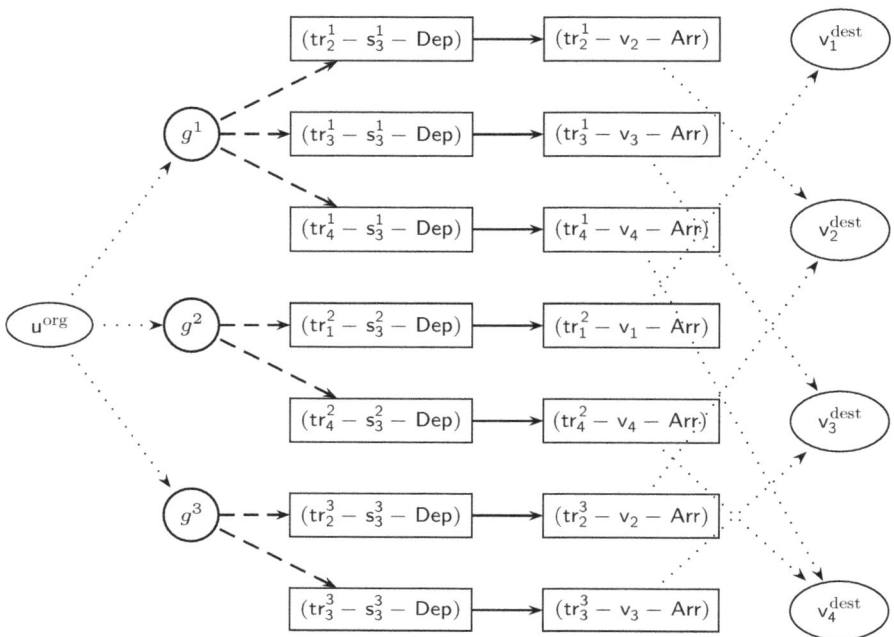

Fig. 3.3 Example for the construction of the EAN in the proof of Theorem 3.4.1

- a set of stations $S = \{s_1\} \cup \{s_2^j, s_3^j, s_4^j, s_5^j : j = 1, \ldots, n\} \cup \{v_i : i = 1, \ldots, m\}$,
- a set of trains $\text{TR} = \{(\text{tr}^j)^1, (\text{tr}^j)^2, (\text{tr}^j)^3 : j = 1, \ldots, n\} \cup \{\text{tr}_i^j : p_i \in q^j\}$ with $l((\text{tr}^j)^1) = (s_2^j, s_5^j)$, $l((\text{tr}^j)^2) = (s_1, s_2^j, s_3^j, s_4^j, s_5^j)$, $l((\text{tr}^j)^3) = (s_1, s_4^j)$ and $l(\text{tr}_i^j) = (s_3^j, v_i)$.
- connections

$$\mathcal{C} = \{((\text{tr}^j)^2, (\text{tr}^j)^1, s_2^j), ((\text{tr}^j)^1, (\text{tr}^j)^2, s_5^j), ((\text{tr}^j)^3, (\text{tr}^j)^2, s_4^j) : j = 1, \ldots, n\}$$
$$\cup \{((\text{tr}^j)^2, \text{tr}_i^j, s_3^j) : p_i \in q^j\}$$

- a lower bound $l_{((\text{tr}^j)^1 - s_2^j - \text{Dep}, (\text{tr}^j)^1 - s_5^j - \text{Arr})} = 1$ on driving activities $((\text{tr}^j)^1 - s_2^j - \text{Dep}, (\text{tr}^j)^1 - s_5^j - \text{Arr})$ for $j = 1, \ldots, n$ and lower bounds $l_a := 0$ on all other activities. The upper bounds are $u_a = 1$ for

$$a \in \{((\text{tr}^j)^1 - s_2^j - \text{Dep}, (\text{tr}^j)^1 - s_5^j - \text{Arr}),$$
$$((\text{tr}^j)^2 - s_2^j - \text{Dep}, (\text{tr}^j)^2 - s_3^j - \text{Arr}),$$
$$((\text{tr}^j)^2 - s_4^j - \text{Dep}, (\text{tr}^j)^2 - s_5^j - \text{Arr}) : j = 1, \ldots, n\}$$

and $u_a := 0$ otherwise.

3.4 Complexity of Timetabling with Routing

- the set of OD-pairs is $\mathcal{OD} = \{(u, v^j) : j = 1, \ldots, n\} \cup \{(u, v_i) : i = 1, \ldots, m\}$ with $u := s_1$ and $v^j := s_5^j$. The weights are $w_{uv^j} = 1$ and $w_{uv_i} = n + 1$.

We show that there is a solution $(\pi, \mathcal{R})$ to the constructed instance I with $c(\pi, \mathcal{R}) \leq K$, if and only if there is a subset Q' solving the Set Cover problem.

1. As a first step we show that given a solution $(\pi, \mathcal{R})$ to I with $c(\pi, \mathcal{R}) \leq K$, we can construct a solution $Q(\pi, \mathcal{R})$ to (P, Q) with $|Q(\pi, \mathcal{R})| \leq c(\pi, \mathcal{R}) \leq K$:

   Let (π, R) be a solution to I with $c(\pi, \mathcal{R}) \leq K$. Since $w_{uv_i} = n + 1$, this implies that $c(\pi, P_i) = 0$, for the path P_i defined in $\mathcal{R}$ for (u, v_i), since otherwise,

$$c(\pi, \mathcal{R}) \geq w_{uv_i} c(\pi, P_i) \geq n + 1 > K.$$

Having a look at the possible paths in the EAN N constructed for our instance I (see Figs. 3.2 and 3.3), we conclude that for every i there is at least one j such that $\pi_{((\text{tr}^j)^2 - s_3^j - \text{Arr})} - \pi_{((\text{tr}^j)^2 - s_2^j - \text{Dep})} = 0$ and such that there exists a train tr_i^j.

With

$$Q(\pi, \mathcal{R}) := \{q^j : \pi_{((\text{tr}^j)^2 - s_3^j - \text{Arr})} - \pi_{((\text{tr}^j)^2 - s_2^j - \text{Dep})} = 0\},$$

due to the construction of our instance I, this is equivalent to saying that for every p_i, there is at least one $q^j \in Q(\pi, \mathcal{R})$ such that $p_i \in q^j$.

Furthermore, we note that if and only if

$$\pi_{((\text{tr}^j)^2 - s_3^j - \text{Arr})} - \pi_{((\text{tr}^j)^2 - s_2^j - \text{Dep})} = 0$$

it follows that

$$\pi_{((\text{tr}^j)^2 - s_5^j - \text{Arr})} - \pi_{((\text{tr}^j)^2 - s_4^j - \text{Dep})} = 1,$$

due to the bounds on activity $((\text{tr}^j)^1 - s_2^j - \text{Dep}, (\text{tr}^j)^1 - s_5^j - \text{Arr})$. For every j with $\pi_{((\text{tr}^j)^2 - s_5^j - \text{Arr})} - \pi_{((\text{tr}^j)^2 - s_4^j - \text{Dep})} = 1$, we see that $c(\pi, P^j) = 1$ for the path P^j for OD-pair (u, v^j). This leads to

$$|Q(\pi, \mathcal{R})| = |\{q^j : \pi_{((\text{tr}^j)^2 - s_3^j - \text{Arr})} - \pi_{((\text{tr}^j)^2 - s_2^j - \text{Dep})} = 0\}|$$

$$= |\{j : \pi_{((\text{tr}^j)^2 - s_5^j - \text{Arr})} - \pi_{((\text{tr}^j)^2 - s_4^j - \text{Dep})} = 1\}|$$

$$\leq \sum_{(u, v^j) \in \mathcal{OD}} c(\pi, P^j)$$

$$\leq c(\pi, \mathcal{R})$$

$$\leq K.$$

2. Now we show that given a solution Q' to (P, Q), we can construct a solution $(\pi(Q'), \mathcal{R}(Q'))$ to I with $c(\pi(Q'), \mathcal{R}(Q')) = |Q'|$:
   Let Q' be a solution to (P, Q, K). We set

$$\pi(Q')_{((\text{tr}^j)^2 - s_1 - \text{Dep})} := 0 \text{ and}$$

$$\pi(Q')_{((\text{tr}^j)^2 - s_3^j - \text{Dep})} := \begin{cases} 0 \text{ if } q^j \in Q' \\ 1 \text{ otherwise.} \end{cases} \quad (3.9)$$

This uniquely defines a timetable $\pi(Q')$ on N.

- For $i = 1, \ldots, m$, we choose a j_i such that $p_i \in q^{j_i}$ and $q^{j_i} \in Q'$ and define the train-route $P_i(Q')$ for (u, v_i) to go via gadget g^{j_i}. This defines $P_i(Q')$ uniquely. Then,

$$c(\pi(Q'), P_i(Q')) = \pi(Q')_{(\text{tr}_i^{j_i} - v_i - \text{Arr})} - \pi(Q')_{(\text{tr}^{j_i})^2 - s_1 - \text{Dep})} = 0.$$

For $j = 1, \ldots, n$, we choose the train-route $P^j(Q')$ defined by taking first $(\text{tr}^j)^3$ and then changing to $(\text{tr}^j)^2$ in s_4^j. Then

$$c(\pi(Q'), P^j(Q')) = \pi(Q')_{((\text{tr}^j)^2 - s_5^j - \text{Arr})} - \pi(Q')_{((\text{tr}^j)^1 - s_1 - \text{Dep})}$$

$$= \begin{cases} 1 - 0 = 1 \text{ if } \pi(Q')_{((\text{tr}^j)^1 - s_1 - \text{Dep})} = 1 \\ 1 - 1 = 0 \text{ otherwise.} \end{cases}$$

$$= \begin{cases} 1 \text{ if } \pi(Q')_{((\text{tr}^j)^2 - s_3 - \text{Arr})} = 1 \\ 0 \text{ otherwise.} \end{cases}$$

Hence

$$c(\pi(Q'), \mathcal{R}(Q')) = \sum_{P_i \in \mathcal{R}} w_{uv_i} c(\pi(Q'), P_i(Q')) + \sum_{P^j \in \mathcal{R}} w_{uv^j} c(\pi(Q'), P^j(Q'))$$

$$= 0 + |\{j : \pi(Q')_{((\text{tr}^j)^2 - s_3 - \text{Arr})} = 1\}|$$

$$= |Q'|.$$

□

3.4.2 Inapproximability of Timetabling with Routing

The proof of Theorem 3.4.1 can be extended to show an inapproximability result for TTwR. To this end, we consider an optimization version of Set Cover. An instance (P, Q) of the optimization version of *Set Cover* consists of a set $P = \{p_1, \ldots, p_m\}$ and a set of subsets $Q = \{q^1, \ldots, q^n\}$ with $q^j \subset P$. The task is to find a subset $Q' \subset Q$ of minimal size $|Q'|$ such that for every $p_i \in P$, there is at least one $q^j \in Q'$ such that p_i is contained in q^j [Fei98].

3.4 Complexity of Timetabling with Routing

There exist many inapproximability results for Set Cover; see, e.g., [BGLR93, RS97, Fei98, AMS06] and references therein. Due to the one-to-one correspondence between solutions and objective values of the optimization version of Set Cover and the instances of TTwR constructed in the proof of Theorem 3.4.1, any approximation algorithm that solves TTwR with a certain approximation ratio could be used to solve Set Cover with a similar approximation ratio. Thus, inapproximability results for Set Cover can easily be transferred to TTwR.

To give an example how this is done, in Theorem 3.4.2, we transfer a result from [BGLR93] which says that there is no constant-factor polynomial-time approximation algorithm for Set Cover unless P = NP to TTwR. To do this, we use a reduction which is very similar to the one in Theorem 3.4.1.

Theorem 3.4.2. *There is no constant-factor polynomial-time approximation algorithm for TTwR unless* P = NP.

Proof. Let A be a polynomial-time approximation algorithm for TTwR. For a given instance I of TTwR, let $(\pi^A, \mathcal{R}^A)$ be the solution returned by the algorithm, and let $(\pi^*, \mathcal{R}^*)$ be an optimal solution. Suppose that there is a constant $C \geq 1$ such that for all instances I of TTwR, it holds that

$$\frac{c(\pi^A, \mathcal{R}^A)}{c(\pi^*, \mathcal{R}^*)} \leq C. \tag{3.10}$$

We show that in this case we can use the algorithm A to solve Set Cover with approximation ratio C in polynomial time.

Let (P, Q) be an instance of Set Cover with $P = \{p_1, \ldots, p_m\}$ and $Q = \{q^1, \ldots, q^n\}$. For P and Q, we construct an instance I of TTwR as described in the proof of Theorem 3.4.1, changing only the passenger weights for OD-pairs (u, v_i) to $w_{uv_i} := Cn + 1$ (instead of $w_{uv_i} = n + 1$ as chosen in the proof of Theorem 3.4.1) for $i = 1, \ldots, m$.

1. As a first step we show that algorithm A always returns a solution $(\pi^A, \mathcal{R}^A)$ with $c(\pi^A, \mathcal{R}^A) \leq n$:

    - We can uniquely define a feasible timetable π' for I setting $\pi'_{((\text{tr}^j)^2 - s_1 - \text{Dep})} := 0$ and $\pi'_{((\text{tr}^j)^2 - s_3^j - \text{Dep})} := 0$ for all $j = 1, \ldots, n$. (This corresponds to the timetable $\pi(Q')$ for $Q' = Q$ we defined in (3.9) in the proof of Theorem 3.4.1.)
    - In this case, for any routing $\mathcal{R}' = \{P'_i : i = 1, \ldots, m\} \cup \{(P^j)' : i = 1, \ldots, n\}$, we have

    $$c(\pi', P'_i) = 0 \text{ and } c(\pi', (P')^j) = 1,$$

    hence

    $$c(\pi', \mathcal{R}') = \sum_{P'_i \in \mathcal{R}'} w_{uv_i} c(\pi', P'_i) + \sum_{(P')^j \in \mathcal{R}'} c(\pi', (P')^j) = n.$$

- Suppose that A finds a solution $(\pi^A, \mathcal{R}^A)$ such that there is an $(u, v_i) \in \mathcal{OD}$ with $c(\pi^A, P_i^A) \neq 0$ for the path P_i^A defined by $\mathcal{R}^A$ for (u, v_i). Then $c(\pi^A, P_i^A) \geq w_{uv_i} c(\pi^A, P_i^A) = Cn + 1$, and it follows that

$$C \geq \frac{c(\pi^A, \mathcal{R}^A)}{c(\pi^*, \mathcal{R}^*)} \geq \frac{c(\pi^A, \mathcal{R}^A)}{c(\pi', \mathcal{R}')} \geq \frac{Cn+1}{n} > C$$

which is a contradiction. Hence, for the solution $(\pi^A, \mathcal{R}^A)$ returned by A, it holds that $c(\pi^A, P_i^A) = 0$ for $i = 1, \ldots, m$.
- Since $c(\pi, P^j) \in \{0, 1\}$ for any choice of timetable π and path P^j for (u, v^j), this implies that

$$c(\pi^A, \mathcal{R}^A) = \sum_{P^j \in \mathcal{R}^A} w_{uv^j} c(\pi, P^j) \leq n.$$

2. Analogously to step (1) in the proof of Theorem 3.4.1, we see that given a solution $(\pi, \mathcal{R})$ to I with $c(\pi, \mathcal{R}) \leq n$, we can construct a solution $Q(\pi, \mathcal{R})$ to (P, Q) with $|Q(\pi, \mathcal{R})| \leq c(\pi, \mathcal{R})$ (setting K:=n). In particular, due to (1) this implies that for the solution $(\pi^A, \mathcal{R}^A)$ to I found by algorithm A, we can construct a solution $Q(\pi^A, \mathcal{R}^A)$ to (P, Q) with $|Q(\pi^A, \mathcal{R}^A)| = c(\pi^A, \mathcal{R}^A)$.
3. Analogously to step (2) in the proof of Theorem 3.4.1 (setting K:=n), we see that given a solution Q to (P, Q), we can construct a solution $(\pi(Q'), \mathcal{R}(Q'))$ to I with $c(\pi(Q'), \mathcal{R}(Q')) = |Q'|$. In particular this implies that for an optimal solution Q^* to (P, Q) and an optimal solution $(\pi^*, \mathcal{R}^*)$ to I, we have $c(\pi^*, \mathcal{R}^*) \leq |Q^*|$.
4. We now apply A to I and obtain a solution $(\pi^A, \mathcal{R}^A)$ which we transfer to a solution $Q(\pi^A, \mathcal{R}^A)$ in polynomial time as described in (2). Then, due to (3),

$$\frac{|Q(\pi^A, \mathcal{R}^A)|}{|Q^*|} \leq \frac{c(\pi^A, \mathcal{R}^A)}{c(\pi^*, \mathcal{R}^*)} \leq C.$$

Hence, using A we can construct a polynomial-time C-approximation algorithm for Set Cover which contradicts the inapproximability results for Set Cover. □

In the instance I of TTwR, constructed from an instance (P, Q) of Set Cover in the proof of Theorem 3.4.2, we use lower bounds of 0 on the driving activities in order to obtain the same objective value as in the Set Cover problem. Hence, the question may arise whether the inapproximability is only due to this choice of zero bounds. In this case, confining the class of instances to such with strictly positive bound could cause a significant improvement in the approximation ratio. However a similar result can be obtained setting $l_a := c$ and $u_a := c$ (where $c \in \mathbb{N}$ is an arbitrary constant) for all activity bounds previously set to 0 in the proof of Theorem 3.4.2 and scaling the other bounds on the activity lengths to numbers that are much higher.

3.4 Complexity of Timetabling with Routing

We notice that the inapproximability result in Theorem 3.4.2 seems to be caused by the high relative difference of the upper and lower bounds on some activities. Indeed, we prove in Theorem 3.4.4 that the output of any (reasonable) approximation algorithm must be bounded by the maximum relative difference of upper and lower bounds.

We make use of the following notation:

Notation 3.4.3. *Given an instance* $(\mathcal{N}, l, u, \mathcal{OD})$ *of TTwR, for every OD-pair* $(u_k, v_k) \in \mathcal{OD}$, *let*

- P_k^{low} *be a shortest path from* u_k^{org} *to* v_k^{dest} *in* $\mathcal{N}$ *with respect to arc lengths* $L_a := l_a$ $\forall a \in \mathcal{A}_{op}$ *and* $L_a := 0$ $\forall a \in \mathcal{A}_{org} \cup \mathcal{A}_{dest}$, *and let* l_k^{low} *denote its length.*
- P_k^{upp} *be a shortest path from* u_k^{org} *to* v_k^{dest} *in* $\mathcal{N}$ *with respect to arc lengths* $L_a := u_a$ $\forall a \in \mathcal{A}_{op}$ *and* $L_a := 0$ $\forall a \in \mathcal{A}_{org} \cup \mathcal{A}_{dest}$, *and let* l_k^{upp} *denote its length.*
- l_k^{opt} *denote the optimal objective value of the instance* $(\mathcal{N}, l, u, \{(u_k, v_k)\})$ *with passenger weight* $w_k := 1$ *of TTwR.*

In Theorem 3.6.4 in the next section we will see that for every OD-pair (u_k, v_k), we can determine l_k^{opt} in polynomial time.

In the next theorem we show three bounds on the quality of a solution $(\pi', \mathcal{R}(\pi'))$ defined by a timetable π'. These bounds differ not only in the time that is needed to calculate them but also in their tightness.

Theorem 3.4.4. *Let* $I = (\mathcal{N}, l, u, \mathcal{OD})$ *be an instance of TTwR, and let* $(\pi^*, \mathcal{R}^*)$ *denote an optimal solution to* I. *Then for any timetable* π' *on* $\mathcal{N}$ *and* $\mathcal{R}' := \mathcal{R}(\pi')$, *it holds that*

$$\frac{c(\pi', \mathcal{R}')}{c(\pi^*, \mathcal{R}^*)} \leq \frac{\sum_{(u_k, v_k) \in \mathcal{OD}} w_k l_k^{upp}}{\sum_{(u_k, v_k) \in \mathcal{OD}} w_k l_k^{opt}} \leq \frac{\sum_{(u_k, v_k) \in \mathcal{OD}} w_k l_k^{upp}}{\sum_{(u_k, v_k) \in \mathcal{OD}} w_k l_k^{low}} \leq \max_{a \in \mathcal{A}_{op}} \frac{u_a}{l_a}.$$

To prove Theorem 3.4.4, we need the following lemma.

Lemma 3.4.5. *For* $a_1, a_2, \ldots, a_n, b_1, b_2, \ldots, b_n > 0$

$$\frac{a_1 + a_2 + \cdots + a_n}{b_1 + b_2 + \cdots + b_n} \leq \max_{i=1,\ldots,n} \left\{\frac{a_i}{b_i}\right\}.$$

Proof. The lemma is proven inductively: for $n = 1$, the statement is obvious. Suppose now that the statement has been shown for $n = k - 1$ with $k > 1$. Denote $A_{k-1} = a_1 + a_2 + \cdots + a_{k-1}$ and $B_{k-1} = b_1 + b_2 + \cdots + b_{k-1}$. Now assume that

$$\frac{A_{k-1} + a_k}{B_{k-1} + b_k} > \max\left\{\frac{A_{k-1}}{B_{k-1}}, \frac{a_k}{b_k}\right\}.$$

Multiplying both inequalities with the denominators yields

$$A_{k-1} B_{k-1} + a_k B_{k-1} > A_{k-1} B_{k-1} + A_{k-1} b_k$$
$$A_{k-1} b_k + a_k b_k > a_k B_{k-1} + a_k b_k,$$

and if we add up the inequalities, we obtain

$$A_{k-1}B_{k-1} + A_{k-1}b_k + a_k B_{k-1} + a_k b_k > A_{k-1}B_{k-1} + A_{k-1}b_k + a_k B_{k-1} + a_k b_k$$

which is clearly a contradiction. $\square$

Now we can prove Theorem 3.4.4.

Proof. 1. For every routing $\mathcal{R} := \{P_k : (u_k, v_k) \in \mathcal{OD}\}$, we have $c(\pi^*, P_k) \geq l_k^{\text{opt}}$; hence

$$c(\pi^*, \mathcal{R}^*) \geq \sum_{(u_k, v_k) \in \mathcal{OD}} w_k l_k^{\text{opt}}.$$

On the other hand, since $\mathcal{R}' := \mathcal{R}(\pi')$, we have $l(P_k') \leq l_k^{\text{upp}}$ for all $P_k' \in \mathcal{R}'$; thus, we obtain

$$c(\pi', \mathcal{R}') \leq \sum_{(u_k, v_k) \in \mathcal{OD}} w_k l_k^{\text{upp}}.$$

This gives us the first inequality.

2. For every OD-pair $(u_k, v_k) \in \mathcal{OD}$, we have $l_k^{\text{low}} \leq l_k^{\text{opt}}$, which implies the second inequality.
3. Using Lemma 3.4.5 several times, we obtain

$$\frac{\sum_{(u_k, v_k) \in \mathcal{OD}} w_k l_k^{\text{upp}}}{\sum_{(u_k, v_k) \in \mathcal{OD}} w_k l_k^{\text{low}}} = \frac{\sum_{(u_k, v_k) \in \mathcal{OD}} w_k \sum_{a \in P_k^{\text{upp}}} u_a}{\sum_{(u_k, v_k) \in \mathcal{OD}} w_k \sum_{a \in P_k^{\text{low}}} l_a}$$

$$\leq \frac{\sum_{(u_k, v_k) \in \mathcal{OD}} w_k \sum_{a \in P_k^{\text{low}}} u_a}{\sum_{(u_k, v_k) \in \mathcal{OD}} w_k \sum_{a \in P_k^{\text{low}}} l_a}$$

$$\leq \max_{(u_k, v_k) \in \mathcal{OD}} \frac{w_k \sum_{a \in P_k^{\text{low}}} u_a}{w_k \sum_{a \in P_k^{\text{low}}} l_a}$$

$$= \max_{(u_k, v_k) \in \mathcal{OD}} \frac{\sum_{a \in P_k^{\text{low}}} u_a}{\sum_{a \in P_k^{\text{low}}} l_a}$$

$$\leq \max_{(u_k, v_k) \in \mathcal{OD}} \max_{a \in P_k^{\text{low}}} \frac{u_a}{l_a}$$

$$\leq \max_{a \in \mathcal{A}_{\text{op}}} \frac{u_a}{l_a}.$$

$\square$

3.5 Timetabling with Routing Between Events

In this section we describe and investigate the problem timetabling with routing between events (TTwRE). Though at first glance TTwRE looks quite similar to TTwR, it turns out to be much easier to solve. TTwRE motivates an alternative formulation for TTwR which provides us with a first solution approach in Sect. 3.6 and, in Sect. 3.7.2, an alternative integer programming formulation.

In TTwRE we assume that instead of knowing only the origin and destination station of each passenger, we are also given the information about the train a passenger wants to board at the origin station and the train he/she wants to arrive with at his/her destination. We then obtain an OD-set consisting of departure events as origins and arrival events as destinations.

Definition 3.5.1. Given an EAN $\mathcal{N}$, an *operational OD-pair* is a pair (i, j) with $i \in \mathcal{E}_{\text{dep}}$ and $j \in \mathcal{E}_{\text{arr}}$. We denote the set of *operational OD-pairs* as $\mathcal{OD}_{\text{op}}$ and the weights for the operational OD-pairs as w_{ij} for $(i, j) \in \mathcal{OD}_{\text{op}}$.

We then restrict our routings to contain only paths which start at a passenger's origin station and end at his or her destination station.

Definition 3.5.2. Let $\mathcal{N}$ be an EAN and $\mathcal{OD}_{\text{op}}$ a set of operational OD-pairs in $\mathcal{N}$.

- A path for $(i, j) \in \mathcal{OD}_{\text{op}}$ is a path from i to j in $\mathcal{N}$.
- A *routing* for $\mathcal{OD}_{\text{op}}$ is defined as $\mathcal{R}_{\text{op}} := \{P_{ij} : (i, j) \in \mathcal{OD}_{\text{op}}\}$ where P_{ij} is a path from i to j in $\mathcal{N}$.
- Like in (3.3), also in TTwRE, for a given timetable π and path P_{ij}, we denote by $c(\pi, P_{ij}) := \pi_j - \pi_i$ the *travel time along* P_{ij}.
- Analogously to (3.4), for a given timetable π and a routing $\mathcal{R}_{\text{op}}$ for $\mathcal{OD}_{\text{op}}$, the travel time in TTwRE is calculated as $c(\pi, \mathcal{R}_{\text{op}}) := \sum_{P_{ij} \in \mathcal{R}_{\text{op}}} w_{ij} c(\pi, P_{ij})$

We define the problem of TTwRE as follows.

Definition 3.5.3. An instance $(\mathcal{N}, l, u, \mathcal{OD}_{\text{op}})$ of timetabling with routing between events (TTwRE) consists of an EAN $\mathcal{N} = (\mathcal{E}, \mathcal{A})$: upper and lower bounds u, l on the operational activities: and a set of operational OD-pairs $\mathcal{OD}_{\text{op}}$. The task is to find a feasible timetable π and a routing $\mathcal{R}_{\text{op}}$ for $\mathcal{OD}_{\text{op}}$ which minimize the overall travel time.

In [Nac98], Nachtigall derives a similar problem from a periodic timetabling problem. He formulates the problem as an integer program using variables for the slack and the passenger flows and solves it by a decomposition in two subproblems to determine an optimal passenger flow and an optimal assignment of slack times separately by linear programming. The approach from [SS10b, SS12b] presented here is similar but allows us to see the problem from a somewhat different perspective, such that, in the subsequent sections, we are able to develop an exact algorithm and an integer program formulation based on the results obtained here.

Note that there may be several paths in $\mathcal{N}$ which, for an operational OD-pair (i, j), connect departure event i to arrival event j. However, the choice of the routing is irrelevant for the objective value.

Lemma 3.5.4. *Given an instance* $(\mathcal{N}, l, u, \mathcal{OD}_{op})$ *of TTwRE and a timetable* π *in* $\mathcal{N}$, *it holds that*

$$c(\pi, \mathcal{R}_{op}^1) = c(\pi, \mathcal{R}_{op}^2) = \sum_{(i,j) \in \mathcal{OD}_{op}} w_{ij} (\pi_j - \pi_i)$$

for every pair of routings $\mathcal{R}_{op}^1, \mathcal{R}_{op}^2$ *for* $\mathcal{OD}_{op}$.

Proof. For a given timetable π, we have

$$c(\pi, \mathcal{R}_{op}^1) = \sum_{P_{ij} \in \mathcal{R}_{op}^1} w_{ij} c(\pi, P_{ij})$$

$$= \sum_{(i,j) \in \mathcal{OD}_{op}} w_{ij} (\pi_j - \pi_i) = \sum_{P_{ij} \in \mathcal{R}_{op}^2} w_{ij} c(\pi, P_{ij}) = c(\pi, \mathcal{R}_{op}^2).$$

□

Hence, for every OD-pair $(i, j) \in \mathcal{OD}_{op}$, we could choose an arbitrary path P_{ij} from i to j as a representative for the class of paths from i to j, set $\mathcal{R}_{op} := \{P_{ij} : (i, j) \in \mathcal{OD}\}$, and concentrate on calculating an optimal timetable. We could then interpret a given instance $(\mathcal{N}, l, u, \mathcal{OD}_{op})$ of TTwRE as an instance $(\mathcal{N}, l, u, \mathcal{R}')$ of sTT where we define $\mathcal{R}'$ to consist of the paths P_{ij} adding origin and destination activities at the stations belonging to departure event i and arrival event j.

However, instead of choosing one of the existing paths between two nodes i and j, we represent the demand between these events by adding a virtual activity (i, j) and choose this activity as a path.

Definition 3.5.5. Given an instance $(\mathcal{N}, l, u, \mathcal{OD}_{op})$ of TTwRE, we define the set of *virtual activities* to be $\mathcal{A}_{virt} := \{(i, j) : (i, j) \in \mathcal{OD}_{op}\}$ and set the activity weight w_{ij} to the passenger weight w_{ij} of (i, j).

Note that due to the same notation for directed arcs and operational OD-pairs, we are now using the same notation for an operational OD-pair (i, j) and the corresponding virtual activity. However, virtual activities should be understood to be just a representation of the operational OD-pairs in the EAN, so we could indeed write $\mathcal{OD}_{op} = \mathcal{A}_{virt}$.

In Fig. 3.4 an example for the concept of virtual activities is given. The OD-pair $(tr_1 - s_1 - Dep, tr_2 - s_4 - Arr)$ is represented by the dotted virtual activity with weight $w_{(tr_1 - s_1 - Dep, tr_2 - s_4 - Arr)}$. Note that three different paths in the EAN can be chosen for operational OD-pair $(tr_1 - s_1 - Dep, tr_2 - s_4 - Arr)$, but that, given a timetable, the travel time of $(tr_1 - s_1 - Dep, tr_2 - s_4 - Arr)$ does not depend on the path choice.

3.6 An Exact Algorithm for Timetabling with Routing

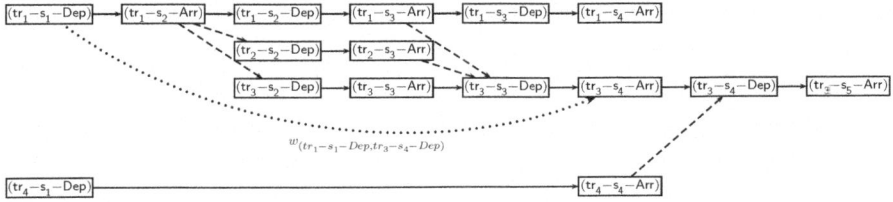

Fig. 3.4 Example for the implementation of virtual activities

We have seen in Lemma 3.5.4 that in TTwRE, the overall travel time does not depend on the routing any more. That is, given a timetable, the overall travel time is $\sum_{(i,j) \in \mathcal{OD}_{\text{op}}} w_{ij} \cdot (\pi_j - \pi_i)$. Hence, similarly to the approach for sTT in Sect. 3.3, we can solve TTwRE by linear programming. Again, we use variables

$$\pi_i := \text{scheduled time at event } i \quad \forall \, i \in \mathcal{E}_{\text{op}}.$$

Then the integer program reads as follows

$$\min \sum_{(i,j) \in \mathcal{OD}_{\text{op}}} w_{ij} \cdot (\pi_j - \pi_i), \tag{3.11}$$

$$\text{s.t.} \quad \pi_h - \pi_g \geq l_{gh} \quad \forall (g,h) \in \mathcal{A}_{\text{op}}, \tag{3.12}$$

$$\pi_h - \pi_g \leq u_{gh} \quad \forall (g,h) \in \mathcal{A}_{\text{op}}, \tag{3.13}$$

$$\pi_g \in \mathbb{Z} \quad \forall g \in \mathcal{E}_{\text{op}}, \tag{3.14}$$

where (3.11) is the overall travel time and (3.12) and (3.13) are the feasibility constraints for the timetable. We have the same totally unimodular coefficient matrix as in the IP formulation (3.5)–(3.8) of sTT.

Theorem 3.5.6. *TTwRE can be solved in polynomial time by linear programming.*

3.6 An Exact Algorithm for Timetabling with Routing

In this section we turn back to the problem TTwR and use the result from Theorem 3.5.6 to solve TTwR exactly.

Definition 3.6.1. Let $\mathcal{N}$ be an EAN and $\mathcal{OD} = \{(u_k, v_k) : k \in K\}$ a set of OD-pairs. We define the set of *departure events for* (u_k, v_k) as

$$\mathcal{E}_{\text{dep}}^k := \{e \in \mathcal{E}_{\text{dep}} : (u_k^{\text{org}}, (\text{tr} - u_k - \text{Dep})) : l(\text{tr}) \ni u_k\}$$

and the set of *arrival events for* (u_k, v_k) as

$$\mathcal{E}_{\text{arr}}^k := \{e \in \mathcal{E}_{\text{arr}} : ((\text{tr} - v_k - \text{Arr}, v_k^{\text{dest}}) : l(\text{tr}) \ni v_k\}.$$

The idea of the approach from [SS12b] is to solve TTwRE for every possible combination of these arrival and destination events for all OD-pairs.

Definition 3.6.2. Let $(u_k, v_k) \in \mathcal{OD}$ be an OD-pair. We define
- the set of *reasonable virtual activities representing* (u_k, v_k)

$$\mathcal{A}_{\text{virt}}^k := \{(i^k, j^k) : i^k \in \mathcal{E}_{\text{dep}}^k, j^k \in \mathcal{E}_{\text{arr}}^k; \exists \text{ a directed path } P_{i^k j^k} \text{ from } i^k \text{ to } j^k \text{ in } \mathcal{N};$$

$$\text{and} \sum_{a \in P_{i^k j^k}} l_a \leq l_k^{\text{upp}}\},$$

- the set of *reasonable virtual activities* $\mathcal{A}_{\text{virt}} = \bigcup_{(u_k, v_k) \in \mathcal{OD}} \mathcal{A}_{\text{virt}}^k$, and
- the set of *reasonable sets of operational OD-pairs representing* $\mathcal{OD}$

$$\mathcal{V}(\mathcal{OD}) := \{(i^1, j^1), (i^2, j^2), \ldots (i^{|\mathcal{OD}_{\text{op}}|}, j^{|\mathcal{OD}_{\text{op}}|})\} : (i^k, j^k) \in \mathcal{A}_{\text{virt}}^k\}.$$

Lemma 3.6.3. *The following definition is equivalent to Definition 3.2.6. An instance* $(\mathcal{N}, l, u, \mathcal{OD})$ *of TTwR consists of an EAN* $\mathcal{N} = (\mathcal{E}, \mathcal{A})$; *upper and lower bounds* u, l *on the operational activities; and a set of OD-pairs* $\mathcal{OD}$.

The task is to find a feasible timetable π *and a set of* operational OD-pairs $\mathcal{OD}_{op} \in \mathcal{V}(\mathcal{OD})$ *such that*

$$\overline{c}(\pi, \mathcal{OD}_{op}) := \sum_{(i,j) \in \mathcal{OD}_{op}} \pi_j - \pi_i$$

is minimal.

Proof. Let $I = (\mathcal{N}, l, u, \mathcal{OD})$ be an instance of TTwR.
- Let $(\pi, \mathcal{OD}_{op})$ be a feasible solution to TTwR as defined in Lemma 3.6.3. Due to the definition of $\mathcal{A}_{\text{virt}}^k$, for $(i^k, j^k) \in \mathcal{A}_{\text{virt}}^k$, there exists a path from u_k^{org} to v_k^{dest} with the first departure node i and last arrival node j. Hence, for every set of operational OD-pairs $\mathcal{OD}_{op}$, we can find a corresponding routing $\mathcal{R}(\mathcal{OD}_{op})$ for TTwR.
- Let an optimal solution $(\pi^*, \mathcal{R}^*)$ to TTwR as defined in Definition 3.2.6 be given. For every $(u_k, v_k) \in \mathcal{OD}$, there is a directed path $P_k \in \mathcal{R}^*$ from u_k^{org} to v_k^{dest}. Let i^k denote the first and j^k the last operational event on P_k, and let $P_{i^k j^k}$ denote the path P_k without origin and destination activity. Suppose that there is an OD-pair $(u_k, v_k) \in \mathcal{OD}$ such that $(i^k, j^k) \notin \mathcal{A}_{\text{virt}}^k$, i.e., $\sum_{a \in P_{i^k j^k}} l_a > l_k^{\text{upp}}$. Then

$$c(\pi^*, P_k) = \sum_{(i,j) \in P_k} (\pi_j - \pi_i) \geq \sum_{(i,j) \in P_k} l_{ij} > l_k^{\text{upp}} \geq c(\pi^*, P_k^{\text{upp}}).$$

Thus, exchanging P_k into P_k^{upp} yields a better objective value. This is a contradiction to the optimality of $(\pi^*, \mathcal{R}^*)$. Hence, every optimal routing $\mathcal{R}$ defines a set of operational OD-pairs $\mathcal{OD}_{\text{op}}(\mathcal{R})$.

- Furthermore, given a timetable π, for every such pair of associated routing $\mathcal{R} := \{P_k : (u_k, v_k) \in \mathcal{OD}\}$ and set of operational OD-pairs $\mathcal{OD}_{\text{op}}$, we have

$$c(\pi, \mathcal{R}) = \sum_{P_k \in \mathcal{R}} (\pi_{j^k} - \pi_{i^k}) = \sum_{(i^k, j^k) \in \mathcal{OD}_{\text{op}}} (\pi_{j^k} - \pi_{i^k}) = \bar{c}(\pi, \mathcal{OD}_{\text{op}}),$$

where i^k denotes the first departure node on P^k and j^k denotes the last arrival node. □

Hence, solving the LP-relaxation of IP formulation (3.11)–(3.14) for all $\mathcal{OD}_{\text{op}} \in \mathcal{V}(\mathcal{OD})$, we find an optimal solution to TTwR.

Theorem 3.6.4. *An instance $(\mathcal{N}, l, u, \mathcal{OD})$ of TTwR can be solved by solving every different instance $(\mathcal{N}, l, u, \mathcal{OD}_{op})$ with $\mathcal{OD}_{op} \in \mathcal{V}(\mathcal{OD})$ of TTwRE and comparing the objective values. In particular we obtain a polynomial-time algorithm in the following cases:*

1. *If for every $(u_k, v_k) \in \mathcal{OD}$, it holds that $|\mathcal{E}_{\text{dep}}^k| = |\mathcal{E}_{\text{arr}}^k| = 1$ an optimal solution to TTwR can be found by solving one linear program (3.11)–(3.14).*
2. *For a fixed maximal number of OD-pairs m, TTwR can be solved in polynomial time by solving $O(\prod_{(u_k, v_k) \in \mathcal{OD}} |\mathcal{E}_{\text{dep}}^k| \cdot |\mathcal{E}_{\text{arr}}^k|) = O(n_{\mathcal{N}}^{2m})$ linear programs, where $n_{\mathcal{N}}$ denotes the number of nodes in the considered EAN $\mathcal{N}$.*
3. *In particular, if there is only one OD-pair (u_1, v_1), an optimal solution to TTwR can be found by solving at most $|\mathcal{E}_{\text{dep}}^1| \cdot |\mathcal{E}_{\text{arr}}^1|$ linear programs.*

In Fig. 3.5 we observe that there are five possible paths from s_1^{org} to s_4^{dest}. However, these paths can be represented using only three virtual activities a_1, a_2, and a_3. Hence, we can solve TTwR by solving three instances of TTwRE and comparing the objective values.

3.7 Integer Programming Formulations

In Sect. 3.5 we saw that an instance of TTwR can be solved by solving many instances of TTwRE. However, this method can only be applied efficiently if there are a small number of OD-pairs or if for every OD-pair, there is a very limited set of choices at which event to start and at which event to end the journey. This is usually not the case in practical applications of timetabling.

In the following, we provide two integer programming formulations for TTwR in Sects. 3.7.1 and 3.7.2. Both formulations rely on the formulation (3.5)–(3.8) used for solving sTT, where the routing was part of the input. To include the routing decisions into the optimization process, in the formulation in Sect. 3.7.1, passengers

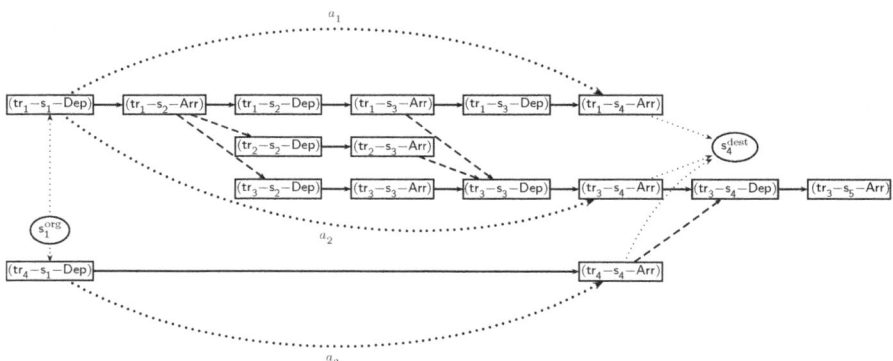

Fig. 3.5 Solving TTwR by solving TTwRE: virtual activities for OD-pair (s_1, s_4)

are modeled as flow in the network, similar to the integer programming approaches for scLPwR, cLPwR, and delay management with routing (DMwR) presented in Sects. 2.6.2 and 4.5. The formulation in Sect. 3.7.2, based on the insights from Sect. 3.5, introduces a variable for every virtual activity instead.

3.7.1 Flow-Based Formulation

Let $I = (\mathcal{N}, l, u, \mathcal{OD})$ be an instance of TTwR. Our first approach is to model the routing constraints using a multi-commodity flow formulation, similar to the one used in the line planning problems in Sect. 2.6.2. To this end, in addition to the timetable variables π_i already used in the formulation (3.5)–(3.8) of sTT, we introduce flow variables q_a^k specifying whether OD-pair (u_k, v_k) uses activity a or not

$$q_a^k := \begin{cases} 1 & \text{if activity } a \text{ is used by OD-pair } (u_k, v_k), \\ 0 & \text{otherwise} \end{cases} \quad \forall a \in \mathcal{A}.$$

Furthermore, for every OD-pair (u_k, v_k), we use variables t_{org}^k and t_{dest}^k specifying the departure time in the origin u_k and the arrival time in the destination v_k.

If the network $\mathcal{N}_{\text{op}}$ is not connected, in order to avoid difficulties in calculating the "big-M"-values needed by our IP formulation, for the calculation of the timetable, we connect the different connected components by *artificial activities* as described in the following.

Notation 3.7.1. *Suppose that $\mathcal{N}_{\text{op}}$ is not connected, but consists of m connected components $\mathcal{N}_{op}^1, \ldots, \mathcal{N}_{op}^m$. Let $\mathcal{N}'$ denote the network built from $\mathcal{N}_{\text{op}}$ by adding a changing activity (j_i, k_{i+1}) from an arbitrary arrival node j_i in $\mathcal{N}_{op}^i$ to an arbitrary*

3.7 Integer Programming Formulations

departure node k_{i+1} in $\mathcal{N}_{op}^{i+1}$ for $i = 1, \ldots, m-1$ with lower bounds $l_{j_i k_{i+1}} := 0$ and $u_{j_i k_{i+1}} := 0$.

That is, $\mathcal{N}'$ is defined as $\mathcal{N}' := (\mathcal{E}', \mathcal{A}')$ with

$$\mathcal{E}' := \mathcal{E}_{op} \text{ and } \mathcal{A}' := \mathcal{A}_{op} \cup \{(j_i, k_{i+1}) : i = 1, \ldots, m-1\}.$$

However, since the artificial activities (j_i, k_{i+1}) do not represent travel possibilities, they are not considered in the flow constraints (3.18)–(3.20).

The flow-based integer program is given as

$$\min \sum_{(u_k, v_k) \in OD} w_k (t_{\text{dest}}^k - t_{\text{org}}^k), \tag{3.15}$$

$$\text{s.t.} \quad \pi_j - \pi_i \geq l_{ij} \qquad \forall (i,j) \in \mathcal{A}', \tag{3.16}$$

$$\pi_j - \pi_i \leq u_{ij} \qquad \forall (i,j) \in \mathcal{A}', \tag{3.17}$$

$$\sum_{a \in \delta^-(u_k^{\text{org}})} q_a^k = 1 \qquad \forall (u_k, v_k) \in \mathcal{OD}, \tag{3.18}$$

$$\sum_{a \in \delta^-(i)} q_a^k - \sum_{a \in \delta^+(i)} q_a^k = 0 \qquad \forall i \in \mathcal{E}_{op}, (u_k, v_k) \in \mathcal{OD}, \tag{3.19}$$

$$\sum_{a \in \delta^+(v_k^{\text{dest}})} q_a^k = 1 \qquad \forall (u_k, v_k) \in \mathcal{OD}, \tag{3.20}$$

$$t_{\text{org}}^k \leq \pi_i + M_{\text{org}}^k (1 - q_{(u_k^{\text{org}}, i)}^k) \qquad \forall i \in \mathcal{E}_{\text{org}}^k, \tag{3.21}$$

$$t_{\text{dest}}^k \geq \pi_i - M_{\text{dest}}^k (1 - q_{(i, v_k^{\text{dest}})}^k) \qquad \forall i \in \mathcal{E}_{\text{dest}}^k, \tag{3.22}$$

$$q_a^k \in \{0, 1\} \qquad \forall (u_k, v_k) \in \mathcal{OD}, a \in \mathcal{A}, \tag{3.23}$$

$$\pi_i \in \mathbb{Z} \qquad \forall i \in \mathcal{E}_{op}, \tag{3.24}$$

$$t_{\text{org}}^k, t_{\text{dest}}^k \in \mathbb{Z} \qquad \forall (u_k, v_k) \in \mathcal{OD}. \tag{3.25}$$

Like in the integer program (3.5)–(3.8) for sTT, constraints (3.16) and (3.17) ensure the feasibility of the timetable and link the timetables of the different connected components of $\mathcal{N}_{op}$. The constraints (3.18)–(3.20) are flow constraints, ensuring that every passenger leaves its origin, travels until its destination, and stops traveling there. Constraints (3.21) and (3.22) together with the objective function (3.15) set the variables t_{org}^k and t_{dest}^k to the corresponding departure and arrival times.

A formal proof of correctness is given in the following Lemmas 3.7.2 and 3.7.3 where we denote $\pi = (\pi_i)_{i \in \mathcal{E}_{op}}$, $q = (q_a^k)_{a \in \mathcal{A}, (u_k, v_k) \in \mathcal{OD}}$, $t_{\text{org}} = (t_{\text{org}}^k)_{(u_k, v_k) \in \mathcal{OD}}$ and $t_{\text{dest}} = (t_{\text{dest}}^k)_{(u_k, v_k) \in \mathcal{OD}}$:

Lemma 3.7.2. *Let* $I = (\mathcal{N}, l, u, \mathcal{OD})$ *be an instance of TTwR. Then for any set* $\{M_{org}^k, M_{dest}^k : (u_k, v_k) \in \mathcal{OD}\} \in \mathbb{N}_0^{2|\mathcal{OD}|}$, *it holds that a feasible solution* $(\pi, q, t_{org}, t_{dest})$ *to the IP (3.15)–(3.25) defines a feasible solution* $(\pi', \mathcal{R}')$ *to instance* I *of TTwR with*

$$c(\pi', \mathcal{R}') \leq \sum_{(u_k, v_k) \in \mathcal{OD}} w_k (t_{dest}^k - t_{org}^k).$$

Proof. Let $(\pi, q, t_{org}, t_{dest})$ be a feasible solution to the IP formulation (3.15)–(3.25), i.e., $(\pi, q, t_{org}, t_{dest})$ fulfills constraints (3.16)–(3.25). We define a timetable π' setting $\pi_i' := \pi_i$ for all $i \in \mathcal{E}_{op}$. π' is feasible due to constraints (3.16) and (3.17). For every $(u_k, v_k) \in \mathcal{OD}$, q^k defines a routing $\mathcal{R}'$ due to constraints (3.18)–(3.20). Thus, every feasible solution $(\pi, q, t_{org}, t_{dest})$ to (3.15)–(3.25) defines a feasible solution (π', R') to the instance I of TTwR. For every $(u_k, v_k) \in \mathcal{OD}$, let j_k denote the last arrival event on $P_k \in \mathcal{R}'$. Then $q_{j_k v_k^{dest}}^k = 1$, hence $t_{dest}^k \geq \pi_{j_k}$ due to constraint (3.22). Just the same, (3.21) forces $t_{org}^k \leq \pi_{i_k}$ for the first departure event i_k on P_k. Hence

$$c(\pi', P_k) = \pi_{j_k} - \pi_{i_k} \leq t_{dest}^k - t_{org}^k$$

and consequently

$$c(\pi', \mathcal{R}') \leq \sum_{(u_k, v_k) \in \mathcal{OD}} w_k (t_{dest}^k - t_{org}^k).$$

$\square$

Lemma 3.7.3. *Let* $I = (\mathcal{N}, l, u, \mathcal{OD})$ *be an instance of TTwR. Let* $\mathcal{E}_{dep}^k, \mathcal{E}_{arr}^k$ *be defined as in Definition 3.6.1. Let for every OD-pair* $(u_k, v_k) \in \mathcal{OD}$ *the numbers* M_{org}^k *and* M_{dest}^k *be chosen in such a way that*

$$M_{org}^k \geq \pi_j - \pi_i \ \forall (i, j) \in \mathcal{E}_{dep}^k \times \mathcal{E}_{dep}^k \text{ and every feasible timetable } \pi \text{ for } \mathcal{N}'$$
(3.26)

and

$$M_{dest}^k \geq \pi_j - \pi_i \ \forall (i, j) \in \mathcal{E}_{arr}^k \times \mathcal{E}_{arr}^k \text{ and every feasible timetable } \pi \text{ for } \mathcal{N}'.$$
(3.27)

Then a solution $(\pi', \mathcal{R}')$ *defined by an optimal solution* $(\pi, q, t_{org}, t_{dest})$ *to the IP (3.15)–(3.25) is an optimal solution to instance* I *of TTwR and*

$$c(\pi', \mathcal{R}') = \sum_{(u_k, v_k) \in \mathcal{OD}} w_k (t_{dest}^k - t_{org}^k).$$

3.7 Integer Programming Formulations

Proof. 1. Let $(\pi', \mathcal{R}')$ be a feasible solution to the instance I of TTwR. Let $\mathcal{N}_{\text{op}}^1, \ldots, \mathcal{N}_{\text{op}}^m$ with $\mathcal{N}_{\text{op}}^i = (\mathcal{E}_{\text{op}}^i, \mathcal{A}_{\text{op}}^i)$ denote the connected components of $\mathcal{N}_{\text{op}}$, and let j_i, k_i be defined as in Notation 3.7.1. We define π as follows:

- We set $\pi_j := \pi'_j$ for all $j \in \mathcal{E}_{\text{op}}^1$. Then we iteratively proceed the connected components as follows:
- For $i = 2, \ldots, m$, we set $\pi_{k_i} = \pi_{j_{i-1}}$.
- We now build the timetable π in $\mathcal{N}_{\text{op}}^i$ setting $\pi_j = \pi'_j - \pi_{k_i}$.

Then for $(i, j) \in \mathcal{A}_{\text{op}}^i$, we have

$$\pi_j - \pi_i = (\pi'_j - \pi_{k_i}) - (\pi'_i - \pi_{k_i}) = \pi'_j - \pi'_i \in [l_{ij}, u_{ij}],$$

since π' is a feasible timetable on $\mathcal{N}$. Since additionally $\pi_{k_i} = \pi_{j_{i-1}}$ for every $i = 2, \ldots, m$, constraints (3.16) and (3.17) are fulfilled.

For every $(u_k, v_k) \in \mathcal{OD}$ denote by P_k the path for OD-pair (u_k, v_k) chosen by $\mathcal{R}'$, and let i_k, j_k be the first departure and last arrival node of P_k. Note that P_k is a path in $\mathcal{N}$; hence i_k and j_j are in the same connected component. We define

$$q_a^k := \begin{cases} 1 & \text{if } a \in P_k \\ 0 & \text{otherwise} \end{cases} \quad \forall a \in \mathcal{A}, (u_k, v_k) \in \mathcal{OD},$$
$$t_{\text{org}}^k := \pi_{i_k} \qquad \forall (u_k, v_k) \in \mathcal{OD},$$
$$t_{\text{dest}}^k := \pi_{j_k} \qquad \forall (u_k, v_k) \in \mathcal{OD}.$$

We show that the so-defined solution fulfills the constraints (3.18)–(3.25).

Due to the definition of q, constraints (3.18)–(3.20) are fulfilled. For the first departure node i_k on P_k, we have $t_{\text{org}}^k = \pi_{i_k}$ by definition. For $i \in \mathcal{E}_{\text{dep}}^k \setminus \{i_k\}$, it holds that $q_{u_k^{\text{org}} i}^k = 0$ due to the definition of q. Hence,

$$t_{\text{org}}^k = \pi_{i_k} = \pi_i + (\pi_{i_k} - \pi_i) \leq \pi_i + M_{\text{org}}^k = \pi_i + M_{\text{org}}^k (1 - q_{u_k^{\text{org}} i}^k),$$

and (3.21) is fulfilled. Analogously, we see that (3.22) is fulfilled. Furthermore, due to the definition of t_{org} and t_{dest} and since for every k, i_k and j_k are in the same connected component, we have

$$\sum_{(u_k, v_k) \in OD} w_k (t_{\text{dest}}^k - t_{\text{org}}^k) = \sum_{(u_k, v_k) \in OD} w_k (\pi_{j_k} - \pi_{i_k})$$
$$= \sum_{(u_k, v_k) \in OD} w_k (\pi'_{j_k} - \pi'_{i_k})$$
$$= c(x', \mathcal{R}'),$$

where j_k denotes the last arrival node on P_k.

2. Let $(\pi, q, t_{\text{org}}, t_{\text{dest}})$ be an optimal solution to (3.15)–(3.25). Due to Lemma 3.7.2, there is a solution $(\pi', \mathcal{R}')$ to I such that

$$c(\pi', \mathcal{R}') \leq \sum_{(u_k, v_k) \in \mathcal{OD}} w_k(t_{\text{dest}}^k - t_{\text{org}}^k).$$

Suppose that for an optimal solution $(x^*, \mathcal{R}^*)$ to TTwR, it holds that

$$c(\pi^*, \mathcal{R}^*) < \sum_{(u_k, v_k) \in \mathcal{OD}} w_k(t_{\text{dest}}^k - t_{\text{org}}^k).$$

Then according to (1), there is a solution $(\pi^*, q^*, t_{\text{org}}^*, t_{\text{dest}}^*)$ to (3.15)–(3.25) with

$$\sum_{(u_k, v_k) \in \mathcal{OD}} w_k((t_{\text{dest}}^*)^k - (t_{\text{org}}^*)^k) = c(x^*, \mathcal{R}^*) < \sum_{(u_k, v_k) \in \mathcal{OD}} w_k(t_{\text{dest}}^k - t_{\text{org}}^k).$$

This is a contradiction to the optimality of $(\pi, q, t_{\text{org}}, t_{\text{dest}})$. □

The minimal values which the numbers $\{M_{\text{org}}^k, M_{\text{dest}}^k : (u_k, v_k) \in \mathcal{OD}\}$ are allowed to take according to Lemma 3.7.3 such that (3.26) and (3.27) are fulfilled with regard to the connected network $\mathcal{N}' = (\mathcal{E}, \mathcal{A}')$ can be rewritten as follows.

$$M_{\text{org}}^k := \max_{\pi \text{ feasible for } \mathcal{N}'} \left(\max_{(i,j) \in \mathcal{E}_{\text{dep}}^k \times \mathcal{E}_{\text{dep}}^k} (\pi_j - \pi_i) \right)$$

and

$$M_{\text{dest}}^k := \max_{\pi \text{ feasible for } \mathcal{N}'} \left(\max_{(i,j) \in \mathcal{E}_{\text{arr}}^k \times \mathcal{E}_{\text{arr}}^k} (\pi_j - \pi_i) \right).$$

These values can be determined in polynomial time. To that end, we solve the linear program

$$\max M_{\text{org}}^k(i, j) := \pi_j - \pi_i, \tag{3.28}$$

$$s.t. \ \pi_h - \pi_g \geq l_{gh} \qquad \forall (g, h) \in \mathcal{A}', \tag{3.29}$$

$$\pi_h - \pi_g \leq u_{gh} \qquad \forall (g, h) \in \mathcal{A}', \tag{3.30}$$

$$\pi_g \in \mathbb{Z} \qquad \forall g \in \mathcal{E}_{\text{op}}, \tag{3.31}$$

for all $(i, j) \in \mathcal{E}_{\text{dep}}^k \times \mathcal{E}_{\text{dep}}^k$ (see Definition 3.6.1). Since the matrix defined by the constraints (3.29) and (3.30) is the same as in formulation (3.5)–(3.8) and hence totally unimodular and the bounds are integer, we obtain an optimal integer solution by linear programming. Then we can set

3.7 Integer Programming Formulations

$$M_{\text{org}}^k := \max_{(i,j) \in \mathcal{E}_{\text{dep}}^k \times \mathcal{E}_{\text{dep}}^k} M_{\text{org}}^k(i,j).$$

Analogously, we can compute M_{dest}^k.

We summarize our result in the following lemma.

Lemma 3.7.4. *The minimal values for $\{M_{\text{org}}^k, M_{\text{dest}}^k : (u_k, v_k) \in \mathcal{OD}\}$ such that (3.26) and (3.27) are fulfilled are*

$$M_{\text{org}}^k := \max_{\pi \text{ feasible for } \mathcal{N}'} \left(\max_{(i,j) \in \mathcal{E}_{\text{dep}}^k \times \mathcal{E}_{\text{dep}}^k} (\pi_j - \pi_i) \right)$$

and

$$M_{\text{dest}}^k := \max_{\pi \text{ feasible for } \mathcal{N}'} \left(\max_{(i,j) \in \mathcal{E}_{\text{dep}}^k \times \mathcal{E}_{\text{dep}}^k} (\pi_j - \pi_i) \right).$$

These values can be calculated in polynomial time.

Although the calculation time for the numbers $\{M_{\text{org}}^k, M_{\text{dest}}^k : (u_k, v_k) \in \mathcal{OD}\}$ is polynomial, it may be very long. The following value for $\{M_{\text{org}}^k, M_{\text{dest}}^k : (u_k, v_k) \in \mathcal{OD}\}$ can be calculated much faster.

Lemma 3.7.5. *Let $I = (\mathcal{N}, l, u, \mathcal{OD})$ be an instance of TTwR. We define $M := \sum_{a \in \mathcal{A}_{op}} u_a$ and set $M_{\text{org}}^k := M$ and $M_{\text{dest}}^k := M$. Then (3.26) and (3.27) are fulfilled.*

Proof. For $(i', j') \in \mathcal{E}_{\text{dep}}^k \times \mathcal{E}_{\text{dep}}^k$, let $P_{i'j'}$ denote an undirected path from i' to j' in $\mathcal{N}'$ (such a path exists since $\mathcal{N}'$ is connected). Then

$$\pi_{j'} - \pi_{i'} \leq \sum_{(i,j) \in P_{i'j'}} \pi_j - \pi_i = \sum_{(i,j) \in P_{i'j'} \cap \mathcal{A}_{op}} \pi_j - \pi_i$$

$$\leq \sum_{(i,j) \in P_{i'j'} \cap \mathcal{A}_{op}} u_{ij} \leq \sum_{a \in \mathcal{A}_{op}} u_a = M.$$

The same holds for $(i', j') \in \mathcal{E}_{\text{arr}}^k \times \mathcal{E}_{\text{arr}}^k$. □

An alternative formulation for integrating routing in timetabling problems using flow constraints is followed by [Lüb09, SG13] for periodic timetabling. There (translated to our notation), variables taking the role of $q_{ij}^k \cdot (\pi_j - \pi_i)$ are introduced for every $(i, j) \in \mathcal{A}_{op}$ and every $(u_k, v_k) \in \mathcal{OD}$. The weighted sum of these variables serves as objective function. The "big-Ms" required to set the new variables to the claimed values are of the size of the upper bounds on the activity lengths and thus significantly smaller than the ones presented in this book; however, the number of variables and constraints is larger than in our formulation.

3.7.2 Virtual-Activity-Based Formulation

Instead of using a flow formulation which introduces a variable for every activity and every OD-pair, it is also a natural idea to introduce a variable for every path an OD-pair might choose. In many routing problems, this approach has the inconvenience that it leads to an exponential number of variables, because of an exponential number of paths between origin and destination node of every OD-pair.

However, it has been shown in [SS12b] that we can exploit the insights gained in Sect. 3.5 to reduce the number of variables significantly. Instead of introducing a variable for every path P_k from u_k^{org} to v_k^{dest} for every OD-pair $(u_k, v_k) \in \mathcal{OD}$, we introduce a variable for every virtual activity $(i, j) \in \mathcal{A}_{\text{virt}}^k$:

$$z_{ij}^k := \begin{cases} 1 & \text{if virtual activity } (i,j) \text{ is used for OD-pair } (u_k, v_k) \\ 0 & \text{otherwise.} \end{cases}$$

Note that for every OD-pair (u_k, v_k), there might be an exponential number of paths from u_k^{org} to v_k^{dest}, but there are at most $O(|\mathcal{E}|^2)$ reasonable virtual activities—and usually much less in practical applications.

We introduce variables t^k which take the value of the travel time of an OD-pair (u_k, v_k) for the timetable and the routing chosen by the integer program. Furthermore we use again the timetable variables π_i for $i \in \mathcal{E}_{\text{op}}$ of formulation (3.5)–(3.8) of sTT.

For a better comparability between the flow-based and the virtual-activity-based IP formulation, also in our second IP for TTwR, in case of an unconnected network we calculate a feasible timetable on $\mathcal{N}'$ instead of $\mathcal{N}_{\text{op}}$.

The virtual-activity-based integer program reads

$$\min \sum_{(u,v) \in \mathcal{OD}} w_k t^k, \qquad (3.32)$$

$$\text{s.t. } \pi_j - \pi_i \geq l_{ij} \qquad \forall (i,j) \in \mathcal{A}', \quad (3.33)$$

$$\pi_j - \pi_i \leq u_{ij} \qquad \forall (i,j) \in \mathcal{A}', \quad (3.34)$$

$$\sum_{(i,j) \in \mathcal{A}_{\text{virt}}^k} z_{ij}^k = 1 \qquad \forall (u_k, v_k) \in \mathcal{OD}, \quad (3.35)$$

$$t^k \geq \pi_j - \pi_i - M^k(1 - z_{ij}^k) \quad \forall (i,j) \in \mathcal{A}_{\text{virt}}^k, (u_k, v_k) \in \mathcal{OD}, \quad (3.36)$$

$$z_{ij}^k \in \{0,1\} \qquad \forall (i,j) \in \mathcal{A}_{\text{virt}}^k, (u_k, v_k) \in \mathcal{OD}, \quad (3.37)$$

$$\pi_i \in \mathbb{Z} \qquad \forall i \in \mathcal{E}_{\text{op}}, \quad (3.38)$$

$$t^k \in \mathbb{Z} \qquad \forall (u_k, v_k) \in \mathcal{OD}. \quad (3.39)$$

3.7 Integer Programming Formulations

Like in the IP (3.5)–(3.8) for sTT, the constraints (3.33) and (3.34) ensure feasibility of the timetable π on $\mathcal{N}' \supset \mathcal{N}$. Constraints (3.35) ensure that exactly one virtual activity is chosen for every OD-pair, and constraints (3.36) set t^k to the travel time of OD-pair (u_k, v_k) if M^k is chosen big enough for every $(u_k, v_k) \in \mathcal{OD}$.

We use the equivalent characterization of TTwR provided by Lemma 3.6.3 to show that this IP formulation for TTwR is correct in the following Lemmas 3.7.6 and 3.7.7.

Lemma 3.7.6. *Let* $I = (\mathcal{N}, l, u, \mathcal{OD})$ *be an instance of TTwR and consider the corresponding IP formulation (3.32)–(3.39). For any set* $\{M^k : (u_k, v_k) \in \mathcal{CD}\} \in \mathbb{N}_0^{|\mathcal{OD}|}$, *a feasible solution* (π, z, t) *to (3.32)–(3.39) defines a feasible solution* $(\pi', \mathcal{OD}'_{op})$ *to I such that*

$$\overline{c}(\pi', \mathcal{OD}'_{op}) \le \sum_{(u_k, v_k) \in \mathcal{OD}} w_k t^k.$$

Proof. Let (π, z, t) be a feasible solution to the IP formulation (3.32)–(3.39). We define a timetable π' setting $\pi'_i := \pi_i$ for all $i \in \mathcal{E}_{\text{org}}$. Due to constraints (3.33) and (3.34), π' is a feasible timetable. Furthermore, z defines a set of operational OD-pairs $\mathcal{OD}'_{op} = \bigcup_{(u_k, v_k) \in \mathcal{OD}} \{(i, j) : z^k_{ij} = 1\}$. Since $z_{i_k j_k} = 1$ for $(i_k, j_k) \in \mathcal{OD}'_{op}$, it follows from (3.36) that

$$\overline{c}(\pi', \mathcal{OD}'_{op}) = \sum_{(i_k, j_k) \in \mathcal{OD}'_{op}} w_k (\pi_{j_k} - \pi_{i_k}) \le \sum_{(u_k, v_k) \in \mathcal{OD}} w_k t^k.$$

□

Lemma 3.7.7. *Let* $I = (\mathcal{N}, l, u, \mathcal{OD})$ *be an instance of TTwR, and let for every OD-pair* $(u_k, v_k) \in \mathcal{OD}$ *the number M^k be chosen in such a way that*

$$M^k \ge (\pi_{j_1} - \pi_{i_1}) - (\pi_{j_2} - \pi_{i_2})$$

$$\forall \, (i_1, j_1), (i_2, j_2) \in \mathcal{A}^k_{virt} \text{ and every feasible timetable } \pi \text{ for } \mathcal{N}'. \tag{3.40}$$

Then a solution $(\pi', \mathcal{OD}'_{op})$, *defined by an optimal solution* (π, z, t) *to the IP (3.32)–(3.39), is an optimal solution to the instance I of TTwR and*

$$\overline{c}(\pi', \mathcal{OD}'_{op}) = \sum_{(u_k, v_k) \in \mathcal{OD}} w_k t^k.$$

Proof. 1. Let $(\pi', \mathcal{R}')$ be a feasible solution to I. Let $\mathcal{N}^1_{op}, \ldots, \mathcal{N}^m_{op}$ with $\mathcal{N}^i_{op} = (\mathcal{E}^i_{op}, \mathcal{A}^i_{op})$ denote the connected components of $\mathcal{N}_{op}$, and let j_i, k_i be defined as in Notation 3.7.1. Like in Lemma 3.7.3, we define π as follows:

- We set $\pi_j := \pi'_j$ for all $j \in \mathcal{E}^1_{\text{op}}$. Then we proceed the connected components iteratively as follows:
- For $i = 2, \ldots, m$, we set $\pi_{k_i} = \pi_{j_{i-1}}$.
- We now build the timetable π in $\mathcal{N}^i_{\text{op}}$ setting $\pi_j = \pi'_j - \pi_{k_i}$.

Then for $(i, j) \in \mathcal{A}^i_{\text{op}}$, we have

$$\pi_j - \pi_i = (\pi'_j - \pi_{k_i}) - (\pi'_i - \pi_{k_i}) = \pi'_j - \pi'_i \in [l_{ij}, u_{ij}],$$

since π' is a feasible timetable on $\mathcal{N}$. Hence, constraints (3.33) and (3.34) are fulfilled. Note that since $\pi_{k_i} = \pi_{j_{i-1}}$ for every $i = 2, \ldots, m$, π is a feasible timetable on $\mathcal{N}'$. Denote by (i_k, j_k) the activity in $\mathcal{OD}'_{\text{op}} \cap \mathcal{A}^k_{\text{virt}}$. We define

$$z^k_{ij} := \begin{cases} 1 & \text{if } (i, j) = (i_k, j_k) \\ 0 & \text{otherwise} \end{cases} \quad \forall (u_k, v_k) \in \mathcal{OD}, (i, j) \in \mathcal{A}^k_{\text{virt}},$$

$$t^k := \pi'_{j_k} - \pi'_{i_k} \quad \forall (u_k, v_k) \in \mathcal{OD}.$$

Constraints (3.35) are fulfilled because $\mathcal{OD}'_{\text{op}}$ is a set of operational OD-pairs. For $(i_k, j_k) \in \mathcal{OD}'_{\text{op}}$, constraints (3.36) are fulfilled by definition. For $(i, j) \in \mathcal{A}^k_{\text{virt}} \setminus (i_k, j_k)$, we have

$$t^k \geq \pi_{j_k} - \pi_{i_k} = \pi_j - \pi_i - [(\pi_j - \pi_i) - (\pi_{j_k} - \pi_{i_k})]$$

$$\geq \pi_j - \pi_i - M^k = \pi_j - \pi_i - M^k(1 - z^k_{ij}).$$

Due to the definition of t^k, $\sum_{(u,v) \in \mathcal{OD}} w_k t^k = \overline{c}(\pi', \mathcal{OD}'_{\text{op}})$.

2. Let (π, z, t) be an optimal solution to (3.32)–(3.39). Due to Lemma 3.7.6, there is a solution $(\pi', \mathcal{OD}'_{\text{op}})$ with

$$\overline{c}(\pi', \mathcal{OD}'_{\text{op}}) \leq \sum_{(u_k, v_k) \in \mathcal{OD}} w_k t^k.$$

Suppose that for an optimal solution $(\pi^*, \mathcal{OD}^*_{\text{op}})$ to TTwR, it holds that

$$\overline{c}(\pi^*, \mathcal{OD}^*_{\text{op}}) < \sum_{(u_k, v_k) \in \mathcal{OD}} w_k t^k.$$

Due to (1), there is a solution (π^*, z^*, t^*) to (3.32)–(3.39) with

$$\sum_{(u_k, v_k) \in \mathcal{OD}} w_k t^*_k = \overline{c}(\pi^*, \mathcal{OD}^*_{\text{op}}) < \sum_{(u_k, v_k) \in \mathcal{OD}} w_k t^k.$$

This contradicts the optimality of (π, z, t). $\square$

3.7 Integer Programming Formulations

The minimal values which the numbers $\{M^k : (u_k, v_k) \in \mathcal{OD}\}$ are allowed to take such that (3.40) is satisfied can be rewritten as

$$\max_{\pi \text{ feasible for } \mathcal{N}'} \max_{((i_1,j_1),(i_2,j_2)) \in \mathcal{A}^k_{\text{virt}} \times \mathcal{A}^k_{\text{virt}}} (\pi_{j_1} - \pi_{i_1}) - (\pi_{j_2} - \pi_{i_2}).$$

Again, these values can be determined in polynomial time by solving a series of linear programs.

$$\max \ M^k((i_1, j_1), (i_2, j_2)) := (\pi_{j_1} - \pi_{i_1}) - (\pi_{j_2} - \pi_{i_2}), \tag{3.41}$$

$$\text{s.t. } \pi_h - \pi_g \geq l_{gh} \qquad \forall (g,h) \in \mathcal{A}', \tag{3.42}$$

$$\pi_h - \pi_g \leq u_{gh} \qquad \forall (g,h) \in \mathcal{A}', \tag{3.43}$$

$$\pi_g \in \mathbb{Z} \qquad \forall g \in \mathcal{E}_{\text{op}}, \tag{3.44}$$

for all $((i_1, j_1), (i_2, j_2)) \in \mathcal{A}^k_{\text{virt}} \times \mathcal{A}^k_{\text{virt}}$. Since the matrix defined by the constraints (3.42) and (3.43) is totally unimodular and we assume the bounds to be integer, we obtain a solution by linear programming. Then we can set

$$M^k := \max_{((i_1,j_1),(i_2,j_2)) \in \mathcal{A}^k_{\text{virt}} \times \mathcal{A}^k_{\text{virt}}} M^k((i_1, j_1), (i_2, j_2)).$$

We summarize our result in the following lemma.

Lemma 3.7.8. *The minimal values for $\{M^k : (u_k, v_k) \in \mathcal{OD}\}$ such that (3.26) and (3.27) are satisfied are*

$$M^k := \max_{\pi \text{ feasible for } \mathcal{N}'} \max_{((i_1,j_1),(i_2,j_2)) \in \mathcal{A}^k_{\text{virt}} \times \mathcal{A}^k_{\text{virt}}} (\pi_{j_1} - \pi_{i_1}) - (\pi_{j_2} - \pi_{i_2}).$$

These values can be calculated in polynomial time.

Like in Lemma 3.7.5, a bound that is faster to calculate can be obtained summing up the upper bounds on all operational activities.

Lemma 3.7.9. *Given an instance $(\mathcal{N}, l, u, \mathcal{OD})$ of TTwR, we set $M^k := \sum_{a \in \mathcal{A}_{op}} u_a$. Then (3.40) is satisfied.*

Proof. For every $(i, j) \in \mathcal{A}^k_{\text{virt}}$, there exists a (directed) path P_{ij} from i to j in $\mathcal{N}_{\text{op}}$. Hence

$$(\pi_{j_1} - \pi_{i_1}) - (\pi_{j_2} - \pi_{i_2}) = \sum_{(i,j) \in P_{i_1 j_1}} (\pi_j - \pi_i) - \sum_{(i,j) \in P_{i_2 j_2}} (\pi_j - \pi_i)$$

$$\leq \sum_{(i,j) \in P_{i_1 j_1}} (\pi_j - \pi_i)$$

$$\leq \sum_{(i,j)\in P_{i_1 j_1}} u_{ij}$$
$$\leq \sum_{a\in \mathcal{A}_{\mathrm{op}}} u_a.$$

□

Numerical evaluations and a comparison of the two integer programming formulations can be found in [Anh12, SS12b]. There, the virtual-activity-based formulation appears to be superior in almost all tested cases, which may be due to the lower number of constraints and variables used.

Chapter 4
Delay Management

4.1 Introduction to Delay Management

4.1.1 Delay Management Problems

Delay management deals with the short-term adaption of a given timetable to small delays as they occur in daily train operations. The main decisions to make are the *wait-depart decisions*: *Should a connecting train wait if a feeder train arrives with a delay?* If it does not wait, transferring passengers lose their connection; however, if it waits it will also receive a delay.

In this section we define the delay management problem relying on the definitions from the preceding chapters, in particular from Sects. 2.1 and 3.1. Like in the timetabling problem, we are given a public transportation network $\mathcal{PTN} = (S, E)$, a set of trains TR with specified lines in the PTN, a set of planned connections $\mathcal{C}$, and lower bounds l on the driving, waiting, and changing times. Furthermore, we are given a timetable π which specifies the arrival and departure times of the trains at the stations.

Delays and Delay Propagation

Delay management becomes relevant whenever delays in the train operations occur. When a train arrives late at a station, in most cases it will also depart later, arrive later at the following station, and so on, until the delay is absorbed by the slack times, i.e., the time buffers in the timetable. If a connecting train waits for passengers from a delayed feeder train, it will also receive a delay. Furthermore, delays may be transferred due to capacity restrictions, e.g., if a regional train has to stop to allow a delayed high-speed train to overtake or if trains have to wait outside stations until a platform becomes available.

We refer to the first occurrence of a delay as a *source delay*. That is, a source delay is one that is not caused by some other delay in the way described above. Source delays can be classified into two groups. We call a source delay *operational* if it arises because an operation (driving or waiting) takes longer than planned. Operational source delays are additive, i.e., if a train already has a delay of 5 min when it receives an additional operational source delay of 6 min, it will have a total delay of 11 min. If, on the other hand, delay is caused because an arrival or a departure cannot take place at the time it was initially scheduled because of e.g., the delayed arrival of a train driver or of a passengers from outside of the considered system (see, e.g., [GJPW07, KTZ11]), we speak of *external* source delays. In this case, if a train has a delay of 5 min and its departure is externally source-delayed by 6 min, it will have a total delay of 6 min.

Some delay management models can deal with both operational and external delays [DPP07, SS08, Bau10, Sch10], while others consider only operational [Dv01, GS07, CDPP12] or only external [GJPS05, Sch06, GJPW07, KTZ11, DHSS12] delays.

The spreading of the delay through the public transportation system depends on the slack times, the wait-depart decisions, and the capacity restrictions. It can be modeled in the framework of max-plus algebras [dDD98, Hd01, Gov05] or incorporated in the event-activity network (EAN) [Sch01, SS08, DSS11, DHSS12, Sch13] or similar graph models [DPP07, HGL08, CDPP12]. This is explained in more detail in Sect. 4.2.

Capacity Restrictions

If trains are rescheduled due to delays, capacity conflicts on tracks and in stations can arise. Hence, in addition to the wait-depart decisions, the planner has to make *rescheduling decisions*. While the delay management problem focuses on the wait-depart decisions in order to minimize passengers' travel time or delay, the *train rescheduling problem* investigates the question how trains should be rescheduled in order to find a feasible disposition timetable. To this end, the railway infrastructure is modeled on a microscopic level including blocks and signals, while most delay management models rely on the macroscopic PTN model as given in Definition 1.3.11.

Most delay management models consider only wait-depart decisions. Schachtebeck and Schöbel [SS08] and the articles [Sch09, Bau10, SS10a, Sch10, Bau10] include priority decisions on the tracks into the macroscopic delay management model. An additional reassignment of trains to platforms is done in [DSS11]. In [DPP07, CDPP12], a more detailed model is used for train rescheduling. Dollevoet et al. [DCDH12] iterate between macroscopic delay management and microscopic train rescheduling steps. Adenso-Díaz et al. [ADOGT99] consider a train rescheduling problem where it is additionally possible to cancel or reassign services.

4.1 Introduction to Delay Management

Online and Offline Delay Management

Since delays in many cases happen unforeseen, delay management is often regarded as an *online* problem [BHLS07, GJPW07, Bau10, KTZ11, KS11], i.e., information about delays is only revealed gradually.

However, it is also interesting to study delay management as an *offline* problem, i.e., to assume that all delays are known in advance, as it is mostly done in the literature. A possible application of offline delay management is to predict secondary delays when source delays are known. Thus, it is a valuable tool for short-time adaption of timetables when delays are known in advance, e.g., because of construction works. Also in an online framework, solution methods for offline delay management can be applied to successively re-optimize the wait-depart decisions whenever new delays get known [Bau10,KS11]. Furthermore, offline delay management is in NP, while PSPACE-hardness was shown for an online delay management problem allowing free choices of trains in [BHLS07].

Disposition Timetable

When delays occur, the initial timetable π cannot be operated anymore and is hence modified to a *disposition timetable x*. The disposition timetable takes into account both the source delays and the lower bounds on the durations of the driving and waiting of trains.

When a train arrives late at a station due to delays, it is possible that connecting trains have already departed; thus, it may not be possible anymore to transfer to the connecting train. We say that the connection is *dropped*, otherwise, it is *maintained*.

Demand and Passengers' Behavior

In train rescheduling [DPP07, CDPP12] and early delay management models [dDD98,Dv01,Hd01], passengers' demand is not taken into account in the modeling and optimization process.

Some delay management models model demand as *passenger loads* on activities, i.e., it is assumed that for every waiting, driving, and changing activities it is known how many passengers perform this activity (see, e.g., [GS07, SS08, Sch10, DSS11]).

Schöbel [Sch01] proposes a model where passengers' demand is given in form of a *routing* in the EAN, i.e., the passengers' paths in the PTN as well as the sequence of trains the passengers take are specified. It is assumed that the underlying timetable is periodic, i.e., that it repeats itself after a fixed time period T, and that whenever the route chosen for an OD-pair cannot be realized because of dropped connections, passengers continue their journey one period later. In the following we refer to this problem as *delay management with penalties*. It is studied in [GGJ+04, GJPS05, HGL08, Sch06, Sch07] as well.

However, there are also some models which assume that demand is given in form of *OD-pairs*. While in some delay management problem, like in line planning and timetabling, only origins and destinations are specified [GGJ+04], other models assume that also *start times* are given [DHSS09, DH11, DCDH12, DHSS12, Sch13]. The start times can be interpreted as the times the passengers arrive at the stations. Hence, the latter models assume that passengers learn about the delays when they arrive at the station.

In a model proposed in [BHLS07] for online delay management, a routing is given in the input, and passengers are not allowed to change the corresponding paths in the PTN; however, they can freely choose among the trains going along these paths. In [GGJ+04, DHSS09, DH11, DCDH12, DHSS12, Sch13], passengers are allowed to adapt their routes to the disposition timetable freely.

Objectives

Delay management aims at finding a trade-off between the inconvenience caused by dropped connections and the additional delay spreading in the system if connections are maintained.

To this end, in [dDD98, Hd01, Dv01], *train delays* and *penalties* for dropping connections are accumulated. In [Sch06, GS07, CDPP12], the delay management problem is considered as a bicriteria problem aiming at minimizing both the delay of the vehicles and the weighted number of missed connections.

Passenger-oriented models aim at minimizing *passengers' delay* or *travel time*. Since an originally planned travel route cannot be followed anymore in case that a connection on the route is canceled, some models assume that passengers can be rerouted and minimize the rerouted travel time [DHSS09, DH11, DCDH12, DHSS12, Sch13]. Instead of rerouting the passengers, other delay management models (see, e.g., [Sch01, GGJ+04, GJPS05, Sch06, Sch07, BHLS07, GJPW07, HGL08]) assume that the underlying timetable is periodic and back on time in the next period and hence estimate the travel time by imposing a *penalty* of the period length in case that a connection is dropped. In [Sch06, Sch07, SS08, Sch10, DSS11], a further simplification of the travel-time objective is investigated. Demand is no longer given in form of routings but in form of *passenger loads* on activities, and a penalty of T multiplied with the activity weight is applied whenever a connection a passenger wanted to use is dropped.

Krumke et al. [KTZ11] maximize the *profit* of the railway company, assuming a refund system for delayed passengers.

4.1.2 Literature Overview

Modeling Framework

Various models for delay management can be found in the literature. The EAN used in this book was introduced into the public transportation context for modeling timetabling problems (see references in Sect. 3.1.2) and used to model the delay management problem for the first time in [Sch01]. Besides the EAN-based model, there are models in the context of max-plus algebra and control theory [dDD98, Dv01, Hd01] and simulation approaches, see, e.g., [SM99, SMBG01, SBK01]. Train rescheduling is sometimes modeled as a job-shop scheduling problem [DPP07, CDPP12] using an alternative graph formulation which is similar to the EAN.

Computational Complexity of Offline Delay Management

In [Sch06], Schöbel shows that if all source delays have the same amount and all slack times are equal to 0, delay management with constant weights can be solved in polynomial time by linear programming. The complexity of (unrestricted) delay management with constant weights is still open.

In general, delay management with penalties is strongly NP-hard. (Strong) NP-hardness is shown in [GGJ$^+$04, GJPS05, Gat07] for many special cases of the problem. The complexity analysis accomplished in these publications also includes algorithms for polynomially solvable subcases. The special cases considered are generated by imposing restrictions on the network structure, the number of connections used per OD-pair, the slack time, the delay scenario, and the numbers of stations a train can visit. In most problems investigated in these publications, external source delays are considered which are brought into the system by passengers arriving from routes outside the system.

Delay management with routing (DMwR) is proven to be strongly NP-hard using a reduction from 3-SAT in [GGJ$^+$04]. More generally speaking, in [GJPS05], it is claimed that every special case considered in the complexity analysis of this paper which is (strongly) NP-hard for the model with fixed passenger routes is also (strongly) NP-hard for delay management with dynamic route choices. In [Sch13], DMwR is shown to be strongly NP-complete even if there is only one passenger. However, under a realistic assumption on the problem instances, delay management for one OD-pair can be solved in polynomial time [DHSS12, Sch13].

In [GS07, Sch06] it is shown that the bicriteria problem of minimizing the delay of the vehicles and the weighted number of missed connections simultaneously is NP-hard by reduction from the knapsack problem.

If additionally to the wait-depart decisions also decisions concerning the limited track capacities have to be taken, delay management with penalties is NP-hard, even in the special case where the wait-depart decisions are already made [CS07].

Solution Approaches in Offline Delay Management

When wait-depart decisions are fixed, the initial timetable can be updated in polynomial time to a disposition timetable by the critical path method [Sch06]. Variants of the delay management problem which can be solved in polynomial time by cut-based approaches and dynamic programming are studied in [Gat07].

However, no polynomial algorithm for delay management is known. Therefore, in most publications the problem is solved by integer programming or integer-programming-based heuristics.

When modeling delay management as an integer program, variables to construct the disposition timetable as well as variables keeping track of the wait-depart decisions are needed in most models.

To update the given timetable to a disposition timetable, in [GS07, SS08, SS10a, Sch10], variables for every departure and every arrival are introduced, and the conditions on the disposition timetable are established as linear constraints in the model. In [Sch01, Sch06, Sch07], the update to a disposition timetable is only implicitly modeled by propagating the delays through the network using delay variables and equivalent constraints. In [HGL08], this model is developed further such that variables neither for the disposition timetable nor for the delays are needed anymore.

To calculate the objective value of delay management with penalties, a straightforward approach is to introduce a variable for every route in the routing, which indicates whether all connections on the route are maintained or not. In the former case, the delay at the destination is counted, while in the latter a penalty of T is added to the objective function. A quadratic formulation and a linearization for this approach are proposed in [Sch01, Sch07, Sch06], and in [HGL08] a similar linear model is treated. Various integer programming formulations for *delay management with penalties* have been developed, analyzed, and compared in [Sch01, Sch07, Sch06, HGL08]. In [Sch06], Schöbel proves upper and lower bounds and network reduction results and develops a branch-and-bound algorithm for delay management with penalties based on this result. Heilporn et al. [HGL08] propose a branch-and-cut procedure and a constraint-generation approach to solve the problem. Furthermore, in [Sch06, Sch07] a nonlinear model with three types of variables based on the activities of the EAN is developed. The first type of variables indicates for every activity how many passengers use it. The second type of variables is introduced only for changing activities and indicates whether a changing activity is maintained or not. Thirdly for every changing activity and every route in the routing which uses the changing activity, a variable indicates whether the changing activity is reached without any anterior missed connection. Then the weighted delay on every maintained activity and the weighted penalty for dropped connections are combined to the objective function. A simplification of this nonlinear program, which fixes the passenger numbers to the values they would take without missed connections, yields a formulation for delay management with constant weights (see [Sch06, Sch07]), which is further used in [SS08, SS10a, Sch10]. Delay management with constant weights in general *overestimates* the overall delay [Sch06, Sch07].

4.1 Introduction to Delay Management

However, in a setting where every node of the EAN is influenced by at most one source delay (this is called the *never-meet property*), the simplified model is equivalent to the original one. Delay management with constant weights is solved by a branch-and-bound procedure or, in case that the never-meet property holds, by decomposing the problem into subproblems, in [Sch06].

In [Sch06, GS07, CDPP12] delay management is modeled as a bicriteria problem minimizing simultaneously the sum over the delays and the number of missed connections. Also in these publications, variables based on changing activities are introduced to count the number of missed connections. An algorithm to find all efficient solutions is given.

Dollevoet et al. [DHSS09, DHSS12] propose and solve an integer programming formulation for DMwR. Since integer programming is impractical for large problem instances, in [DH11], Dollevoet and Huisman test various heuristic solution approaches for DMwR.

Delay management including priority decisions is analyzed in [Sch10, SS10a, SS08, Sch09]. Solution approaches include integer programming and variable-fixing heuristics. In [DPP07], a microscopic model for a train rescheduling problem is formulated as a job-shop problem and solved using branch-and-bound. Dollevoet et al. [DCDH12] iterate between delay management and train scheduling to use a delay management problem including capacity constraints on a microscopic level.

Online Delay Management

Since delays in many cases happen unforeseen, delay management is often regarded as an online problem. A competitive analysis for online delay management on a single line is done in [GJPW07, Gat07, KTZ11]. Different strategies for online delay management have been compared numerically in [Bau10, KS11]. In [BHLS07] it is shown that the considered online delay management problem is PSPACE-hard and a simulation platform to compare different heuristics for online delay management is presented.

Related Problems

When rescheduling trains in case of delays, it is also possible to skip services or to replace them by other train units. This extended rescheduling problem is studied, e.g., in [ADOGT99].

Other approaches to deal with delays in the daily railway operations are to give delay-resistant travel advice to passengers [GKMH$^+$13, GHMH$^+$13] or to construct delay-resistant timetables (see references in Sect. 3.1.2). In particular the concept of recoverable robust timetabling [CDS$^+$09, LLMS09, Sch10] or recoverable robust timetable information [GHMH$^+$13], where the possibility to reroute passengers is explicitly taken into account, is related to the delay management problem.

Anderegg et al. [APW02] consider a problem where buses arrive with an unknown delay at an origin station and the waiting time at this station has to be minimized.

4.1.3 The Delay Management Problem Considered in This Book

In this section we define delay management with routing (DMwR) in the context of the PTN relying on the definitions and notations from the preceding chapters.

Delays

In the remainder of this book we assume *external* source delays which occur at arrivals and departures of trains.

Definition 4.1.1. Given a PTN $\mathcal{PTN} = (S, E)$ and a set of trains TR, to every departure and every arrival of a train tr $\in$ TR at a station $s \in S$, an (external) *source delay* $d_{(tr-s-Dep)} \in \mathbb{N}_0$ (or $d_{(tr-s-Arr)} \in \mathbb{N}_0$, respectively) is assigned.

Note that in realistic instances, for most events the associated source delay is 0. If $d_{(tr-s-Dep)} > 0$ (or $d_{(tr-s-Arr)} > 0$), we say that the departure (or arrival) of train tr at station s is *delayed*. Otherwise, it is *not delayed*.

Disposition Timetable

Like in timetabling, we denote the lower bounds on driving, waiting, and changing times by vectors l of nonnegative integers.

Given a PTN $\mathcal{PTN} = (S, E)$, a set of trains TR, lower bounds l on the driving and waiting times, a timetable π, and source delays d, a *disposition timetable* x is a timetable (as defined in Definition 3.1.1) that satisfies

$$x_{(tr-s-Dep)} \geq \pi_{(tr-s-Dep)} + d_{(tr-s-Dep)} \ \forall \text{ departures of trains tr at stations } s,$$

$$x_{(tr-s-Arr)} \geq \pi_{(tr-s-Arr)} + d_{(tr-s-Arr)} \ \forall \text{ arrivals of trains tr at stations } s.$$

and

$$x_{(tr-s-Dep)} - x_{(tr-s'-Arr)} \geq l_{(tr-s-Dep,tr-s'-Arr)}$$
$$\forall \text{ driving times of train tr on edge } \{s, s'\},$$
$$x_{(tr-s-Arr)} - x_{(tr-s-Dep)} \geq l_{(tr-s-Arr,tr-s-Dep)}$$
$$\forall \text{ waiting times of train tr at station } s.$$

4.1 Introduction to Delay Management

We say that a connection $(\text{tr}, \text{tr}', s) \in \mathcal{C}$ is *maintained* under disposition timetable x if

$$x_{(\text{tr}'-s-\text{Dep})} - x_{(\text{tr}-s-\text{Arr})} \geq l_{(\text{tr}-s-\text{Dep},\text{tr}'-s-\text{Arr})}$$

for the lower bound $l_{(\text{tr}-s-\text{Dep},\text{tr}'-s-\text{Arr})}$ on connection $(\text{tr}, \text{tr}', s)$. Otherwise, $(\text{tr}, \text{tr}', s)$ is *dropped*.

Demand, Passenger Routes, and Travel Time

Different to our assumptions in the considered line planning and timetabling problems, in delay management, we assume that passengers are already at the origin station when they are informed about delays. Thus, other than in Definition 2.1.3, in delay management, the OD-pairs include the start times. That is, the *passengers' demand* is given in form of an *origin*, a *destination*, and a *start time* $(u, v, \sigma) \subset S \times S \times \mathbb{Z}$. For every so-defined *OD-pair* (u, v, σ), the *number of passengers* or *passenger weight* $w_{uv\sigma}$ is specified.

We restrict the train-routes which a passenger can take to the ones starting not earlier than the passenger's arrival at the origin station.

Definition 4.1.2. Given a timetable π, a *start-time-train-route (ST-train-route)* $P_{uv\sigma}$ for an OD-pair $(u, v, \sigma) \in \mathcal{OD}$ is a train-route for (u, v) as defined in Definition 3.1.2, satisfying $\pi_{(\text{tr}-u-\text{Dep})} \geq \sigma$ for the train tr chosen for the departure at the origin.

Like in the preceding chapters, when the set of OD-pairs is indexed, e.g., $\mathcal{CD} = \{(u_k, v_k, \sigma_k) : k \in K\}$ for an index set K, for the sake of simplicity, we also use the notation $w_k := w_{u_k v_k \sigma_k}$ and $P_k := P_{u_k v_k \sigma_k}$.

Due to dropped connections, it is possible that some ST-train-routes are not realizable anymore. An ST-train-route $P_{uv\sigma}$ is *feasible for a disposition timetable* x if the connections from one train to another, defined by $P_{uv\sigma}$, are maintained. We can regard the feasibility from a different perspective, i.e., require the disposition timetable to maintain all connections on a given ST-train-route: A disposition timetable x is *feasible for an ST-train-route* $P_{uv\sigma}$ if the connections from one train to another, defined by $P_{uv\sigma}$, are maintained.

These feasibility decisions are equivalent in the sense that if and only if an ST-train-route $P_{uv\sigma}$ is feasible for a disposition timetable x, x is feasible for $P_{uv\sigma}$.

Since we assume that a passenger arrives at a given time σ at his/her origin station, his/her travel time along an ST-train-route $P_{uv\sigma}$ is calculated as the time difference between the start time σ and the arrival time at the destination, according to the disposition timetable:

$$c(x, P_{uv\sigma}) := x_{(\text{tr}(v,\text{Arr})-v-\text{Arr})} - \sigma,$$

where $\text{tr}(v, \text{Arr})$ is the train chosen for the arrival at station v in $P_{uv\sigma}$. See Lemma 3.1.3 for the different definition in timetabling.

An *ST-train-routing* $\mathcal{R}$ is a set containing exactly one ST-train-route for every OD-pair $(u, v, \sigma) \in \mathcal{OD}$, i.e., $\mathcal{R} := \{P_{uv\sigma} : (u, v, \sigma) \in \mathcal{OD}\}$. An ST-train-routing $\mathcal{R}$ is *feasible for a disposition timetable* x if every ST-train-route $P_{uv\sigma} \in \mathcal{R}$ is feasible for x. A disposition timetable x is *feasible for an ST-train-routing* $\mathcal{R}$ if it is feasible for every ST-train-route $P_{uv\sigma} \in \mathcal{R}$. Again, both feasibility definitions are equivalent in the sense that ST-train-routing $\mathcal{R}$ is feasible for disposition timetable x if and only if x is feasible for $\mathcal{R}$.

Also in delay management our objective is to minimize the passengers' overall *travel time* $c(x, \mathcal{R}) := \sum_{P_{uv\sigma} \in \mathcal{R}} w_{uv\sigma} c(x, P_{uv\sigma})$ over all feasible combinations of disposition timetables x and ST-train-routings $\mathcal{R}$.

Delay Management with Routing

We are now ready to define the problem DMwR from [DHSS09, DHSS12].

Definition 4.1.3. An instance $(\mathcal{PTN}, \mathrm{TR}, \mathcal{C}, l, \pi, \mathcal{OD}, d)$ of delay management with routing (DMwR) consists of a public transportation network $\mathcal{PTN}$, a set of trains TR with specified lines in the PTN, a set of planned connections $\mathcal{C}$, lower bounds on driving, waiting, and changing times l, a timetable π, a set of OD-pairs $\mathcal{OD}$, and source delays d. The task is to find a disposition timetable x and a feasible ST-train-routing $\mathcal{R}$ for x that minimize the overall travel time $c(x, \mathcal{R})$.

Given a connection $(\mathrm{tr}, \mathrm{tr}', s) \in \mathcal{C}$, the question whether $(\mathrm{tr}, \mathrm{tr}', s)$ is maintained in the disposition timetable or not is referred to as a *wait-depart decision*.

In many papers, delay management is regarded from the perspective of the wait-depart decisions which relies on the insight that, given a set of maintained connections, a disposition timetable and an ST-train-routing which are optimal for the chosen set of maintained connections can easily be constructed (see, e.g., [Sch06]). Hence, the task of delay management can be reformulated as the decision about which connections should be maintained and which connections should be dropped. This idea is explained in more detail in Sect. 4.3.

4.1.4 Outline of the Delay Management Chapter

We use EANs for modeling DMwR. The slight modification in the definition of an EAN compared to Sect. 3.2 that is necessary in the definition of an EAN for DMwR is explained in the next section, and DMwR is formulated in terms of the EAN in Definition 4.2.5.

In Sect. 4.3 we investigate how a given disposition timetable, a given set of maintained connections, or a given routing define a solution to DMwR.

Section 4.4 comprises a complexity analysis of DMwR. In Sects. 4.4.1 and 4.4.2 we concentrate on instances of DMwR with a single OD-pair. This problem is shown to be strongly NP-hard in Theorem 4.4.1. However, under some assumptions on the network structure, Algorithm 1 presented in Sect. 4.4.2 solves DMwR to optimality. Section 4.4.2 in addition contains further analysis of Algorithm 1, namely, it is shown that even if the provided solution is not optimal, it yields a lower bound as well as a two-approximation on the optimal objective value.

In Sect. 4.4.3 we investigate whether Algorithm 1 can be extended to find an optimal solution for instances of DMwR in which all OD-pairs have the same origin under the mentioned assumption on the network structure. Then, in Sect. 4.4.4, we show that for general instances of DMwR there can be no approximation algorithm, but that for instances in which the delay is bounded with respect to the travel time an easy approximation can be given.

Finally, in Sect. 4.5 we extend an integer program for delay management with constant weights to deal with DMwR.

4.2 Modeling Delay Management with Routing Using Event-Activity Networks

EANs as used in Chap. 3 are also used to model delay management problems (see, e.g., [Sch06, Sch07, DHSS09, SS10a, DHSS12]). The construction of these networks from a public transportation network $\mathcal{PTN}$, a set of trains TR, a set of planned connections $\mathcal{C}$, and a set of OD-pairs $\mathcal{OD} = \{(u_k, v_k) : k \in K\}$ is explained in detail in Sect. 3.2. Therefore, in this section we limit ourselves to point out the small difference of the EAN used in DMwR to the EAN used in TTwR which is due to the additional information of the start time in the set of OD-pairs $\mathcal{OD} = \{(u_k, v_k, \tau_k) : k \in K\}$ that we are given in DMwR. To this end, let $(\mathcal{PTN}, \text{TR}, \mathcal{C}, l, \pi, \mathcal{OD}, d)$ be an instance of DMwR.

To include the start time we define a mapping org' which maps an OD-pair (u, v, σ) to an *origin node* $u^{\text{org}}(\sigma) := \text{org}'(u, v, \sigma)$.

Definition 4.2.1. We define

- the set of *origin events*

$$\mathcal{E}'_{\text{org}} := \text{org}'(\mathcal{OD}) = \{u^{\text{org}}(\sigma) : (u, v, \sigma) \in \mathcal{OD}\}$$

- and the set of *origin activities* $\mathcal{A}'_{\text{org}} \subset \mathcal{E}'_{\text{org}} \times \mathcal{E}_{\text{dep}}$

$$\mathcal{A}'_{\text{org}} := \{(u^{\text{org}}(\sigma), \text{tr}-u-\text{Dep}) : u^{\text{org}}(\sigma) \in \mathcal{E}'_{\text{org}}, (\text{tr}-u-\text{Dep}) \in \mathcal{E}_{\text{dep}},$$

$$\pi_{(\text{tr}-u-\text{Dep})} \geq \sigma\}.$$

In other words, compared to Definitions 3.2.2 and 3.2.4, we add the start time to the description of the origin nodes and connect each origin node $u_k^{\text{org}}(\sigma)$ only to departure event $(\text{tr} - u - \text{Dep})$ at station u if the departure of train tr at station u is scheduled after the OD-pair's start time. Hence, a passenger cannot leave the station earlier than planned. Operational and destination events and activities are defined as in Definitions 3.2.1–3.2.4.

In this chapter the EAN is given by $\mathcal{N} := (\mathcal{E}, \mathcal{A})$ with

$$\mathcal{E} := \mathcal{E}_{\text{op}} \cup \mathcal{E}'_{\text{org}} \cup \mathcal{E}_{\text{dest}} \quad \text{and} \quad \mathcal{A} := \mathcal{A}_{\text{op}} \cup \mathcal{A}'_{\text{org}} \cup \mathcal{A}_{\text{dest}}.$$

Note that in models where the a routing or passenger loads are fixed in the input (see, e.g., [Sch06]), the smaller operational EAN $\mathcal{N}_{\text{op}} = (\mathcal{E}_{\text{op}}, \mathcal{A}_{\text{op}})$ is used and referred to as "EAN".

We reinterpret the source delays as assignments $d_i \in \mathbb{N}_0$ to the events $i \in \mathcal{E}_{\text{op}}$ and transfer the definition of disposition timetables to EANs.

Definition 4.2.2. Given an EAN $\mathcal{N} = (\mathcal{E}, \mathcal{A})$ with lower bounds l_a on the operational activities $a \in \mathcal{A}_{\text{op}}$, a timetable π, and source delays d_e on the events $e \in E$, a *disposition timetable* x is a timetable that satisfies

$$x_i \geq \pi_i + d_i \ \forall i \in \mathcal{E} \quad \text{and} \quad x_j \geq x_i + l_{ij} \ \forall (i,j) \in \mathcal{A}_{\text{drive}} \cup \mathcal{A}_{\text{wait}}.$$

In delay management the driving and waiting activities impose operational restrictions on the vehicles, whereas a changing activity does not imply that a train has to wait in case of a delay of another train. In fact, the wait-depart decision can be regarded as the main decision we want to take during the optimization process. Given a timetable π, we denote by

$$\mathcal{A}_{\text{fix}}(x) := \{(i,j) \in \mathcal{A}_{\text{change}} : x_j - x_i \geq l_{ij}\} \tag{4.1}$$

the set of *maintained changing activities*.

Analogously to Sects. 2.2 and 3.2 every ST-train-route in the PTN represents a path in the EAN and vice versa. Hence, for a path $P_k := (u_k^{\text{org}}(\sigma_k), i_1^k, \ldots, i_{n_k}^k, v_k^{\text{dest}})$ in $\mathcal{N}$ and a disposition timetable x, we calculate the travel time as

$$c(x, P_k) = x_{i_{n_k}^k} - \sigma.$$

A *routing* in the EAN is a collection $\mathcal{R} := \{P_k : (u_k, v_k, \sigma_k) \in \mathcal{OD}\}$ with P_k being a path from $u_k^{\text{org}}(\sigma_k)$ to v_k^{dest} in the EAN. For a timetable π and a routing $\mathcal{R}$ in the EAN, we can interpret the overall *travel time* as

$$c(x, \mathcal{R}) = \sum_{(u_k, v_k, \sigma_k) \in \mathcal{OD}} w_k c(x, P_k).$$

The feasibility definitions from Sect. 4.1.1 can now be translated to the EAN.

Definition 4.2.3. A routing $\mathcal{R}$ is *feasible for a disposition timetable x* (or, equivalently, a *disposition timetable x is feasible for a routing $\mathcal{R}$*) if and only if

$$P_k \cap \left(\mathcal{A}_{\text{change}} \setminus \mathcal{A}_{\text{fix}}(x)\right) = \emptyset \quad \forall P_k \in \mathcal{R}.$$

4.2 Modeling Delay Management with Routing Using Event-Activity Networks

Hence, a routing is feasible for a disposition timetable x if and only if the corresponding ST-train-routing is feasible for x. (And a disposition timetable x is feasible for a routing if and only if it is feasible for the corresponding ST-train-routing.)

Let x be a disposition timetable. To illustrate the paths which can be chosen for the OD-pairs and their lengths, we define a network that excludes all dropped changing activities and takes as arc lengths the time distance between the incident events:

Definition 4.2.4. Given an EAN $\mathcal{N}$ and a disposition timetable x, we define the *routing network* $\mathcal{N}(x) := (\mathcal{E}, \mathcal{A}(x))$ with

$$A(x) := \mathcal{A}_{\text{drive}} \cup \mathcal{A}_{\text{wait}} \cup \mathcal{A}_{\text{fix}}(x) \cup \mathcal{A}'_{\text{org}} \cup \mathcal{A}_{\text{dest}}.$$

The length of $(i, j) \in \mathcal{A}(x)$ is

$$L_{ij} := \begin{cases} \pi_j - \pi_i & \text{if } (i, j) \in \mathcal{A}_{\text{op}} \cap \mathcal{A}(x), \\ \pi_j - \sigma_k & \text{if } j \in \mathcal{E}_{\text{op}} \text{ and } i = u_k^{\text{org}}(\sigma_k) \text{ for an OD-pair } (u_k, v_k, \sigma_k) \in \mathcal{OD}, \\ 0 & \text{if } i \in \mathcal{E}_{\text{op}} \text{ and } j = v_k^{\text{org}} \text{ for an OD-pair } (u_k, v_k, \sigma_k) \in \mathcal{OD}. \end{cases}$$

We can now redefine DMwR as stated in Definition 4.1.3.

Definition 4.2.5. An instance $I = (\mathcal{N}, l, \pi, \mathcal{OD}, d)$ of DMwR consists of an EAN $\mathcal{N}$, lower bounds on the lengths of the operational activities l, a timetable π, a set of OD-pairs $\mathcal{OD}$, and a set of source delays d. The task is to find a disposition timetable x and a feasible routing $\mathcal{R}$ for x in $\mathcal{N}(x)$ that minimize the overall travel time $c(x, \mathcal{R})$.

Since they describe the same problem, in the following sections we equivalently use Definitions 4.1.3 and 4.2.5 and the associated definitions and notations for representing DMwR as a problem in the PTN or the EAN.

Note that the objective of DMwR can equivalently be described as the minimization of delays instead of travel times. Since this alternative point of view is useful in some explanations and proofs in the remainder of this chapter, we introduce it formally. To this end, let $I = (\mathcal{N}, l, \pi, \mathcal{OD}, d)$ be an instance of DMwR. For every $(u_k, v_k, \sigma_k) \in \mathcal{OD}$, let P_k denote a path from $u_k^{\text{org}}(\sigma_k)$ to v_k^{dest} in $\mathcal{N}$ with last operational node j_k that minimizes $SP_k := \pi_{j_k} - \sigma_k$. (Such a path can easily be found by shortest path calculations in $\mathcal{N}(\pi)$.) We call SP_k the *nominal travel time for OD-pair* (u_k, v_k, σ_k).

The overall *delay* of a feasible solution $(x, \mathcal{R})$ to I is

$$c^d(x, \mathcal{R}) := \sum_{P_k \in \mathcal{R}} w_k(c(x, P) - SP_k) \tag{4.2}$$

and the *delay in a node* $e \in \mathcal{E}_{\text{op}}$ is defined as $d(x, e) = x_e - \pi_e$.

Naturally, the overall delay can be calculated from the delays in the last arrival nodes of every OD-pair's path.

Lemma 4.2.6. *Let $(x, \mathcal{R})$ be a feasible solution to I with $\mathcal{R} = \{P_k : (u_k, v_k, \sigma_k) \in \mathcal{OD}\}$. Denote by j_k the last arrival node on a shortest path from $u_k^{org}(\sigma_k)$ to v_k^{dest} in $N(\pi)$. Let j_k' denote the last node on $P_k \in \mathcal{R}$. Then*

$$c^d(x, \mathcal{R}) = \sum_{P_k \in \mathcal{R}} w_k(x_{j_k'} - \pi_{j_k}).$$

Proof.

$$c^d(x, \mathcal{R}) := \sum_{P_k \in \mathcal{R}} w_k(c(x, P_k) - SP_k) = \sum_{P_k \in \mathcal{R}} w_k(c(x, P_k) + \sigma_k - (SP_k + \sigma_k))$$

$$= \sum_{(u_k, v_k, \sigma_k) \in \mathcal{OD}} w_k(x_{j_k'} - \pi_{j_k}).$$

□

Since SP_k is constant for every $(u_k, v_k, \sigma_k) \in \mathcal{OD}$, the overall delay c^d is an alternative objective function for DMwR:

Lemma 4.2.7. *Let I be an instance of DMwR and $(x, \mathcal{R})$ a solution to I. $(x, \mathcal{R})$ minimizes c if and only if $(x, \mathcal{R})$ minimizes c^d.*

4.3 Solving Delay Management When Part of the Solution Is Fixed

In this section we show that solutions to DMwR can be defined in a sensible way by the specification of a disposition timetable x, a set of maintained changing activities $\mathcal{A}_{\text{fix}}$, or a routing $\mathcal{R}$. Furthermore, we prove some results about the so-defined solutions which we use in the proofs in the following sections and give some more formal prerequisites for the next sections.

First, suppose that a disposition timetable x is given. The set of corresponding maintained changing activities $\mathcal{A}_{\text{fix}}(x)$ is given as

$$\mathcal{A}_{\text{fix}}(x) := \{(i, j) \in \mathcal{A}_{\text{change}} : x_j - x_i \geq l_{ij}\}$$

according to (4.1).

We now define the corresponding routings:

Definition 4.3.1. *Let $I = (\mathcal{N}, l, \pi, \mathcal{OD}, d)$ be an instance of DMwR and let x be a disposition timetable for $\mathcal{N}$. If there exists a feasible routing for x, for every OD-pair (u_k, v_k, σ_k), we define $\mathcal{P}_{u_k v_k \sigma_k}(x) = \mathcal{P}_k(x)$ to be the set of shortest paths $P_k(x)$ from $u_k^{org}(\sigma_k)$ to v_k^{dest} in the routing network $\mathcal{N}(x)$.*

4.3 Solving Delay Management When Part of the Solution Is Fixed

- The set of *routings corresponding to* x is

$$\text{OPT}(x) := \{\mathcal{R}(x) : \mathcal{R}(x) = \{P_1(x), \ldots, P_{|\text{OD}|}(x)\} \text{ with } P_k(x) \in \mathcal{P}_k(x)\}.$$

- The set of *solutions corresponding to* x is defined as $\{(x, \mathcal{R}(x)) : \mathcal{R}(x) \in \text{OPT}(x)\}$.

Since $\mathcal{R}(x)$ is defined as a shortest-path routing, Lemma 4.3.2 follows directly:

Lemma 4.3.2. *Let* $I = (\mathcal{N}, l, \pi, \mathcal{OD}, d)$ *be an instance of DMwR, and let* x *be a disposition timetable for* $\mathcal{N}$. *If there is a feasible routing for* x, *for all* $\mathcal{R}(x) \in \text{OPT}(x)$, *it holds that*

$$c(x, \mathcal{R}(x)) \leq c(x, \mathcal{R}')$$

for all routings R' *for* $\mathcal{OD}$ *that are feasible for* x.

In order to define a solution corresponding to a given set of maintained changing activities $\mathcal{A}_{\text{fix}}$, we consider the *delay management problem with fixed connections* defined in [Sch06]:

Definition 4.3.3 ([Sch06]). An instance $(\mathcal{N}, l, \pi, \mathcal{A}_{\text{fix}}, w, d)$ of *delay management with fixed connections (DMwFC)* consists of an EAN $\mathcal{N}$, lower bounds l on the operational activities, a timetable π, a set of maintained changing activities $\mathcal{A}_{\text{fix}}$, weights $w := \{w_e \in \mathbb{N} : e \in \mathcal{E}_{\text{op}}\}$ for the events, and source delays d. The task is to find a disposition timetable x such that all changing activities in $\mathcal{A}_{\text{fix}}$ are maintained and $\sum_{e \in \mathcal{E}} w_e(x_e - \pi_e)$ is minimized.

The result from [Sch06], which we cite in Lemma 4.3.5, states that the solution to this problem can be calculated, independently of the weights w_e for $e \in \mathcal{E}_{\text{op}}$, as follows:

Definition 4.3.4. Consider an EAN $\mathcal{N}$ with lower bounds l_a on the operational activities $a \in \mathcal{A}_{\text{op}}$, a timetable π on $\mathcal{N}$ and source delays d_e on the operational events $e \in \mathcal{E}_{\text{op}}$. Given a set of maintained changing activities $\mathcal{A}_{\text{fix}} \subset \mathcal{A}_{\text{change}}$, we define the disposition timetable $x(\mathcal{A}_{\text{fix}})$ as

$$x(\mathcal{A}_{\text{fix}})_j := \max\{\pi_j + d_j, \max_{i:(i,j) \in \mathcal{A}_{\text{drive}} \cup \mathcal{A}_{\text{wait}} \cup \mathcal{A}_{\text{fix}}} \{x(\mathcal{A}_{\text{fix}})_i + l_{ij}\}\}.$$

This method to generate a new disposition timetable is called *critical path method* [Sch06].

Note that in general, for a given set of maintained changing activities $\mathcal{A}'_{\text{fix}}$

$$\mathcal{A}_{\text{fix}}(x(\mathcal{A}'_{\text{fix}})) = \{(i, j) \in \mathcal{A}_{\text{change}} : x(\mathcal{A}'_{\text{fix}})_j - x(\mathcal{A}'_{\text{fix}})_i \geq l_{ij}\} \supseteq \mathcal{A}'_{\text{fix}}.$$

Lemma 4.3.5 ([Sch06]). *Let* $I = (\mathcal{N}, l, \pi, \mathcal{A}_{\text{fix}}, w, d)$ *be an instance of DMwFC. The disposition timetable* $x(\mathcal{A}_{\text{fix}})$ *is an optimal solution to* I.

This implies the following result for DMwR.

Definition 4.3.6. Let $I = (\mathcal{N}, l, \pi, \mathcal{OD}, d)$ be an instance of DMwR and let $\mathcal{A}_{\text{fix}}$ be a subset of the changing activities.

- The *disposition timetable corresponding to* $\mathcal{A}_{\text{fix}}$ is $x(\mathcal{A}_{\text{fix}})$ as defined in Definition 4.3.4.
- The set of *routings corresponding to* $\mathcal{A}_{\text{fix}}$ is $\text{OPT}(x(\mathcal{A}_{\text{fix}}))$.
- The set of *solutions corresponding to* $\mathcal{A}_{\text{fix}}$ is

$$\{(x(\mathcal{A}_{\text{fix}}), \mathcal{R}(x(\mathcal{A}_{\text{fix}}))) : \mathcal{R}(x(\mathcal{A}_{\text{fix}})) \in \text{OPT}(x(\mathcal{A}_{\text{fix}}))\}.$$

Lemma 4.3.7. *Let $I = (\mathcal{N}, l, \pi, \mathcal{OD}, d)$ be an instance of DMwR and let $\mathcal{A}_{fix}$ be a subset of the changing activities. Then for every $\mathcal{R}(x(\mathcal{A}_{fix})) \in OPT(x(\mathcal{A}_{fix}))$*

$$c(x(\mathcal{A}_{fix}), \mathcal{R}(x(\mathcal{A}_{fix}))) \le c(x', \mathcal{R}')$$

for every disposition timetable x' for which $\mathcal{A}_{fix} \subseteq \mathcal{A}_{fix}(x')$ and every routing $\mathcal{R}'$ in $\mathcal{N}(x')$.

Proof. Consider $e \in \mathcal{E}$. Setting weights

$$w_i^e := \begin{cases} 1 & \text{if } i = e \\ 0 & \text{otherwise.} \end{cases}$$

Lemma 4.3.5 yields $x(\mathcal{A}_{\text{fix}})_e \le x'_e$. It follows that

$$\begin{aligned}
c(x(\mathcal{A}_{\text{fix}}), \mathcal{R}(x(\mathcal{A}_{\text{fix}}))) &= \sum_{P_k \in \mathcal{R}(x(\mathcal{A}_{\text{fix}}))} w_k c(x(\mathcal{A}_{\text{fix}}), P_k) \\
&\le \sum_{P'_k \in \mathcal{R}'} w_k c(x(\mathcal{A}_{\text{fix}}), P'_k) \\
&= \sum_{P'_k \in \mathcal{R}'} w_k \left(x(\mathcal{A}_{\text{fix}})_{i_k} - \sigma_k \right) \\
&\le \sum_{P'_k \in \mathcal{R}'} w_k \left(x'_{i_k} - \sigma_k \right) \\
&= c(x', \mathcal{R}'),
\end{aligned}$$

where i_k denotes the last arrival event on path P'_k. □

The following two observations on disposition timetables will be used in the following sections.

Lemma 4.3.8. *Let $\mathcal{N}$ be an EAN with timetable π and lower bounds l. Let $\mathcal{A}_{fix}$ be a subset of the changing activities. For $i', j' \in \mathcal{E}_{op}$ with $i' \le_{\mathcal{N}(x(\mathcal{A}_{fix}))} j'$ denote by $\mathcal{P}(i', j', \mathcal{A}_{fix})$, the set of paths from i' to j' in $\mathcal{N}(x(\mathcal{A}_{fix}))$, with $\mathcal{P}(j', j', \mathcal{A}_{fix}) := \{\emptyset\}$. Then*

4.3 Solving Delay Management When Part of the Solution Is Fixed

$$x(\mathcal{A}_{fix})_{j'} = \max_{i' \in \mathcal{E}_{op}: i' \leq_{\mathcal{N}(x(\mathcal{A}_{fix}))} j'} \max_{P \in \mathcal{P}(i',j',\mathcal{A}_{fix})} \left\{ \pi_{i'} + d_{i'} + \sum_{(i,j) \in P} l_{ij} \right\}.$$

This follows by induction from Definition 4.3.6.

Corollary 4.3.9. *Let $\mathcal{N}$ be an EAN with timetable π and lower bounds l. Let $\mathcal{A}_{fix}$ be a subset of the changing activities. Then for every $j \in \mathcal{E}_{op}$*

$$x(\mathcal{A}_{fix})_j \leq \pi_j + \max_{i \in \mathcal{E}_{op}} d_i.$$

Other than, e.g., in *delay management with penalties* or in *delay management with constant weights* (see, e.g., [Sch06] or other papers referenced in Sect. 4.1.2), in DMwR for every OD-pair a path in the EAN has to be found. Hence, the routing defines a set of maintained changing activities and a timetable:

Definition 4.3.10. *Let $I = (\mathcal{N}, l, \pi, \mathcal{OD}, d)$ be an instance of DMwR and let $\mathcal{R}$ be a routing for $\mathcal{OD}$.*

- The set of *maintained changing activities corresponding to $\mathcal{R}$* is

$$\mathcal{A}_{\text{fix}}(\mathcal{R}) := \bigcup_{P_k \in \mathcal{R}} \left(P_k \cap \mathcal{A}_{\text{change}} \right).$$

- The *disposition timetable corresponding to $\mathcal{R}$* is $x(\mathcal{R})$ where $x(\mathcal{R}) := x(\mathcal{A}_{\text{fix}}(\mathcal{R}))$ with $\mathcal{A}_{\text{fix}}(\mathcal{R})$ as defined above.
- The *solution corresponding to $\mathcal{R}$* is $(x(\mathcal{R}), \mathcal{R})$.

Corollary 4.3.11. *Let $I = (\mathcal{N}, l, \pi, \mathcal{OD}, d)$ be an instance of DMwR and let $\mathcal{R}$ be a routing for $\mathcal{OD}$. Then*

$$c(x(\mathcal{R}), \mathcal{R}) \leq c(x', \mathcal{R})$$

for every disposition timetable x' for which $\mathcal{A}_{fix}(\mathcal{R}) \subseteq \mathcal{A}_{fix}(x')$.

Proof.

$$c(x(\mathcal{R}), \mathcal{R}) = c(x(\mathcal{A}_{\text{fix}}(\mathcal{R})), \mathcal{R}) \leq c(x', \mathcal{R}),$$

according to Lemma 4.3.7. □

In this sense, the specification of a disposition timetable, a set of maintained changing activities, or a routing determines a solution to DMwR.

Notation 4.3.12. *Let $I = (\mathcal{N}, l, \pi, \mathcal{OD}, d)$ be an instance of DMwR. In the following we sometimes describe a solution to I stating only*

- *the disposition timetable x referring to a corresponding full solution $(x, \mathcal{R}(x))$ with $\mathcal{R}(x)$ being an arbitrary routing in $OPT(x)$*

- the set of maintained changing activities $\mathcal{A}_{fix}$, referring to a corresponding solution $(x(\mathcal{A}_{fix}), \mathcal{R}(x(\mathcal{A}_{fix})))$ with $\mathcal{R}(x(\mathcal{A}_{fix}))$ being an arbitrary routing in $OPT(x(\mathcal{A}_{fix}))$,
- or the routing $\mathcal{R}$, referring to $(x(\mathcal{R}), \mathcal{R})$.

As specified in Definition 4.3.6, every set $\mathcal{A}_{\text{fix}} \subset \mathcal{A}_{\text{change}}$ defines routings $\mathcal{R}(x(\mathcal{A}_{\text{fix}}))$ as a collection of shortest paths in $\mathcal{N}(x)$. However, if there is only one OD-pair, for some sets $\mathcal{A}_{\text{fix}}$, it is possible to define another path which represents the structure of $\mathcal{A}_{\text{fix}}$ in a better way. Since we make use of this concept in the next sections, we define it here.

Lemma 4.3.13. *Let $\mathcal{N} = (\mathcal{E}, \mathcal{A})$ be an EAN, $e_1, e_2 \in \mathcal{E}$ and $\mathcal{A}_{fix} \subset \mathcal{A}_{change}$. There is at most one path P from e_1 to e_2 such that $P \cap \mathcal{A}_{change} = \mathcal{A}_{fix}$.*

Proof. Note that such a path P exists if and only if $\mathcal{A}_{\text{fix}} = \{(j_i, k_i) : i = 1, \ldots, l\}$ can be sorted in such a way that $j_i <_{\mathcal{N}} j_{i+1}$ and $\text{tr}(j_1) = \text{tr}(e_1)$, $\text{tr}(j_{i+1}) = \text{tr}(k_i)$ and $\text{tr}(k_l) = \text{tr}(e_2)$ for all $i = 1, \ldots, l-1$. Since the network is acyclic, this sorting is unique if it exists.

Since in $(\mathcal{E}_{\text{op}}, \mathcal{A}_{\text{drive}} \cup \mathcal{A}_{\text{wait}})$ the path between two nodes $j, k \in \mathcal{E}_{\text{op}}$ is unique (if it exists, i.e., if $\text{tr}(j) = \text{tr}(k)$), the specification of e_1, e_2 and the above sorting determine P uniquely. □

Notation 4.3.14. *Let $\mathcal{A}_{fix}$ be a set of maintained changing activities and $e_1, e_2 \in \mathcal{E}$. If there exists a path P from e_1 to e_2, defined by the equality*

$$P \cap \mathcal{A}_{change} = \mathcal{A}_{fix}$$

as described in Lemma 4.3.13, we write $P_{e_1 e_2}^{\mathcal{A}_{fix}} := P$.

Let $(\mathcal{N}, l, \pi, \mathcal{OD}, d)$ be an instance of DMwR with $\mathcal{N} = (\mathcal{E}, \mathcal{A})$, $e_1, e_2 \in \mathcal{E}$ and $\mathcal{A}_{\text{fix}} \subset \mathcal{A}_{\text{change}}$ such that $P_{e_1 e_2}^{\mathcal{A}_{\text{fix}}}$ exists. Let $P_{e_1 e_2}(\mathcal{A}_{\text{fix}})$ denote a shortest path from e_1 to e_2 in $\mathcal{N}(x(\mathcal{A}_{\text{fix}}))$. Note that in general

$$P_{e_1 e_2}^{\mathcal{A}_{\text{fix}}} \neq P_{e_1 e_2}(\mathcal{A}_{\text{fix}}).$$

4.4 Complexity of Delay Management with Routing

In this section we summarize the results on the computational complexity of DMwR from [DHSS09, DHSS12, Sch13]. First, in Sect. 4.4.1, we see that DMwR is strongly NP-hard, even for the case of only one OD-pair. Then, in Sect. 4.4.2, we develop a polynomial-time algorithm for DMwR for one OD-pair which provides us with a lower bound on the objective value for general DMwR problems. Under the assumption that for the considered OD-pair (u, v, σ) there is no path from $u^{\text{org}}(\sigma)$ to v^{dest} that enters a train more than once, this algorithm can even be guaranteed to find an optimal solution. Finally, in Sect. 4.4.3, we show that when generalizing our

4.4 Complexity of Delay Management with Routing   131

problem to a set of several OD-pairs, it turns out to be strongly NP-hard again, even under the mentioned restriction and if all OD-pairs have the same origin and the same start time.

4.4.1 NP-Hardness of Delay Management with Routing for One OD-Pair

Theorem 4.4.1. *DMwR is strongly NP-hard, even if*

- *there is only one OD-pair, and*
- *there is only one source delay.*

We prove this theorem by reduction from the problem *Hamiltonian Path*, which remains NP-complete when the first and the last node of the path to find are fixed [GJ79]. An instance (G, t_1, t_2) of the problem *Hamiltonian Path with fixed start and end nodes* consists of an undirected graph G and two specified vertices t_1 and t_2. The question is whether there is a path from t_1 to t_2 such that every vertex is contained exactly once in this path.

Proof. Consider an instance (G, t_1, t_2) of the Hamiltonian Path problem with fixed start and end nodes and let n denote the number of nodes in G.
We construct an instance $I = (\mathcal{PTN}, \text{TR}, \mathcal{C}, l, \pi, \mathcal{OD}, d)$ of DMwR in the following way:

- The set of stations is $S := \{s_u, s_v, s^1, s^{n-1}\} \cup \{s_i : i = 2, \ldots, n-1\} \cup \{s_i^j : i = 2, \ldots, n-1, j = 2, \ldots, n-2\}$. The edges and edge lengths of the PTN are characterized implicitly by the description of the trains, assuming a driving time function of $\alpha_l(x) := x$ for all lines.
- The set of trains is $\text{TR} := \{\text{tr}_1, \ldots, \text{tr}_n\}$.
    - Train tr_1 starts at station s_u at time 0 and runs to station s^1 where it arrives at time 1.
    - Train tr_n starts at station s^{n-1} at time $(n-2)(n-3)4 + 8$ and runs to station s_v where it arrives at time $(n-2)(n-3)4 + 9$.
    - The trains tr_i for $i = 2, \ldots, n-1$ start at station s_i at time 0 and go on driving through stations $s^1, s_2^2, s_3^2, \ldots, s_{n-1}^2, s_2^3, s_3^3, \ldots, s_{n-1}^3, \ldots, s_{n-1}^{n-2}, s^{n-1}$ in this order. Note that the trains tr_i do not stop at every station they pass. The stopping pattern and the timetable of the trains $\text{tr}_2, \ldots, \text{tr}_{n-1}$ is as follows:
        · For $i = 2, \ldots, n-1$, if $\{1, i\} \in G$, tr_i stops at station s^1. The arrival time according to π is 3 and the departure time is 4.
        · Train tr_i stops at station s_i^j for every $i = 2, \ldots, n-1$ and $j = 2, \ldots, n-2$. It arrives there at time $(j-2)(n-2)4 + (i-2)4 + 5$ and departs at time $(j-2)(n-2)4 + (i-2)4 + 6$.

- Whenever $\{i,i'\} \in G$ for $i,i' \in \{2,\ldots,n-1\}$, $\text{tr}_{i'}$ stops at station s_i^j. It arrives there at time $(j-2)(n-2)4 + (i-2)4 + 7$ and departs at time $(j-2)(n-2)4 + (i-2)4 + 8$.
- If $\{i,n\} \in G$, tr_i stops at station s^{n-1}. Its arrival time at this station is $(n-2)(n-3)4 + 5$.

• We define the lower bounds on the driving and waiting times such that all slack times for driving and waiting are 0, i.e., $l_{ij} := \pi_j - \pi_i$ for all subsequent arrivals and departures i, j.
• We define

$$\mathcal{C} := \{(\text{tr}_1, \text{tr}_{i'}, s^1) : \{1, i'\} \in G\}$$

$$\cup \bigcup_{j=2}^{n-2} \bigcup_{i=2}^{n-1} \{(\text{tr}_i, \text{tr}_{i'}, s_i^j) : \{i, i'\} \in G\}$$

$$\cup \{(\text{tr}_{i'}, \text{tr}_n, s^{n-1}) : \{i', n\} \in G\}. \tag{4.3}$$

The lower bounds on the changing times are $l_{(\text{tr}-s-\text{Arr},\text{tr}'-s-\text{Dep})} := 2$ for all $(\text{tr}, \text{tr}', s) \in \mathcal{C}$. Consequently, there is a slack time of 1 on each connection.
• A delay of $d_{(\text{tr}_1-s_u-\text{Dep})} := n - 1$ occurs in the departure node at station s_u. This is the only delay.
• The only OD-pair is $(u, v, \sigma) := (s_u, s_v, 0)$.

Let $\mathcal{N}$ be the EAN for instance I. From now on, we regard I as an instance in the EAN, i.e., $I = (\mathcal{N}, l, \pi, \mathcal{OD}, d)$.

An example for this construction is given in Fig. 4.1. Figure 4.1a shows an instance of Hamiltonian Path with fixed start node 1 and fixed end node 5. The EAN constructed for the reduction is shown in Fig. 4.1b. The numbers in the nodes indicate the times the corresponding events are scheduled according to π.

We show that there is a Hamiltonian path in G with start node 1 and end node n if and only if there is a solution $(x, \mathcal{R})$ with

- $c^d(x, \mathcal{R}) = x_{(\text{tr}_n-s_v-\text{Arr})} - \pi_{(\text{tr}_n-s_v-\text{Arr})} = 0$, and
- $P \in \mathcal{R}$ is a path from $u^{\text{org}}(\sigma)$ to v^{dest} in $\mathcal{N}(x)$ to I,

i.e., if and only if it is possible for the OD-pair to arrive in s_v without a delay.

This, of course, is equivalent to showing that if and only if there is a Hamiltonian path in G from 1 to n, there is a solution $(x, \mathcal{R})$ with $c(x, \mathcal{R}) \leq (n-2)4(n-3) + 9$, and hence proves our theorem.

1. First we prove that $\mathcal{R} = \{P\}$ is a solution to I if and only if for every tr_i for $i = 2, \ldots, n$, there is exactly one changing activity $(e, e_i) \in P \cap \mathcal{A}_{\text{change}}$ with $\text{tr}(e_i) = \text{tr}_i$:
   Let P be a path in $\mathcal{N}$ from $u^{\text{org}}(\sigma)$ to v^{dest} and $\mathcal{R} := \{P\}$.

4.4 Complexity of Delay Management with Routing

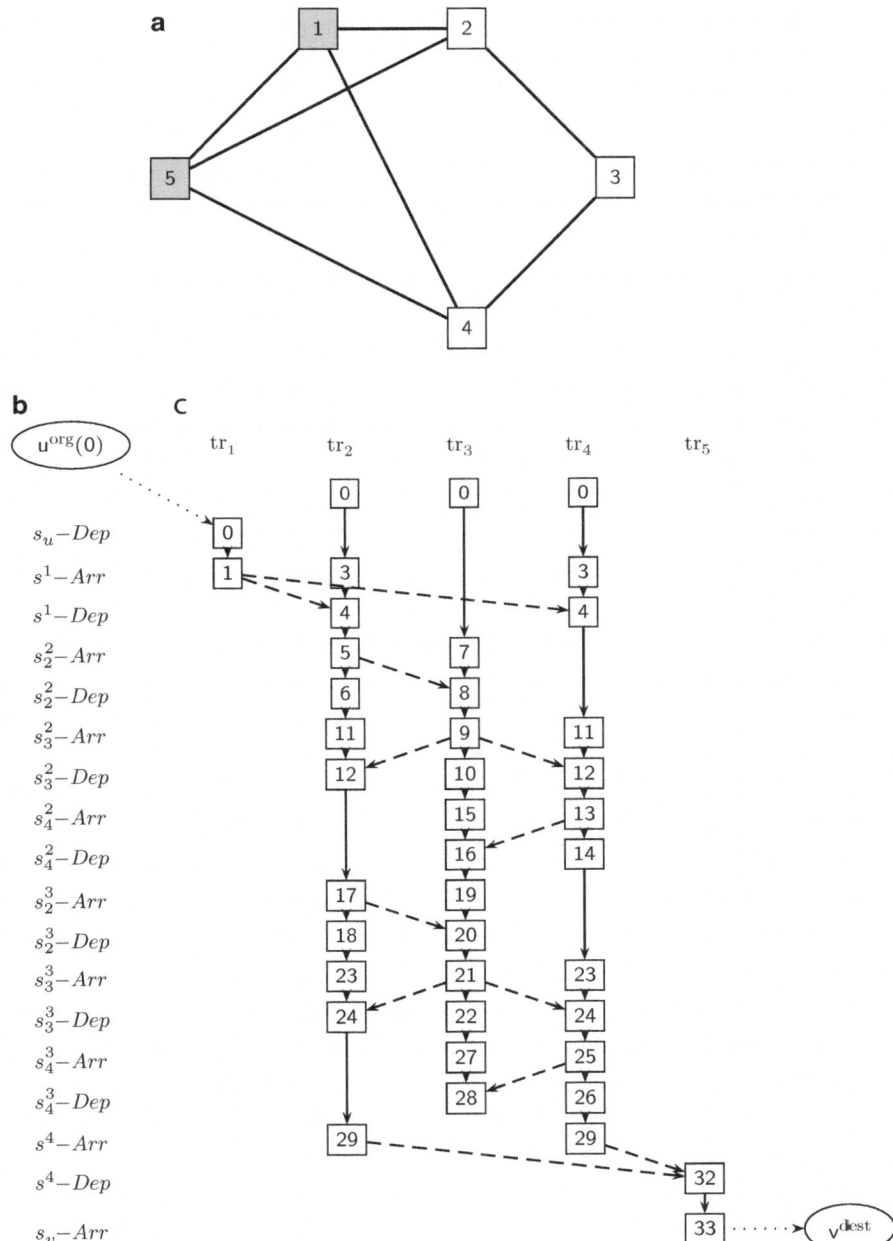

Fig. 4.1 Example for the constructed instance of DMwR in the proof of Theorem 4.4.1. (**a**) Instance of Hamiltonian Path with fixed start node 1 and fixed end node 5. (**b**) EAN and timetable for the instance of DMwR constructed from (**a**) in the proof of Theorem 4.4.1

(a) The delay in the first arrival node of P is

$$d(x(\mathcal{R}), (\text{tr}_1-s_u-\text{Dep})) = x(\mathcal{R})_{(\text{tr}_1-s_u-\text{Dep})} - \pi_{(\text{tr}_1-s_u-\text{Dep})} = d_{(\text{tr}_1-s_u-\text{Dep})}$$
$$= n - 1.$$

(b) Consider $(i, j) \in P \setminus \mathcal{A}_{\text{change}}$. Since there are no slack times on driving and waiting activities, we have $\pi_j = \pi_i + l_{ij}$ and it follows from the definitions in Sect. 4.3 that

$$x(\mathcal{R})_j - \pi_j = \max\{\pi_j, x(\mathcal{R})_i + l_{ij}\} - \pi_j = \max\{0, x(\mathcal{R})_i - \pi_i\} = x(\mathcal{R})_i - \pi_i$$

Thus, given a node $i \in \mathcal{E}_{\text{op}}$, we have $d(x(\mathcal{R}), j) \geq d(x(\mathcal{R}), i)$ for all successors j of i on train $\text{tr}(i)$.

(c) Consider $(i', j) \in P \cap \mathcal{A}_{\text{change}}$ and let i be the direct predecessor of j on $\text{tr}(j)$. Like before we have $\pi_j = \pi_i + l_{ij}$ and furthermore, according to the definitions of the changing times, $\pi_j = \pi_{i'} + l_{i'j} + 1$. Hence, due to Definitions 4.3.10 and 4.3.4, it follows that

$$x(\mathcal{R})_j - \pi_j = \max\{\pi_j, x(\mathcal{R})_i + l_{ij}, x(\mathcal{R})_{i'} + l_{i'j}\} - \pi_j$$
$$= \max\{0, x(\mathcal{R})_i + l_{ij} - \pi_j, x(\mathcal{R})_{i'} + l_{i'j} - \pi_j\}$$
$$= \max\{0, x(\mathcal{R})_i - \pi_i, x(\mathcal{R})_{i'} - \pi_{i'} - 1\}.$$

Using (1b) it follows inductively that

$$x(\mathcal{R})_j - \pi_j$$
$$= \begin{cases} x(\mathcal{R})_i - \pi_i \geq x(\mathcal{R})_{i'} - \pi_{i'} & \text{if tr}(i, j) \text{ was already entered on path } P, \\ x(\mathcal{R})_{i'} - \pi_{i'} - 1 & \text{otherwise.} \end{cases}$$

(d) Hence, for $(i, j) \in P$

$$x(\mathcal{R})_j - \pi_j$$
$$= \begin{cases} x(\mathcal{R})_i - \pi_i & \text{if } (i, j) \in \mathcal{A}_{\text{drive}} \cup \mathcal{A}_{\text{wait}}, \\ x(\mathcal{R})_i - \pi_i - 1 & \text{if } (i, j) \in \mathcal{A}_{\text{change}} \text{ and} \\ & P \cap \{k : \text{tr}(k) = \text{tr}(j), k \leq_{\mathcal{N}} j,\} = \emptyset \\ x(\mathcal{R})_k - \pi_k \geq x(\mathcal{R})_i - \pi_i & \text{for the direct predecessor } k \text{ of } j \text{ on } \text{tr}(j) \\ & \text{otherwise.} \end{cases}$$

Thus,

$$d(x(\mathcal{R}), j) = x(\mathcal{R})_j - \pi_j = x(\mathcal{R})_i - \pi_i - 1 = d(x(\mathcal{R}), i) - 1$$

if and only if (i, j) is a changing activity to a train tr and no predecessor of j in tr is contained in P.

4.4 Complexity of Delay Management with Routing

It follows that $\mathcal{R} = \{P\}$ is a solution to I if and only if for every tr_i for $i = 2, \ldots, n$, there is exactly one changing activity $(e, e_i) \in P \cap \mathcal{A}_{\text{change}}$ with $\text{tr}(e_i) = \text{tr}_i$.

2. We use the result of (1) to show that given a Hamiltonian path P' in G with start node 1 and end node n we can construct a path P from $u^{\text{org}}(\sigma)$ to v^{dest} with $c^d(x(\{P\}), \{P\}) = 0$.

Suppose that $P' = (1, k_2, \ldots, k_{n-1}, n)$ is a Hamiltonian path in G. Similar to Notation 4.3.14, we implicitly define a path $P := P^S_{u^{\text{org}}(\sigma) v^{\text{dest}}}$ from $u^{\text{org}}(\sigma)$ to v^{dest} in $\mathcal{N}$ by the sequence S of changing activities contained in P,

$$S = [(\text{tr}_1 - s^1 - \text{Arr}, \text{tr}_{k_2} - s^1 - \text{Dep})]$$
$$\cup [(\text{tr}_{k_i} - s^i_{k_i} - \text{Arr}, \text{tr}_{k_{i+1}} - s^i_{k_i} - \text{Dep}) : i = 2, \ldots, i-2]$$
$$\cup [(\text{tr}_{k_{n-1}} - s^{n-1} - \text{Arr}, \text{tr}_n - s^{n-1} - \text{Dep})].$$

That is, the corresponding ST-train-route starts by taking tr_1 from s_u to s^1. In s^1 the OD-pair transfers to train tr_{k_2} and uses this train until station $s^2_{k_2}$. Then in $s^2_{k_2}$ the OD-pair transfers to train tr_{k_3}, in $s^3_{k_3}$ the OD-pair transfers to train tr_{k_4}, and so on. Finally, at station s^{n-1}, the OD-pair transfers from tr_{n-1} to tr_n.

Due to the construction of I, such a path exists and it enters every train exactly once. Hence, $\mathcal{R} := \{P\}$ is a solution to I with overall delay

$$c^d(x(\mathcal{R}), \mathcal{R}) = x(\mathcal{R})_{(\text{tr}_n - s_v - \text{Arr})} - \pi_{(\text{tr}_n - s_v - \text{Arr})} = 0.$$

3. Now, using (1), we show that given a solution $(x, \mathcal{R})$ to I with

$$c^d(x, \mathcal{R}) = x_{(\text{tr}_n - s_v - \text{Arr})} - \pi_{(\text{tr}_n - s_v - \text{Arr})} = 0,$$

we can construct a solution to the given instance of Hamiltonian Path with fixed start and end nodes.

Let $\mathcal{R} = \{P\}$ be a routing consisting of a path P from $u^{\text{org}}(\sigma)$ to v^{dest} in the constructed instance of DMwR with delay

$$d(x(\mathcal{R}), (\text{tr}_n - s_v - \text{Arr})) = x(\mathcal{R})_{(\text{tr}_n - s_v - \text{Arr})} - \pi_{(\text{tr}_n - s_v - \text{Arr})} = 0.$$

Consider the sequence of maintained changing activities

$$S = ((e_1, f_1), (e_2, f_2), \ldots, (e_l, f_l)) := \mathcal{A}_{\text{change}} \cap P$$

defined by P. As shown in (1), for every $\text{tr}_i, i = 2, \ldots, n$, S contains exactly one changing activity (e_i, f_i) with $\text{tr}(f_i) = \text{tr}_i$. It follows that $l = n - 1$.

Furthermore, due to (4.3), we observe that

- $(e_1, f_1) = (\text{tr}_1 - s^1 - \text{Dep}, \text{tr}_{k_2} - s^1 - \text{Dep})$ for a train tr_{k_2} with $\{1, k_2\} \in G$,

- for all $i = 2, \ldots, n-1$, we have $(e_i, f_i) = (\text{tr}_{k_i} - s_i^j - \text{Dep}, \text{tr}_{k_{i+1}} - s_i^j - \text{Dep})$ for trains $\text{tr}_{k_i}, \text{tr}_{k_{i+1}}$ with $\{k_i, k_{i+1}\} \in G$ and $j \in \{2, \ldots, n-2\}$,
- $(e_{n-1}, f_{n-1}) = (\text{tr}_{k_{n-1}} - s^{n-1} - \text{Arr}, \text{tr}_n - s^{n-1} - \text{Dep})$ for a train tr_{n-1} with $\{k_{n-1}, n\} \in G$.

Hence, $(1, k_2, k_3, \ldots, k_{n-1}, n)$ is a Hamiltonian path in G. □

4.4.2 An Algorithm for Delay Management with Routing with One OD-Pair

We have seen in Theorem 4.4.1 that DMwR is strongly NP-hard in general, even if there is only one OD-pair. In this section we present a 2-approximation algorithm for DMwR with one OD-pair, which in addition calculates a lower bound on the objective value. Under some quite realistic assumptions on the network structure, the outcome of the algorithm is even guaranteed to be optimal.

Let $I = (\mathcal{N}, l, \pi, \{(u, v, \sigma)\}, d)$ be an instance of DMwR with one OD-pair. For the sake of simplicity, we assume that the passenger weight is $w_{uv\sigma} := 1$. Then the objective value of a solution $(x, \mathcal{R})$ can be calculated as $\tilde{c}(x, \mathcal{R}) := c(x, P) = x_{j'} - \sigma$ for the last arrival node j' on the path $P \in \mathcal{R}$. For any instance of DMwR with one OD-pair with higher passenger weights, we can obtain the solution from $\tilde{c}(x, \mathcal{R})$ by multiplication with the passenger weight.

In order to expose the structure of our algorithm, we extend the concept of DMwR with one OD-pair to deal with OD-pairs that are given as pairs of actual nodes in the EAN $\mathcal{N}$, i.e., elements of the type $(u^{\text{org}}(\sigma), i)$ where $u^{\text{org}}(\sigma)$ is the origin node and $i \in \mathcal{E}$ is an arbitrary successor of $u^{\text{org}}(\sigma)$ in $\mathcal{N}$:

Definition 4.4.2. An instance $(\mathcal{N}, l, \pi, \{(u^{\text{org}}(\sigma), i)\}, d)$ of *extended delay management with routing with one OD-pair (DMwR')* consists of an EAN $\mathcal{N}$, lower bounds l_a on the lengths of the operational activities $a \in \mathcal{A}_{\text{op}}$, a timetable π, a pair of nodes $(u^{\text{org}}(\sigma), i)$ where $u^{\text{org}}(\sigma) \in \mathcal{E}'_{\text{org}}$, $i \in \mathcal{E}_{\text{op}} \cup \mathcal{E}_{\text{dest}}$, and source delays d_e on the operational events $e \in \mathcal{E}_{\text{op}}$. The task is to find a disposition timetable x and a routing $\mathcal{R} := \{P\}$ where P is a path from $u^{\text{org}}(\sigma)$ to i in $\mathcal{N}(x)$ that minimize $c'(x, \mathcal{R}) := x_i - \sigma$.

Also in DMwR', a solution $(x, \mathcal{R})$ is determined by the disposition timetable x, the routing $\mathcal{R}$, or the set of maintained changing activities $\mathcal{A}_{\text{fix}}$, as explained in Sect. 4.3 and summarized in Notation 4.3.12 for DMwR.

Notation 4.4.3. *We sometimes describe a solution $(x, \mathcal{R})$ to DMwR' stating only x, $\mathcal{R}$, or $\mathcal{A}_{\text{fix}}$, where we refer to the corresponding full solutions*

- $(x, \mathcal{R}(x))$ with $\mathcal{R}(x) := \{P(x)\}$ and $P(x)$ is an arbitrary shortest path from $u^{\text{org}}(\sigma)$ to i in $\mathcal{N}(x) = (\mathcal{E}, \mathcal{A}(x))$.
- $(x(\mathcal{R}), \mathcal{R})$ where $x(\mathcal{R}) = x(P \cap \mathcal{A}_{change})$ for the path P in $\mathcal{R}$.
- $(x(\mathcal{A}_{fix}), \mathcal{R}(x(\mathcal{A}_{fix})))$ where

4.4 Complexity of Delay Management with Routing

- $x(\mathcal{A}_{fix})$ is calculated like in Definition 4.3.4, and
- $\mathcal{R}(x(\mathcal{A}_{fix})) := \{P(x(\mathcal{A}_{fix}))\}$, with $P(x(\mathcal{A}_{fix}))$ being an arbitrary shortest path from $u^{org}(\sigma)$ to i in $\mathcal{N}(x(\mathcal{A}_{fix})) = (\mathcal{E}, \mathcal{A}(x(\mathcal{A}_{fix})))$.

Note that every instance $(\mathcal{N}, l, \pi, \{(u, v, \sigma)\}, d)$ of DMwR with one OD-pair can be understood as an instance $(\mathcal{N}, l, \pi, \{(u^{org}(\sigma), v^{dest})\}, d)$ of DMwR'; hence, DMwR' indeed is an extension of DMwR with one OD-pair.

Before we detail our algorithm for DMwR with one OD-pair, we show some properties which we need in order to prove correctness of the algorithm. To this end, let $\mathcal{N}$ be an EAN with lower bounds l, π be a timetable, and d be a vector of source delays.

In Lemma 4.4.4 we observe that the addition of changing activities to a set $A_1 \subset \mathcal{A}_{change}$ does not influence the time for events i that happen before the added activities take place:

Lemma 4.4.4. *Let $A_1 \subset A_2 \subset \mathcal{A}_{change}$ be two subsets of the changing activities. Let $i \in \mathcal{E}_{op}$ be an event with the following property:*

$$x(A_2)_i \leq x(A_2)_j \text{ for all } (i, j) \in A_2 \setminus A_1.$$

Then it holds that $x(A_1)_i = x(A_2)_i$.

Similarly to the NLT property in line planning, we can define the *no-train-twice (NTT) property* which states that there is always a path with minimal arrival time such that the passengers use every train at most once.

Definition 4.4.5. *Let $\mathcal{N}$ be an EAN. A path P has the NTT property if the following holds: If $i, k \in \mathcal{E}_{op} \cap P$ with $i \leq_P k$ and $\operatorname{tr}(i) = \operatorname{tr}(k)$, then $\operatorname{tr}(j) = \operatorname{tr}(i)$ for all j such that $i \leq_P j \leq_P k$.*

Lemma 4.4.6. *Let e be an event in $\mathcal{N}$. For every set $\mathcal{A}_{fix} \subset \mathcal{A}_{change}$ for which there exists a path from the origin node $u^{org}(\sigma)$ to e in $\mathcal{N}(x(\mathcal{A}_{fix}))$, there also exists a path $P = (\mathcal{E}^P, \mathcal{A}^P)$ from $u^{org}(\sigma)$ to e that has the NTT property.*

Proof. Assume that P_0 is a path from $u^{org}(\sigma)$ to e in $\mathcal{N}(x(\mathcal{A}_{fix}))$ and let i, j, k be three events on P_0 such that $\operatorname{tr}(k) = \operatorname{tr}(i) \neq \operatorname{tr}(j)$ and $i <_P j <_P k$. We construct a new path P_1 in the following way: P_1 consists of the same events as P_0 between $u^{org}(\sigma)$ and i. Then P_1 continues on $\operatorname{tr}(i)$ until k is reached. From k to e, P_1 consists of the same events as P_0. Note that P_1 is a path in $\mathcal{N}(\mathcal{A}_{fix})$. Repeating this construction, we obtain a path with the claimed property. □

The following Notation 4.4.7 and Lemmas 4.4.8 and 4.4.9 simplify the calculations of disposition timetables in some situations.

Notation 4.4.7. *Let $\tilde{\pi} := x(\emptyset)$ denote the disposition timetable for an empty set of maintained changing activities calculated as defined in Definition 4.3.4. We set $\tilde{\pi}_{u^{org}(\sigma)} := \sigma$ and $\tilde{\pi}_{v^{dest}} := \sigma$.*

Lemma 4.4.8 makes use of $\tilde{\pi}$ for the calculation of a disposition timetable $x(\mathcal{A}_{\text{fix}})$:

Lemma 4.4.8. *For a given set of maintained changing activities $\mathcal{A}_{\text{fix}} \subset \mathcal{A}_{\text{change}}$, we have*

$$x(\mathcal{A}_{\text{fix}})_e = \max\left\{\tilde{\pi}_e, \max_{i:(i,e)\in\mathcal{A}_{\text{drive}}\cup\mathcal{A}_{\text{wait}}\cup\mathcal{A}_{\text{fix}}} \{x(\mathcal{A}_{\text{fix}})_i + l_{ie}\}\right\}. \quad (4.4)$$

Proof. According to Definition 4.3.4, we have

$$x(\mathcal{A}_{\text{fix}})_e = \max\left\{\pi_e + d_e, \max_{i:(i,e)\in\mathcal{A}_{\text{drive}}\cup\mathcal{A}_{\text{wait}}\cup\mathcal{A}_{\text{fix}}} \{x(\mathcal{A}_{\text{fix}})_i + l_{ie}\}\right\}.$$

We observe that for every set $\mathcal{A}_{\text{fix}} \subset \mathcal{A}_{\text{change}}$ and every event $e \in \mathcal{E}_{\text{arr}} \cup \mathcal{E}_{\text{dep}}$

$$x(\mathcal{A}_{\text{fix}})_e \geq \tilde{\pi}_e \geq \pi_e + d_e.$$

Hence, we can equivalently determine the minimal arrival times in the events e like in (4.4). □

Lemma 4.4.9. *Let P be a path from $u^{org}(\sigma)$ to e in $\mathcal{N}$. Denote $\mathcal{A}_{fix}^P := \mathcal{A}_{change} \cap \mathcal{A}^P$. Then*

$$x(\mathcal{A}_{fix}^P)_e \geq \max\{\tilde{\pi}_e, x(\mathcal{A}_{fix}^P)_j + l_{je}\}$$

for the direct predecessor j of e on the path P. If P has the NTT property, we even have equality:

$$x(\mathcal{A}_{fix}^P)_e = \max\{\tilde{\pi}_e, x(\mathcal{A}_{fix}^P)_j + l_{je}\}.$$

Proof.

$$x(\mathcal{A}_{\text{fix}}^P)_e = \max\left\{\tilde{\pi}_e, \max_{i:(i,e)\in\mathcal{A}_{\text{drive}}\cup\mathcal{A}_{\text{wait}}\cup\mathcal{A}_{\text{fix}}^P} \{x(\mathcal{A}_{\text{fix}}^P)_i + l_{ie}\}\right\}$$

$$\geq \max\{\tilde{\pi}_e, x(\mathcal{A}_{\text{fix}}^P)_j + l_{je}\}.$$

This proves the first statement of the lemma. Now suppose that P has the NTT property.

4.4 Complexity of Delay Management with Routing

- If $(j,e) \in \mathcal{A}_{\text{drive}} \cup \mathcal{A}_{\text{wait}}$, or if $(j,e) \in \mathcal{A}_{\text{fix}}^P$ and (j,e) is the first trip of the train $\text{tr}(e)$, then (j,e) is the only activity terminating in e in $\mathcal{N}(\mathcal{A}_{\text{fix}}^P)$. Thus

$$x(\mathcal{A}_{\text{fix}}^P)_e = \max\left\{\tilde{\pi}_e, \max_{i:(i,e)\in\mathcal{A}_{\text{drive}}\cup\mathcal{A}_{\text{wait}}\cup\mathcal{A}_{\text{fix}}^P}\{x(\mathcal{A}_{\text{fix}}^P)_i + l_{ie}\}\right\}$$
$$= \max\{\tilde{\pi}_e, x(\mathcal{A}_{\text{fix}}^P)_j + l_{je}\}.$$

- Now let $(j,e) \in \mathcal{A}_{\text{fix}}^P$ and $(k,e) \in \mathcal{A}_{\text{wait}}$ with $\text{tr}(k) = \text{tr}(e)$. Since there is no $i \in P$ with $i \leq_P k$ and $\text{tr}(i) = \text{tr}(k)$, we have $x(\mathcal{A}_{\text{fix}}^P)_k = \tilde{\pi}_k$ and

$$\tilde{\pi}_e \geq \tilde{\pi}_k + l_{ke} = x(\mathcal{A}_{\text{fix}}^P)_k + l_{ke}.$$

It follows that

$$x(\mathcal{A}_{\text{fix}}^P)_e = \max\left\{\tilde{\pi}_e, \max_{i:(i,e)\in\mathcal{A}_{\text{drive}}\cup\mathcal{A}_{\text{wait}}\cup\mathcal{A}_{\text{fix}}^P}\{x(\mathcal{A}_{\text{fix}}^P)_i + l_{ie}\}\right\}$$
$$= \max\{\tilde{\pi}_e, x(\mathcal{A}_{\text{fix}}^P)_j + l_{je}, x(\mathcal{A}_{\text{fix}}^P)_k + l_{ke}\}$$
$$= \max\{\tilde{\pi}_e, x(\mathcal{A}_{\text{fix}}^P)_j + l_{je}\}.$$

□

Based on the preceding Lemmas 4.4.4–4.4.9, we now explain Algorithm 1. The procedure is similar to Dijkstra's label setting algorithm for finding a shortest path in a graph. In every iteration k a best path from the origin event $u^{\text{org}}(\sigma)$ to a node $\hat{e}_k$ is determined, using the results about best paths to the preceding nodes.

In other words, given an instance $I = (\mathcal{N}, l, \pi, \{(u,v,\sigma)\}, d)$ of DMwR with one OD-pair, we iteratively solve instances $I_{\hat{e}} = (\mathcal{N}, l, \pi, \{(u^{\text{org}}(\sigma), \hat{e})\}, d)$ of DMwR' for different OD-pairs $(u^{\text{org}}(\sigma), \hat{e})$, thereby constructing the solution to $I_{\hat{e}}$ from the solution to $I_{i_{\hat{e}}}$ for a direct predecessor $i_{\hat{e}}$ of $\hat{e}$ in $\mathcal{N}$.

To do so, in every iteration k we store a *permanent time label* $T[\hat{e}_k]$ and a set of *maintained changing activities* $\mathcal{A}_{\text{fix}}[\hat{e}_k]$ that defines a path $P_{u^{\text{org}}(\sigma)\hat{e}_k}^{\mathcal{A}_{\text{fix}}[\hat{e}_k]}$ in the sense of Lemma 4.3.14.

In an iteration k let PERM(k) be the set of events i for which a solution to DMwR' has been found and the above values $T[i]$ and $\mathcal{A}_{\text{fix}}[i]$ have been determined. For every $e \notin \text{PERM}(k)$ with a direct predecessor $i \in \text{PERM}(k)$ we determine the preliminary time label

$$\tilde{T}[e] = \min_{i\in\text{PERM}(k):(i,e)\in\mathcal{A}}\{\tilde{\pi}_e, T[i] + l_{ie}\}.$$

Like in Dijkstra's algorithm we fix the event $\hat{e}$ with smallest preliminary time label. In order to update the set of fixed changing activities $\mathcal{A}_{\text{fix}}[\hat{e}]$, we distinguish two cases: Let $i_{\hat{e}}$ be the direct predecessor of $\hat{e}$ in the solution of DMwR' for OD-pair $(u^{\text{org}}(\sigma), \hat{e})$. If $\hat{a} = (i_{\hat{e}}, \hat{e}) \in \mathcal{A}_{\text{change}}$, we obtain

Algorithm 1 Modified Dijkstra's algorithm for DMwR with one OD-pair

Require: Instance $I = (\mathcal{N}, l, \pi, \{(u, v, \sigma)\}, d)$ of DMwR.
Ensure: Set $\mathcal{A}_{\text{fix}} \subset \mathcal{A}_{\text{change}}$ and a lower bound lb on the optimal objective value.
1: Define $l_a := 0$ for $a \in \mathcal{A}'_{\text{org}} \cup \mathcal{A}_{\text{dest}}$.
2: Generate $\tilde{\pi}$.
3: Set PERM $:= \{u^{\text{org}}(\sigma)\}$, TEMP $:= \mathcal{E} \setminus \{u^{\text{org}}(\sigma)\}$, $T[u^{\text{org}}(\sigma)] := \sigma$, $\tilde{T}[e] := \infty$ for every $e \in$ TEMP, $\mathcal{A}_{\text{fix}}[u^{\text{org}}(\sigma)] := \emptyset$, $\hat{e}^{\text{old}} := u^{\text{org}}(\sigma)$.
4: **for** $e \in$ TEMP such that $(\hat{e}^{\text{old}}, e) \in \mathcal{A}$ **do**
5:     Set $\tilde{T}[e] := \min\{\tilde{T}[e], \max\{\tilde{\pi}_e, T[\hat{e}^{\text{old}}] + l_{\hat{e}^{\text{old}}e}\}\}$.
6: **end for**
7: Choose $\hat{e} \in \arg\min_{e \in \text{TEMP}} \tilde{T}[e]$. Set PERM $:=$ PERM $\cup \{\hat{e}\}$, TEMP $:=$ TEMP $\setminus \{\hat{e}\}$, $T[\hat{e}] := \tilde{T}[\hat{e}]$.
8: Let $i_{\hat{e}}$ be the corresponding direct predecessor of $\hat{e}$, i.e., either $T[i_{\hat{e}}] = \tilde{T}[\hat{e}] - l_{i_{\hat{e}}\hat{e}}$, or, if no such $i_{\hat{e}}$ exists, $i_{\hat{e}}$ is the direct predecessor of $\hat{e}$ on $\text{tr}(\hat{e})$.
9: **if** $(i_{\hat{e}}, \hat{e}) \in \mathcal{A}_{\text{change}}$ **then**
10:     Set $\mathcal{A}_{\text{fix}}[\hat{e}] := \mathcal{A}_{\text{fix}}[i_{\hat{e}}] \cup \{(i_{\hat{e}}, \hat{e})\}$.
11: **else** Set $\mathcal{A}_{\text{fix}}[\hat{e}] := \mathcal{A}_{\text{fix}}[i_{\hat{e}}]$.
12: **end if**
13: Set $\hat{e}^{\text{old}} := \hat{e}$.
14: **if** $\hat{e} = v^{\text{dest}}$ **then**
15:     Set $\mathcal{A}_{\text{fix}} := \mathcal{A}_{\text{fix}}[v^{\text{dest}}]$ and lb $:= T[v^{\text{dest}}] - \sigma$.
16:     **return** $\mathcal{A}_{\text{fix}}$ and lb.
17: **else** Go to step 4.
18: **end if**

$$\mathcal{A}_{\text{fix}}[\hat{e}] := \mathcal{A}_{\text{fix}}[i_{\hat{e}}] \cup \{(i_{\hat{e}}, \hat{e})\}.$$

Otherwise, we simply set

$$\mathcal{A}_{\text{fix}}[\hat{e}] := \mathcal{A}_{\text{fix}}[i_{\hat{e}}].$$

The algorithm is summarized above.

To prove correctness of the described algorithm in Theorem 4.4.19, we need some insights about the stored time labels $T[i]$ and sets of maintained changing activities $\mathcal{A}_{\text{fix}}[i]$. These are provided in the following.

Let $I = (\mathcal{N}, l, \pi, \{(u, v, \sigma)\}, d)$ be an instance of DMwR to which we apply Algorithm 1.

Lemma 4.4.10. *Let $\hat{e}$ be the node chosen in step 7 of Algorithm 1, let $i_{\hat{e}}$ be its direct predecessor as specified in step 8 and let $\mathcal{A}_{fix}[\hat{e}]$ be defined as in steps 9–12. Then the path $P_{u^{\text{org}}(\sigma)\hat{e}}^{\mathcal{A}_{fix}[\hat{e}]}$ as defined in Notation 4.3.14 exists, and $i_{\hat{e}}$ is the direct predecessor of $\hat{e}$ on $P_{u^{\text{org}}(\sigma)\hat{e}}^{\mathcal{A}_{fix}[\hat{e}]}$.*

Proof. If it exists, $P_{u^{\text{org}}(\sigma)\hat{e}}^{\mathcal{A}_{\text{fix}}[\hat{e}]}$ is uniquely defined by $P_{u^{\text{org}}(\sigma)\hat{e}}^{\mathcal{A}_{\text{fix}}[\hat{e}]} \cap \mathcal{A}_{\text{change}} = \mathcal{A}_{\text{fix}}$ (see Lemma 4.3.13 and Notation 4.3.14). We show the existence of $P_{u^{\text{org}}(\sigma)\hat{e}}^{\mathcal{A}_{\text{fix}}[\hat{e}]}$ inductively:

4.4 Complexity of Delay Management with Routing

- For $u^{\text{org}}(\sigma)$

$$P^{\mathcal{A}_{\text{fix}}[u^{\text{org}}(\sigma)]}_{u^{\text{org}}(\sigma)u^{\text{org}}(\sigma)} \cap \mathcal{A}_{\text{change}} = \emptyset \cap \mathcal{A}_{\text{change}} = \emptyset = \mathcal{A}_{\text{fix}}[u^{\text{org}}(\sigma)].$$

- Assume that $P^{\mathcal{A}_{\text{fix}}[i_{\hat{e}}]}_{u^{\text{org}}(\sigma)i_{\hat{e}}}$ exists.
- If $(i_{\hat{e}}, \hat{e}) \in \mathcal{A}_{\text{drive}} \cup \mathcal{A}_{\text{wait}}$, in step 11, the algorithm sets $\mathcal{A}_{\text{fix}}[\hat{e}] = \mathcal{A}_{\text{fix}}[i_{\hat{e}}]$. Then $P^{\mathcal{A}_{\text{fix}}[\hat{e}]}_{u^{\text{org}}(\sigma)\hat{e}} = P^{\mathcal{A}_{\text{fix}}[i_{\hat{e}}]}_{u^{\text{org}}(\sigma)i_{\hat{e}}} \cup (i_{\hat{e}}, \hat{e})$ since there is no $(i_{\hat{e}}, i) \in \mathcal{A}_{\text{fix}}[\hat{e}]$ and $\text{tr}(i_{\hat{e}}) = \text{tr}(\hat{e})$. If $(i_{\hat{e}}, \hat{e}) \in \mathcal{A}_{\text{change}}$, in step 10, the algorithm sets $\mathcal{A}_{\text{fix}}[\hat{e}] = \mathcal{A}_{\text{fix}}[i_{\hat{e}}] \cup \{(i_{\hat{e}}, \hat{e})\}$. Then $P^{\mathcal{A}_{\text{fix}}[\hat{e}]}_{u^{\text{org}}(\sigma)\hat{e}} = P^{\mathcal{A}_{\text{fix}}[i_{\hat{e}}]}_{u^{\text{org}}(\sigma)i_{\hat{e}}} \cup (i_{\hat{e}}, \hat{e})$ since $(i_{\hat{e}}, \hat{e})$ must be contained in $P^{\mathcal{A}_{\text{fix}}[\hat{e}]}_{u^{\text{org}}(\sigma)\hat{e}}$.

□

Notation 4.4.11. *Let $\mathcal{N}$ be an EAN and $u^{\text{org}}(\sigma)$ an origin node in $\mathcal{N}$. For the sake of an easier notation in the following for any disposition timetable x, we write $x_{u^{\text{org}}(\sigma)} := \sigma$.*

In the next lemma we show that the permanent time labels $T[i]$ are a lower bound on the scheduled time in the events i with respect to disposition timetables that make it possible to reach event i from $u^{\text{org}}(\sigma)$.

Lemma 4.4.12. *Let $\hat{e} \neq v^{\text{dest}}$ be an event chosen in step 7 of an iteration k of Algorithm 1. There is no $A \subset \mathcal{A}_{\text{change}}$ such that $x(A)_{\hat{e}} < T[\hat{e}]$ and such that there is a path from $u^{\text{org}}(\sigma)$ to $\hat{e}$ in $\mathcal{N}(A)$.*

Proof. This lemma is proven inductively:

- For the origin event $u^{\text{org}}(\sigma)$, we have

$$T[u^{\text{org}}(\sigma)] = \sigma = x(\emptyset)_{u^{\text{org}}(\sigma)} = \min_{A \subset \mathcal{A}_{\text{change}}} x(A)_{u^{\text{org}}(\sigma)}.$$

- Suppose that in the iterations $k' = 1, \ldots, k-1$ of the algorithm for all the elements $\hat{e}_{k'}$ added to PERM and all $A \subset \mathcal{A}_{\text{change}}$ it holds that $x(A)_{\hat{e}_{k'}} \geq T[\hat{e}_{k'}]$. It follows that

$$x(A)_e \geq T[e] \quad \forall e \in \text{PERM}. \tag{4.5}$$

- Now let $\hat{e} := \hat{e}_k$ be the event chosen in step 7 of the kth iteration. Suppose that there is a set $A \subset \mathcal{A}_{\text{change}}$ such that

$$x(A)_{\hat{e}} < T[\hat{e}] \tag{4.6}$$

and there is a path $\tilde{P}$ from $u^{\text{org}}(\sigma)$ to $\hat{e}$ in $\mathcal{N}(A)$. Due to Lemma 4.4.6, there also is a path P from $u^{\text{org}}(\sigma)$ to $\hat{e}$ in $\mathcal{N}(A)$, which has the NTT property.

1. If the direct predecessor e_0 of $\hat{e}$ in P is in PERM

$$\begin{aligned}x(A)_{\hat{e}} &= \max\left\{\tilde{\pi}_{\hat{e}}, \max_{i:(i,\hat{e})\in\mathcal{A}_{\text{drive}}\cup\mathcal{A}_{\text{wait}}\cup\mathcal{A}_{\text{fix}}}\{x(\mathcal{A}_{\text{fix}})_i + l_{i\hat{e}}\}\right\} \quad \text{[because of Lemma 4.4.8]}\\ &\geq \max\{\tilde{\pi}_{\hat{e}}, x(A)_{e_0} + l_{e_0\hat{e}}\}\\ &\geq \max\{\tilde{\pi}_{\hat{e}}, T[e_0] + l_{e_0\hat{e}}\} \quad \text{[because of (4.5)]}\\ &\geq \min_{i\in\text{PERM}:(i,\hat{e})\in\mathcal{A}}\max\{\tilde{\pi}_{\hat{e}}, T[i] + l_{i\hat{e}}\}\\ &= T[\hat{e}].\end{aligned}$$

which contradicts (4.6).

2. If the direct predecessor e_0 of $\hat{e}$ in P is in TEMP, let e_1 denote the last event in PERM on the path P. (e_1 exists because $u^{\text{org}}(\sigma) \in$ PERM.) Let $e_2 \in$ TEMP denote the direct successor of e_1 on P. Since $\hat{e} \in \arg\min_{e\in\text{TEMP}} \tilde{T}[e]$ it follows that

$$\begin{aligned}x(A)_{\hat{e}} &\geq x(A)_{e_2}\\ &= \max\left\{\tilde{\pi}_{e_2}, \max_{i:(i,e_2)\in\mathcal{A}_{\text{drive}}\cup\mathcal{A}_{\text{wait}}\cup\mathcal{A}_{\text{fix}}}\{x(\mathcal{A}_{\text{fix}})_i + l_{ie_2}\}\right\} \quad \text{[because of Lemma 4.4.8]}\\ &\geq \max\{\tilde{\pi}_{e_2}, x(A)_{e_1} + l_{e_1 e_2}\}\\ &\geq \max\{\tilde{\pi}_{e_2}, T[e_1] + l_{e_1 e_2}\} \quad \text{[because of (4.5)]}\\ &\geq \min_{i\in\text{PERM}:(i,e_2)\in\mathcal{A}}\max\{\tilde{\pi}_{e_2}, T[i] + l_{ie_2}\}\\ &= \tilde{T}[e_2] \geq \tilde{T}[\hat{e}] = T[\hat{e}]\end{aligned}$$

which contradicts (4.6).

□

We emphasize that in particular, Lemma 4.4.12 holds for the sets $\mathcal{A}_{\text{fix}}[\hat{e}]$ constructed in the algorithm:

Corollary 4.4.13. *In every iteration of Algorithm 1 for every event $e \in \mathcal{E}_{op} \cup \mathcal{E}'_{org}$ that is contained in PERM it holds that*

$$T[\hat{e}] \leq x(\mathcal{A}_{fix}[\hat{e}])_{\hat{e}} \tag{4.7}$$

for the labels $T[\hat{e}]$ and the sets of changing activities $\mathcal{A}_{fix}[\hat{e}]$ calculated by the algorithm.

Proof. Due to the construction of $\mathcal{A}_{\text{fix}}[\hat{e}]$, the path $P^{\mathcal{A}_{fix}[\hat{e}]}_{u^{\text{org}}(\sigma)\hat{e}}$ as defined in Notation 4.3.14 exists. The statement thus follows from Lemma 4.4.12. □

If the path which is constructed by the algorithm has the NTT property, we can even show equality of the permanent time labels and the disposition timetable:

Lemma 4.4.14. *Let $\hat{e} \neq v^{dest}$ be the event chosen in step 7 of iteration k of Algorithm 1. If path $P^{\mathcal{A}_{fix}[\hat{e}]}_{u^{org}(\sigma)\hat{e}}$ as defined in Notation 4.3.14 has the NTT property it holds that*

4.4 Complexity of Delay Management with Routing

$$T[\hat{e}] = x(\mathcal{A}_{fix}[\hat{e}])_{\hat{e}}$$

for the labels $T[\hat{e}]$ and the sets of changing activities $\mathcal{A}_{fix}[\hat{e}]$ calculated by the algorithm.

Proof. We show the lemma inductively.

- For the origin node $u^{org}(\sigma)$, we have

$$T[u^{org}(\sigma)] = \sigma = x(\emptyset)_{u^{org}(\sigma)} = x(\mathcal{A}_{fix}(u^{org}(\sigma)))_{u^{org}(\sigma)}.$$

- Assume that the assumption holds for all predecessors of $\hat{e}$ on $P^{\mathcal{A}_{fix}[\hat{e}]}_{u^{org}(\sigma)\hat{e}}$.

- Let $\tilde{T}^{old}[\hat{e}]$ be the label of $\hat{e}$ at the beginning of step 4 in iteration k and let $\hat{e}^{old}$ be the event that was added to PERM in the $(k-1)$th iteration. Then the new temporary label of $\hat{e}$ is given as

$$\tilde{T}[\hat{e}] = \begin{cases} \tilde{T}^{old}[\hat{e}] & \text{if } (\hat{e}^{old}, \hat{e}) \notin \mathcal{A} \\ \min\{\tilde{T}^{old}[\hat{e}], \max\{\tilde{\pi}_{\hat{e}}, T[\hat{e}^{old}] + l_{\hat{e}^{old}\hat{e}}\}\} & \text{otherwise} \end{cases}$$

$$= \min_{i \in \text{PERM}:(i,\hat{e}) \in \mathcal{A}} \max\{\tilde{\pi}_{\hat{e}}, T[i] + l_{i\hat{e}}\}.$$

Denote $P := P^{\mathcal{A}_{fix}[\hat{e}]}_{u^{org}(\sigma)\hat{e}}$ and let P' be the subpath from $u^{org}(\sigma)$ to the direct predecessor $i_{\hat{e}}$ of $\hat{e}$ in P. Since there is no node i with $i \leq_P \hat{e}$ and $\text{tr}(i) = \text{tr}(\hat{e})$ it follows that

$$x(\mathcal{A}_{fix}[\hat{e}])_{j_{\hat{e}}} = x\left(\left\{P^{\mathcal{A}_{fix}[\hat{e}]}_{u^{org}(\sigma)\hat{e}}\right\}\right)_{j_{\hat{e}}} = \tilde{\pi}_{j_{\hat{e}}} \quad (4.8)$$

for the direct predecessor $j_{\hat{e}}$ of $\hat{e}$ on $\text{tr}(\hat{e})$. Since $\mathcal{A}_{fix}[\hat{e}] = P^{\mathcal{A}_{fix}[\hat{e}]}_{u^{org}(\sigma)\hat{e}} \cap \mathcal{A}_{change}$ we have $\{i : (i, \hat{e}) \in \mathcal{A}_{drive} \cup \mathcal{A}_{wait} \cup \mathcal{A}_{fix}\} = \{i_{\hat{e}}, j_{\hat{e}}\}$. Therefore, and because of (4.8), it follows that

$$\max\{\tilde{\pi}_{\hat{e}}, x(\mathcal{A}_{change} \cap P)_{i_{\hat{e}}} + l_{i_{\hat{e}}\hat{e}}\}$$

$$= \max\{\tilde{\pi}_{\hat{e}}, \tilde{\pi}_{j_{\hat{e}}} + l_{j_{\hat{e}}\hat{e}}, x(\mathcal{A}_{change} \cap P)_{i_{\hat{e}}} + l_{i_{\hat{e}}\hat{e}}\}$$

$$= \max\{\tilde{\pi}_{\hat{e}}, x(\mathcal{A}_{change} \cap P)_{j_{\hat{e}}} + l_{j_{\hat{e}}\hat{e}}, x(\mathcal{A}_{change} \cap P)_{i_{\hat{e}}} + l_{i_{\hat{e}}\hat{e}}\}$$

$$= \max\left\{\tilde{\pi}_{\hat{e}}, \max_{(i,\hat{e}) \in \mathcal{A}_{drive} \cup \mathcal{A}_{wait} \cup \mathcal{A}_{fix}} \{x(\mathcal{A}_{change} \cap P)_i + l_{i\hat{e}}\}\right\}. \quad (4.9)$$

Then

$$
\begin{aligned}
T[\hat{e}] &= \min_{i \in \text{PERM}:(i,\hat{e}) \in \mathcal{A}} \max\{\tilde{\pi}_{\hat{e}}, T[i] + l_{i\hat{e}}\} \\
&= \max\{\tilde{\pi}_{\hat{e}}, T[i_{\hat{e}}] + l_{i_{\hat{e}}\hat{e}}\} \\
&= \max\{\tilde{\pi}_{\hat{e}}, x(\mathcal{A}_{\text{fix}}[i_{\hat{e}}])_{i_{\hat{e}}} + l_{i_{\hat{e}}\hat{e}}\} \qquad \text{[due to the induction hypothesis]} \\
&= \max\{\tilde{\pi}_{\hat{e}}, x(\mathcal{A}_{\text{change}} \cap P')_{i_{\hat{e}}} + l_{i_{\hat{e}}\hat{e}}\} \\
&= \max\{\tilde{\pi}_{\hat{e}}, x(\mathcal{A}_{\text{change}} \cap P)_{i_{\hat{e}}} + l_{i_{\hat{e}}\hat{e}}\} \qquad \text{(due to Lemma 4.4.4)} \\
&= \max\left\{\tilde{\pi}_{\hat{e}}, \max_{(i,\hat{e}) \in \mathcal{A}_{\text{drive}} \cup \mathcal{A}_{\text{wait}} \cup \mathcal{A}_{\text{fix}}}\{x(\mathcal{A}_{\text{change}} \cap P)_i + l_{i\hat{e}}\}\right\} \qquad \text{[see (4.9)]} \\
&= x(\mathcal{A}_{\text{change}} \cap P)_{\hat{e}} \qquad \text{(due to Lemma 4.4.8)} \\
&= x(\mathcal{A}_{\text{fix}}[\hat{e}])_{\hat{e}}.
\end{aligned}
$$
(4.10)

This concludes the proof of the lemma. □

This implies that every time Algorithm 1 treats a node $\hat{e}$ for which the constructed path $P_{u^{\text{org}}(\sigma)\hat{e}}^{\mathcal{A}_{\text{fix}}[\hat{e}]}$ has the NTT property, $T[\hat{e}] - \sigma$ is an optimal solution to the corresponding instance of DMwR':

Corollary 4.4.15. *Let $I' = (\mathcal{N}, l, \pi, \{(u^{org}(\sigma), \hat{e})\}, d)$ be an instance of DMwR' and denote by $\mathcal{A}_{fix}[\hat{e}]$ the set of maintained changing activities for node $\hat{e} \in \mathcal{E}_{op}$ calculated in Algorithm 1 for the instance $I = (\mathcal{N}, l, \pi, \{(u, v, \sigma)\}, d)$ of DMwR. If the path $P_{u^{org}(\sigma)\hat{e}}^{\mathcal{A}_{fix}[\hat{e}]}$ as defined in Notation 4.3.14 has the NTT property, it follows that $(x(\mathcal{A}_{fix}[\hat{e}]), \{P_{u^{org}(\sigma)\hat{e}}^{\mathcal{A}_{fix}[\hat{e}]}\})$ is an optimal solution to I'.*

Proof. This follows from Lemmas 4.4.12 and 4.4.14. □

Now we show that the lower bound property of the permanent time labels proven in Lemma 4.4.12 for operational nodes also holds for the destination node v^{dest}. For a path P from $u^{\text{org}}(\sigma)$ to v^{dest} we abbreviate $c'(P) := c(x(\{P\}), \{P\})$.

Lemma 4.4.16. *Let $\mathcal{P}$ denote the set of paths from $u^{org}(\sigma)$ to v^{dest} in $\mathcal{N}$. It holds that*

$$T[v^{dest}] \leq \min_{P \in \mathcal{P}} c'(P) + \sigma.$$

Proof. Suppose that $T[v^{\text{dest}}] > \min_{P \in \mathcal{P}} c'(P) + \sigma$. Let $P^* \in \operatorname{argmin}_{P \in \mathcal{P}} c'(P)$ and $A^* := P^* \cap \mathcal{A}_{\text{change}}$ and let e_0 be the direct predecessor of v^{dest} on P^*. Then

$$c'(P^*) + \sigma < T[v^{\text{dest}}]. \tag{4.11}$$

1. If $e_0 \in$ PERM

$$\begin{aligned}
&c'(P^*) + \sigma \\
&= \max\{\tilde{\pi}_{v^{\text{dest}}}, x(A^*)_{e_0} + l_{e_0 v^{\text{dest}}}\} \\
&= \min_{i \in \text{PERM}:(i,v^{\text{dest}}) \in \mathcal{A}^*} \max\{\tilde{\pi}_{v^{\text{dest}}}, x(A^*)_i + l_{i v^{\text{dest}}}\} \\
&\geq \min_{i \in \text{PERM}:(i,v^{\text{dest}}) \in \mathcal{A}^*} \max\{\tilde{\pi}_{v^{\text{dest}}}, T[i] + l_{i v^{\text{dest}}}\} \quad \text{[because of Corollary 4.4.13]} \\
&= \min_{i \in \text{PERM}:(i,v^{\text{dest}}) \in \mathcal{A}^*} \max\{\sigma, T[i] + l_{i v^{\text{dest}}}\} \\
&= T[v^{\text{dest}}].
\end{aligned}$$

which contradicts (4.11).

2. If the direct predecessor e_0 of v^{dest} in P^* is in TEMP, let e_1 denote the last event in PERM on the path P^* and $e_2 \in$ TEMP its direct successor. Since $T[v^{\text{dest}}] = \tilde{T}[v^{\text{dest}}] \leq \tilde{T}[e]$ for $e \in$ TEMP:

$$c'(P^*) + \sigma = x(A^*)_{e_0} \geq x(A^*)_{e_2} \geq \max\{\tilde{\pi}_{e_2}, T[e_1] + l_{e_1 e_2}\}$$
$$\geq \tilde{T}[e_2] \geq \tilde{T}[v^{\text{dest}}] = T[v^{\text{dest}}]$$

which contradicts (4.11). □

If $P^{\mathcal{A}_{\text{fix}}[v^{\text{dest}}]}_{u^{\text{org}}(\sigma) v^{\text{dest}}}$ has the NTT property, similar to Lemma 4.4.14 we obtain equality of the permanent time label and the arrival time in v^{dest}:

Lemma 4.4.17. *If the path $P^{\mathcal{A}_{\text{fix}}[v^{\text{dest}}]}_{u^{\text{org}}(\sigma) v^{\text{dest}}}$ as defined in Notation 4.3.14 has the NTT property,*

$$T[v^{\text{dest}}] = c'\left(P^{\mathcal{A}_{\text{fix}}[v^{\text{dest}}]}_{u^{\text{org}}(\sigma) v^{\text{dest}}}\right) + \sigma$$

for the label $T[v^{\text{dest}}]$ and the set $\mathcal{A}_{\text{fix}}[v^{\text{dest}}]$ defined in the last iteration of Algorithm 1.

Proof. Let e_0 be the direct predecessor of v^{dest} on $P^{\mathcal{A}_{\text{fix}}[\hat{e}]}_{u^{\text{org}}(\sigma) v^{\text{dest}}}$ and PERM(k) the set of permanently labeled nodes in Algorithm 1 in the last iteration k.

$$T[v^{\text{dest}}] = T[e_0] = x(\mathcal{A}_{\text{fix}}[e_0])_{e_0} = c'\left(P^{\mathcal{A}_{\text{fix}}[e_0]}_{u^{\text{org}}(\sigma) v^{\text{dest}}}\right) + \sigma = c'\left(P^{\mathcal{A}_{\text{fix}}[v^{\text{dest}}]}_{u^{\text{org}}(\sigma) v^{\text{dest}}}\right) + \sigma.$$

□

We say that *an instance* has the NTT property if for every path for an OD-pair which is not too long in $\mathcal{N}(\pi)$ the NTT property holds (see Definition 4.4.5 for the definition of the NTT property for paths):

Definition 4.4.18. Let $I = (\mathcal{N}, l, \pi, \mathcal{OD}, d)$ with $\mathcal{OD} = \{(u_k, v_k, \sigma_k) : k \in K\}$ be an instance of DMwR. Let $\hat{P}_k$ be a path from $u_k^{\text{org}}(\sigma_k)$ to v_k^{dest} in $\mathcal{N}$ with last arrival node $\hat{j}_k$ that minimizes the nominal arrival time $SP_k = x_{\hat{j}_k} - \sigma_k$.

We say that I has the *NTT property*, if for every $k \in K$ every path P_k from $u_k^{\text{org}}(\sigma_k)$ to v_k^{dest} with last arrival node j_{P_k} and

$$\pi_{j_{P_k}} \leq \pi_{\hat{j}_k} + \max_{i \in \mathcal{E}_{\text{op}}} d_i$$

has the NTT property.

Given Lemmas 4.4.12–4.4.17, we conclude the main theorem of this section:

Theorem 4.4.19. *Let $I = (\mathcal{N}, l, \pi, \{(u, v, \sigma)\}, d)$ be an instance of DMwR and let $(\mathcal{A}_{\text{fix}}, \text{lb})$ be the output returned by Algorithm 1 for I. Let $P_{u^{\text{org}}(\sigma)v^{\text{dest}}}^{\mathcal{A}_{\text{fix}}}$ be defined as in Notation 4.3.14.*

- *lb is a lower bound on the objective value of I.*
- *If I has the NTT property, $(x(\mathcal{A}_{\text{fix}}), \{P_{u^{\text{org}}(\sigma)v^{\text{dest}}}^{\mathcal{A}_{\text{fix}}}\})$ is an optimal solution to I with objective value*

$$c(x(\mathcal{A}_{\text{fix}}), \{P_{u^{\text{org}}(\sigma)v^{\text{dest}}}^{\mathcal{A}_{\text{fix}}}\}) = \text{lb}.$$

The running time of Algorithm 1 is $O(n^2)$ where $n = |\mathcal{E}|$ is the number of activities in the EAN $\mathcal{N}$.

Proof.

- It follows from Lemma 4.4.16 that

$$\text{lb} = T[v^{\text{dest}}] - \sigma \leq \min_{P \in \mathcal{P}} c'(x(\{P\}), \{P\})$$

for the set $\mathcal{P}$ of paths from $u^{\text{org}}(\sigma)$ to v^{dest} in $\mathcal{N}$. Hence, lb is a lower bound on the optimal objective value of the instance $I' = (\mathcal{N}, l, \pi, \{(u^{\text{org}}(\sigma), v^{\text{dest}})\}, d)$ of DMwR'. Therefore, lb is also a lower bound on the objective value of I.
- Like in Definition 4.4.18, let $\hat{P}$ denote a path from $u^{\text{org}}(\sigma)$ to v^{dest} in $\mathcal{N}$ with last arrival node $\hat{j}$ that minimizes the nominal arrival time $SP = x_{\hat{j}} - \sigma$.

  For every node $j \in \mathcal{E}_{\text{op}}$ and every $A \subset \mathcal{A}_{\text{change}}$, due to Corollary 4.3.9, it holds that

$$x(A)_j \leq \pi_j + \max_{i \in \mathcal{E}_{\text{op}}} d_i.$$

Let j' be the last arrival node on $P_{u^{\text{org}}(\sigma)v^{\text{dest}}}^{\mathcal{A}_{\text{fix}}}$. Since $T[v^{\text{dest}}] = \max\{\sigma, T[j']\}$, in particular $T[v^{\text{dest}}] \geq \pi_{j'}$. Thus, due to Lemma 4.4.16,

$$\pi_{j'} \leq T[v^{\text{dest}}] \leq c'(\hat{P}) + \sigma = x(\mathcal{A}_{\text{change}} \cap \hat{P})_{\hat{j}} \leq \pi_{\hat{j}} + \max_{i \in \mathcal{E}_{\text{op}}} d_i.$$

Hence, since I has the NTT property, $P_{u^{\text{org}}(\sigma)v^{\text{dest}}}^{\mathcal{A}_{\text{fix}}}$ has the NTT property. Due to Lemma 4.4.17, it follows that

4.4 Complexity of Delay Management with Routing

$$T[v^{\text{dest}}] = c' \left(P^{\mathcal{A}_{\text{fix}}}_{u^{\text{org}}(\sigma)v^{\text{dest}}} \right) + \sigma.$$

According to Lemma 4.4.16, this implies that $\mathcal{R} := \{P^{\mathcal{A}_{\text{fix}}}_{u^{\text{org}}(\sigma)v^{\text{dest}}}\}$ is an optimal solution to $I' = (\mathcal{N}, l, \pi, \{(u^{\text{org}}(\sigma), v^{\text{dest}})\}, d)$ with optimal objective value $\text{lb} = T[v^{\text{dest}}] - \sigma$. Therefore, $\mathcal{R}$ is also an optimal solution to I with optimal objective value lb.

The generation of the disposition timetable in step 2 of Algorithm 1 is done in time $O(m)$ using the method given in Definition 4.3.4. Inspecting steps 4–6, we note that for the setting of the temporary labels $\tilde{T}$, summing up over all iterations, every activity $a \in \mathcal{A}$ has to be considered at most once. As the steps 7–8 are done in time $O(n)$, steps 7–13 need time $O(m)$. Since step 15 is also in $O(m)$, the running time of the modified Dijkstra's algorithm is $O(m)$. □

In practice, situations like the one constructed in the NP-hardness proof for DMwR in Theorem 4.4.1, where a lot of trains run on the same physical route and constantly overtake each other, are not very likely to happen. In fact, the NTT property, i.e., the assumption that for an OD-pair there is no reasonably short path where a train is entered more than once, under which Algorithm 1 returns optimal solutions for instances of DMwR with one OD-pair, holds for the real-world instances considered in [DHSS09, DHSS12].

However, we show in Theorem 4.4.21 that, even if optimality is not guaranteed, the solution obtained using Algorithm 1 for DMwR with one OD-pair is a 2-approximation of the optimal objective value. To prove the theorem we need the following lemma.

Lemma 4.4.20. *Let $I = (\mathcal{N}, l, \pi, \{(u, v, \sigma)\}, d)$ be an instance of DMwR with one OD-pair. Let $(\mathcal{A}_{\text{fix}}, \text{lb})$ be the output of Algorithm 1 and denote $P := P^{\mathcal{A}_{\text{fix}}}_{u^{\text{org}}(\sigma)v^{\text{dest}}} \cap \mathcal{A}_{op}$ with $P^{\mathcal{A}_{\text{fix}}}_{u^{\text{org}}(\sigma)v^{\text{dest}}}$ as defined in Notation 4.3.14. Then*

$$\text{lb} = \max_{i_{k'} \in P} \left\{ \tilde{\pi}_{i_{k'}} + \sum_{(i,j) \in P : i \geq_P i'} l_{ij} \right\} - \sigma. \quad (4.12)$$

Proof. We first observe how lb is calculated in Algorithm 1. Let $l_{j'}$ be the number of the iteration when for the first time a direct predecessor j' of v^{dest} enters PERM. Then in step 5 of iteration $l_{j'}$ we set $\tilde{T}[v^{\text{dest}}] := \min\{\infty, \max\{T[j'], \sigma\}\} = T[j']$. In step 7 of iteration $l_{j'} + 1$ there is no $j \in \text{TEMP}$ with $\tilde{T}[v^{\text{dest}}] < \tilde{T}[j]$, thus without loss of generality we can assume that $T[v^{\text{dest}}] = T[j']$ is set in step 7 of iteration $l_{j'} + 1$. Then lb is set to $T[v^{\text{dest}}] - \sigma$.

Let $P = (1, 2 \ldots, j')$ and $P_k := (k, k+1, \ldots, j') \subset P$.
We now show inductively that for every $k = 1, \ldots, j'$

$$T[v^{\text{dest}}] = \max\left\{T[k] + \sum_{(i,j)\in P_k} l_{ij}, \max_{k'\in P_{k+1}}\left\{\tilde{\pi}_{k'} + \sum_{(i,j)\in P_{k'}} l_{ij}\right\}\right\}. \quad (4.13)$$

- As we have just seen, $T[v^{\text{dest}}] = T[j']$.
- Assume that (4.13) holds for all $k = \hat{k}, \ldots, j'$.
- Then

$$\begin{aligned}\text{lb} &= T[v^{\text{dest}}] \\ &= \max\left\{T[\hat{k}] + \sum_{(i,j)\in P_{\hat{k}}} l_{ij}, \max_{k'\in P_{\hat{k}+1}}\left\{\tilde{\pi}_{k'} + \sum_{(i,j)\in P_{k'}} l_{ij}\right\}\right\} \\ &= \max\left\{\max\{\tilde{\pi}_{\hat{k}}, T[\hat{k}-1] + l_{\hat{k}-1\hat{k}}\} + \sum_{(i,j)\in P_{\hat{k}}} l_{ij}, \max_{k'\in P_{\hat{k}+1}}\left\{\tilde{\pi}_{k'} + \sum_{(i,j)\in P_{k'}} l_{ij}\right\}\right\} \\ &= \max\left\{\tilde{\pi}_{\hat{k}} + \sum_{(i,j)\in P_{\hat{k}}} l_{ij}, T[\hat{k}-1] + \sum_{(i,j)\in P_{\hat{k}}} l_{ij}, \max_{k'\in P_{\hat{k}+1}}\left\{\tilde{\pi}_{k'} + \sum_{(i,j)\in P_{k'}} l_{ij}\right\}\right\} \\ &= \max\left\{T[\hat{k}-1] + \sum_{(i,j)\in P_{\hat{k}}} l_{ij}, \max_{k'\in P_{\hat{k}}}\left\{\tilde{\pi}_{k'} + \sum_{(i,j)\in P_{k'}} l_{ij}\right\}\right\}.\end{aligned}$$

Since $T[1] = \max\{T[u^{\text{org}}(\sigma)]+0, \tilde{\pi}_1\} = \tilde{\pi}_1$, the statement of the lemma follows. $\square$

We now show that Algorithm 1 is a 2-approximation algorithm for DMwR with one OD-pair.

Theorem 4.4.21. *Let $I = (\mathcal{N}, l, \pi, \{(u, v, \sigma)\}, d)$ be an instance of DMwR with one OD-pair and let $(x^*, \mathcal{R}^*)$ be an optimal solution to I. Let $\mathcal{A}_{\text{fix}} \subset \mathcal{A}_{\text{change}}$ denote the set of maintained changing activities returned by Algorithm 1 for I and $\mathcal{R}(x(\mathcal{A}_{\text{fix}})) \in OPT(x(\mathcal{A}_{\text{fix}}))$. Then it holds that*

$$\frac{c(x(\mathcal{A}_{\text{fix}}), \mathcal{R}(x(\mathcal{A}_{\text{fix}})))}{c(x^*, P^*)} \leq 2.$$

Proof. We introduce some notation that is used in the proof:

- Let $P^{\mathcal{A}_{\text{fix}}}_{u^{\text{org}}(\sigma)v^{\text{dest}}}$ be the path from $u^{\text{org}}(\sigma)$ to v^{dest} defined by $\mathcal{A}_{\text{fix}}$ in the sense of Notation 4.3.14. Let j' denote the last arrival node on $P^{\mathcal{A}_{\text{fix}}}_{u^{\text{org}}(\sigma)v^{\text{dest}}}$. Let P' denote the subpath of $P^{\mathcal{A}_{\text{fix}}}_{u^{\text{org}}(\sigma)v^{\text{dest}}}$ from the first departure node to j', i.e., $P' = P^{\mathcal{A}_{\text{fix}}}_{u^{\text{org}}(\sigma)v^{\text{dest}}} \cap \mathcal{A}_{\text{op}}$. For any $i' \in P'$ denote by $P'_{i'}$ the subpath of P' starting in i'.
- For i, j with $\text{tr}(i) = \text{tr}(j)$ and $i \leq_{\mathcal{N}} j$ let $P^{\text{tr}(j)}_{ij} = P^{\text{tr}(i)}_{ij}$ be the path from i to j entirely contained in $\text{tr}(j)$.
- For $i, j \in \mathcal{E}_{\text{op}}$ with $i \leq_{\mathcal{N}(\mathcal{A}_{\text{fix}})} j$ denote by $\hat{\mathcal{P}}(i, j, \mathcal{A}_{\text{fix}})$ the set of all paths from i to j in $\mathcal{N}(x(\mathcal{A}_{\text{fix}}))$.

4.4 Complexity of Delay Management with Routing

Note that for all $i' \in P'$ it holds that $\pi_{i'} \geq \sigma$. Hence, for all $i' \in P'$

$$\sum_{(i,j)\in P'_{i'}} s_{ij} = \pi_{j'} - \pi_{i'} - \sum_{(i,j)\in P'_{i'}} l_{ij} \leq \pi_{j'} - \sigma. \qquad (4.14)$$

Furthermore, with T denoting the permanent time labels used in Algorithm 1

$$\text{lb} = T[v^{\text{dest}}] - \sigma = T[j'] - \sigma \geq \pi_{j'} - \sigma. \qquad (4.15)$$

We can now bound lb from below.

$\text{lb} + \sigma$

$$= \max_{i' \in P'} \left\{ \tilde{\pi}_{i'} + \sum_{(i,j)\in P'_{i'}} l_{ij} \right\}$$

[due to Lemma 4.4.20]

$$= \max_{i' \in P'} \left\{ \max_{\substack{e:\text{tr}(e)=\text{tr}(i'),\\ e \leq_\mathcal{N} i'}} \left\{ \pi_e + d_e + \sum_{(i,j)\in P^{\text{tr}(i')}_{ei'}} l_{ij} \right\} + \sum_{(i,j)\in P'_{i'}} l_{ij} \right\}$$

[due to Lemma 4.3.8 and the definition of $\tilde{\pi} = x(\emptyset)$]

$$= \max_{i' \in P'} \left\{ \max_{\substack{e:\text{tr}(e)=\text{tr}(i'),\\ e \leq_\mathcal{N} i'}} \left\{ \pi_e + d_e + \sum_{(i,j)\in P^{\text{tr}(i')}_{ei'}} l_{ij} \right\} + \pi_{j'} - \pi_{i'} - \sum_{(i,j)\in P'_{i'}} s_{ij} \right\}$$

$$\left(\text{since } \pi_{j'} = \pi_{i'} + \sum_{(i,j)\in P'_{i'}} (l_{ij} + s_{ij}) \right)$$

$$\geq \max_{i' \in P'} \left\{ \max_{\substack{e:\text{tr}(e)=\text{tr}(i'),\\ e \leq_\mathcal{N} i'}} \left\{ \pi_e + d_e + \sum_{(i,j)\in P^{\text{tr}(i')}_{ei'}} l_{ij} \right\} + \pi_{j'} - \pi_{i'} - (\pi_{j'} - \sigma) \right\}$$

[due to (4.14)]

$$= \max_{i' \in P'} \left\{ \max_{\substack{e:\text{tr}(e)=\text{tr}(i'),\\ e \leq_\mathcal{N} i'}} \left\{ \pi_e + d_e + \sum_{(i,j)\in P^{\text{tr}(i')}_{ei'}} l_{ij} \right\} - \pi_{i'} + \sigma \right\}.$$

Now $x(\mathcal{A}_{\text{fix}})_{j'} - \sigma$ can be bounded by 2lb:

$$x(\mathcal{A}_{\text{fix}})_{j'}$$

$$= \max_{e \in \mathcal{E}_{\text{op}}: e \leq_{\mathcal{N}}(\mathcal{A}_{\text{fix}})j'} \max_{P \in \hat{\mathcal{P}}(e,j',\mathcal{A}_{\text{fix}})} \left\{ \pi_e + d_e + \sum_{(i,j) \in P} l_{ij} \right\}$$

[due to Lemma 4.3.8]

$$= \max_{i' \in P'} \left\{ \max_{\substack{e:\text{tr}(e)=\text{tr}(i') \\ e \leq_{\mathcal{N}} i'}} \left\{ \pi_e + d_e + \sum_{(i,j) \in P_{ei'}^{\text{tr}(i')}} l_{ij} \right\} + \max_{P \in \hat{\mathcal{P}}(i',j',\mathcal{A}_{\text{fix}})} \left\{ \sum_{(i,j) \in P} l_{ij} \right\} \right\}$$

[since $\hat{\mathcal{P}}(e, j', \mathcal{A}_{\text{fix}}) \neq \emptyset \Leftrightarrow \exists i' \in P'$ such that $\text{tr}(i') = \text{tr}(e)$ and $e \leq_{\mathcal{N}} i'$]

$$= \max_{i \in P'} \left\{ \max_{\substack{e:\text{tr}(e)=\text{tr}(i') \\ e \leq_{\mathcal{N}} i'}} \left\{ \pi_e + d_e + \sum_{(i,j) \in P_{ei'}^{\text{tr}(i')}} l_{ij} \right\} + \max_{P \in \hat{\mathcal{P}}(i',j',\mathcal{A}_{\text{fix}})} \left\{ \pi_{j'} - \pi_{i'} - \sum_{(i,j) \in P} s_{ij} \right\} \right\}$$

$$\left[\text{since } \pi_{j'} = \pi_{i'} + \sum_{(i,j) \in P} (l_{ij} + s_{ij}) \right]$$

$$\leq \max_{i \in P'} \left\{ \max_{\substack{e:\text{tr}(e)=\text{tr}(i') \\ e \leq_{\mathcal{N}} i'}} \left\{ \pi_e + d_e + \sum_{(i,j) \in P_{ei'}^{\text{tr}(i')}} l_{ij} \right\} + \pi_{j'} - \pi_{i'} \right\}$$

[since $s_{ij} \geq 0 \ \forall (i,j) \in \mathcal{A}_{\text{op}}$]

$$= \max_{i \in P'} \max_{\substack{e:\text{tr}(e)=\text{tr}(i') \\ e \leq_{\mathcal{N}} i'}} \left\{ \pi_e + d_e + \sum_{(i,j) \in P_{ei'}^{\text{tr}(i')}} l_{ij} \right\} + \pi_{j'} - \pi_{i'} - (\pi_{j'} - \sigma) + (\pi_{j'} - \sigma)$$

$$= \underbrace{\left(\max_{i \in P'} \left\{ \max_{\substack{e:\text{tr}(e)=\text{tr}(i') \\ e \leq_{\mathcal{N}} i'}} \left\{ \pi_e + d_e + \sum_{(i,j) \in P_{ei'}^{\text{tr}(i')}} l_{ij} \right\} - \pi_{i'} + \sigma \right\} \right)}_{\leq \text{lb} + \sigma} + \underbrace{(\pi_{j'} - \sigma)}_{\leq \text{lb}}$$

[due to (4.14) and (4.15)]

$\leq 2\text{lb} + \sigma$.

Due to Theorem 4.4.19, $c(x^*, P^*) \geq \text{lb}$, hence

$$\frac{c(x(\mathcal{A}_{\text{fix}}), \mathcal{R}(x(\mathcal{A}_{\text{fix}})))}{c(x^*, P^*)} = \frac{x(\mathcal{A}_{\text{fix}})_{j'} - \sigma}{c(x^*, P^*)} \leq \frac{2\text{lb}}{\text{lb}} = 2.$$

This concludes the proof. □

Solving DMwR for one OD-pair may be of only theoretical interest since there are many OD-pairs in real-world instances. However, Algorithm 1 can be used to determine a lower bound on the optimal objective value for the general DMwR problem which improves the integer programming formulation for DMwR in Sect. 4.5.

Lemma 4.4.22. *Let* $I = (\mathcal{N}, l, \pi, \mathcal{OD}, d)$ *be an instance of DMwR with* $\mathcal{OD} = \{(u_k, v_k, \sigma_k) : k \in K\}$ *with weights* w_k. *Let* $(x^*, \mathcal{R}^*)$ *be an optimal solution to* I. *For every* $P_k \in \mathcal{R}^*$ *let* j_k *denote the last arrival node on* P_k, *and for every* $(u_k, v_k, \sigma_k) \in \mathcal{OD}$ *let* lb_k *denote the bound calculated by Algorithm 1 for the instance* $I_k := (\mathcal{N}, l, \pi, \{(u_k, v_k, \sigma_k)\}, d)$.

It holds that

$$\text{lb}_k \leq x^*_{j_k} - \sigma$$

for any $(u_k, v_k, \sigma_k) \in \mathcal{OD}$. In particular, we have

$$\sum_{(u_k, v_k, \sigma_k) \in \mathcal{OD}} w_k \text{lb}_k \leq c(x^*, R^*).$$

Moreover, this bound on the objective value can be calculated in $O(|\mathcal{OD}|n^2)$.

4.4.3 NP-Hardness of Delay Management with Routing for Instances with NTT Property

In Sect. 4.4.2 we have seen that for the case of only one OD-pair, instances of DMwR can be solved to optimality in polynomial time by a Dijkstra-type solution algorithm if they have the NTT property. Dijkstra's shortest path algorithm can be generalized to find the shortest paths from an origin to all nodes in the graph. It is hence a natural question, whether such a generalization is also possible for Algorithm 1. Therefore, in this section we consider instances $I = (\mathcal{N}, l, \pi, \mathcal{OD}, d)$ of DMwR where $\mathcal{OD} = \{(u, v_k, \sigma) : k \in K\}$, i.e., all OD-pairs have the same origin u and the same start time σ, which have the NTT property. This generalization of our problem already turns out to be strongly NP-hard.

Theorem 4.4.23. *DMwR is strongly NP-hard, even if*

- *all OD-pairs $(u_k, v_k, \sigma_k) \in \mathcal{OD}$ have the same origin $u_k := u$ and the same start time $\sigma_k := 0$,*
- *all origin and destination events are connected to only one event in the network,*
- *for every $(u, v_k, \sigma) \in \mathcal{OD}$, every path from $u^{org}_k(\sigma)$ to u^{dest}_k has the NTT property, and*
- *there is only one source delay.*

Proof. This theorem is proven by reduction from the NP-complete decision problem Set Cover (see Definition on p. 88). Let (P, Q, K) be an instance of Set Cover with $P = \{p_1, \ldots, p_m\}$ and $Q = \{q^1, \ldots, q^n\}$. Note that the structure of Set Cover can be represented by a matrix $M = (m_{ij})_{i=1,\ldots,m, j=1,\ldots,n}$ with

$$m_{ij} = \begin{cases} 1 & \text{if } p_i \in q^j \\ 0 & \text{otherwise.} \end{cases}$$

We construct an instance I of DMwR in which P corresponds to the OD-pairs and Q corresponds to connections for which we have to decide whether they are maintained or not. We have to cover all OD-pairs (i.e., make sure that all passengers reach their destinations) with a minimal set of maintained connections, since maintaining a connection causes costs (represented as delays) to other passengers. Our construction of an instance of DMwR is the following:

- We transfer the matrix M to a set of stations $S = \{s_i^j : m_{ij} = 1\}$, i.e., whenever $m_{ij} = 1$ for $1 \leq i \leq m$ and $1 \leq j \leq n$ there is a station s_i^j. For every $j = 1, \ldots, n$, we introduce two more stations s_{m+1}^j and s_{v^j}. Finally we add stations s_u and s_0.
- There is a train tr^0 that runs from s_u to s_0 and after leaving s_0 runs through the stations s_{m+1}^j for all $j = 1, \ldots, n$. For every $j = 1, \ldots, n$, there is a train tr^j which starts at s_0 and runs through the existing stations s_i^j in increasing order of i. For every $i = 1, \ldots, m$, there is a train tr_i that runs through the existing stations s_i^j in increasing order of j.
- At the driving activity from s_0 to s_{m+1}^1 we set the slack time to be 1. The slack time on all other driving and waiting activities is set to 0.
- At station s_0 there are changing activities from s_0 to the departure events of the trains tr^j. The stations s_i^j offer the possibility to change from train tr^j to train tr_i, so if and only if $m_{ij} = 1$, it is possible to change from tr^j to tr_i. We allow the passengers to change from tr^0 to tr^j at every station s_{m+1}^j. The slack time for all changing activities is 0.
- Let π and l be an arbitrary pair of timetable and set of lower bounds on the (implicitly) constructed EAN $\mathcal{N}$ which satisfy the requirements on the slack times and $\pi_{s_u} = 0$.
- The set of OD-pairs is $\mathcal{OD} := \{(u, v_i, 0) : i = 1, \ldots, m\} \cup \{(u, v^j, 0) = 1, \ldots, n\}$ with $u := s_u$, $v_i := s_{v_i}$, $v^j := s_{v^j}$ and $w_{uv\sigma} := 1$ for all $(u, v, \sigma) \in \mathcal{OD}$.
- $d_{(\text{tr}^0 - s_0 - \text{Arr})} := 1$. This is the only source delay.

We denote the corresponding EAN by $\mathcal{N}$. Note that in this construction there can be no directed path that enters the same train more than once; hence, I has the NTT property.

In Fig. 4.2 the PTN for an instance of DMwR constructed from an instance of Set Cover with $P = \{p_1, p_2\}$ and $Q = \{\{p_1\}, \{p_2\}, \{p_1, p_2\}\}$ is pictured. There are six trains, tr^0 represented by the thick line, tr^1, tr^2, and tr^3 starting at station s_0 and going from left to right and tr_1 and tr_2 going top down.

The detailed event-activity network $\mathcal{N}$ can be found in Fig. 4.3. Here, the square nodes are the departure and arrival events. The origin and destination events are represented by ovals. The dotted lines are the origin and destination activities, the solid lines represent driving and waiting activities, and changing activities are represented by dashed lines.

In this proof we use the reformulation of the objective function to the minimization of the overall delay instead of the overall travel time. We show that the instance (P, Q, K) of Set Cover has a solution if and only if there is a set $\mathcal{A}_{\text{fix}}$ such that $c^d(x(\mathcal{A}_{\text{fix}}), \mathcal{R}(x(\mathcal{A}_{\text{fix}}))) \leq \tilde{K} := m + K$ for $\mathcal{R}(x(\mathcal{A}_{\text{fix}})) \in \text{OPT}(x(\mathcal{A}_{\text{fix}}))$.

4.4 Complexity of Delay Management with Routing

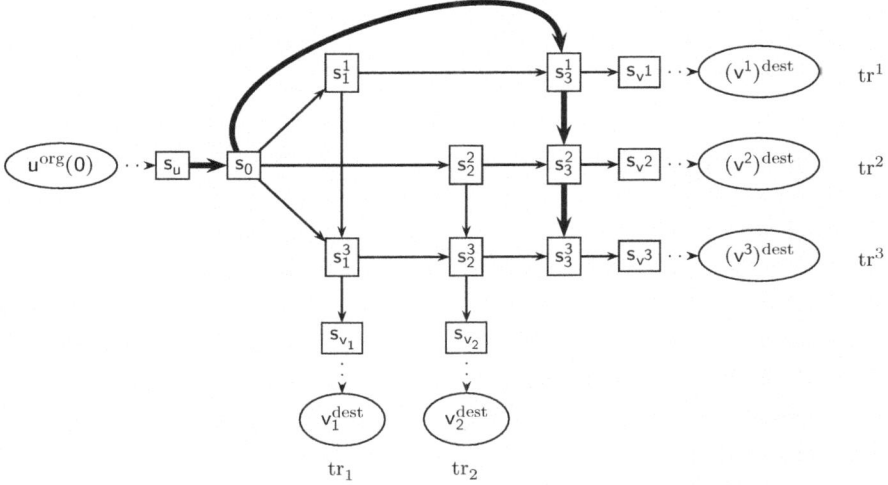

Fig. 4.2 The PTN (indicating the directions of the trains) for the instance of DMwR constructed from an instance of Set Cover with $P = \{p_1, p_2\}$ and $Q = \{\{p_1\}, \{p_2\}, \{p_1, p_2\}\}$

We divide $\mathcal{A}_{\text{change}}$ into three sets:

1. the changing activities at station s_0

$$\mathcal{A}_{\text{change}}(s_0) := \{(\text{tr}^0 - s_0 - \text{Arr}, \text{tr}^j - s_0 - \text{Dep}) : j = 1, \ldots, m\},$$

2. the later changing activities from train tr^0

$$\mathcal{A}_{\text{change}}(\text{tr}^0) := \{(\text{tr}^0 - s_{m+1}^j - \text{Arr}, \text{tr}^j - s_{m+1}^j - \text{Dep})\},$$

3. and the changing activities defined by M

$$\mathcal{A}_{\text{change}}(M) := \{(\text{tr}^j - s_i^j - \text{Arr}, \text{tr}_i - s_i^j - \text{Dep})\}.$$

- Let Q' be a solution to (P, Q, K). We define

$$\mathcal{A}_{\text{fix}}(Q') := \mathcal{A}_{\text{change}}(M) \cup \mathcal{A}_{\text{change}}(\text{tr}^0) \cup \{(\text{tr}^0 - s_0 - \text{Arr}, \text{tr}^j - s_0 - \text{Dep}) : q^j \in Q'\}$$

and set $x(Q') := x(\mathcal{A}_{\text{fix}}(Q'))$. For $(u, v^j, 0) \in \mathcal{OD}$, we choose the following train-route P^j: train tr^0 is entered at station s_u and left at station s_{m+1}^j, where tr^j is entered. Then

$$c^d(x(Q'), P^j) = \begin{cases} 1 & \text{if } (\text{tr}^0 - s_0 - \text{Arr}, \text{tr}^j - s_0 - \text{Dep}) \in \mathcal{A}_{\text{fix}}(Q'), \\ 0 & \text{otherwise.} \end{cases}$$

$$= \begin{cases} 1 & \text{if } q^j \in Q', \\ 0 & \text{otherwise.} \end{cases}$$

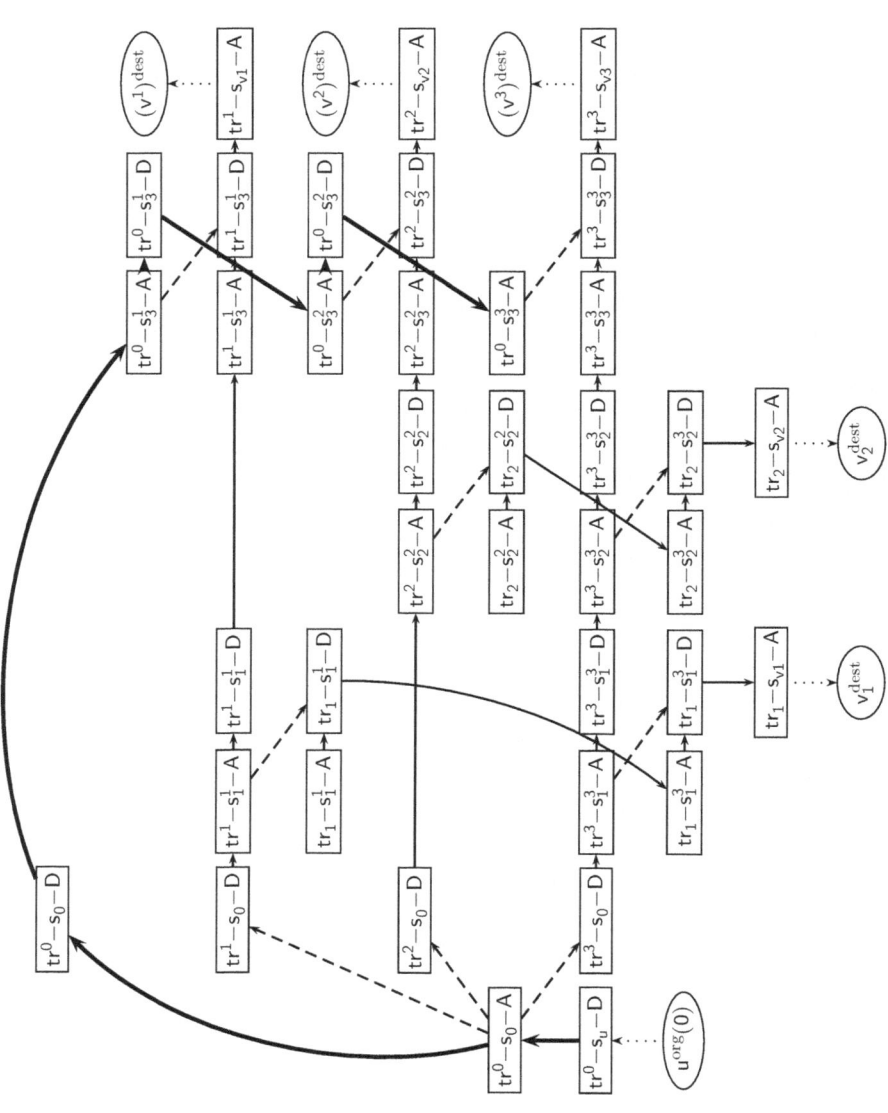

Fig. 4.3 The EAN for the instance of DMwR constructed from an instance of Set Cover with $P = \{p_1, p_2\}$ and $Q = \{\{p_1\}, \{p_2\}, \{p_1, p_2\}\}$

4.4 Complexity of Delay Management with Routing

For $(u, v_i, 0) \in \mathcal{OD}$, we choose the following train-route P_i: train tr^0 is entered at station s_u and left at s_i^j, where tr_i is entered. Hence, $c^d(x(Q'), P_i) = 1$. With $\mathcal{R}(Q') := \{P_i : i = 1, \ldots, m\} \cup \{P^j : j = 1, \ldots, n\}$, it follows that

$$c^d(x(Q'), \mathcal{R}(Q')) = \sum_{i=1}^{m} c^d(x(Q'), P_i) + \sum_{j=1}^{n} c^d(x(Q'), P^j) = m + |Q'| \leq m + K.$$

- Let x be a timetable for I with $\text{OPT}(x) \neq \emptyset$ and $c(x, \mathcal{R}(x)) \leq m + K$ for any $\mathcal{R}(x) \in \text{OPT}(x)$. We set $Q(x) := \{q^j : (\text{tr}^0 - s_0 - \text{Arr}, \text{tr}^j - s_0 - \text{Dep}) \in \mathcal{A}_{\text{fix}}(x)\}$. Consider $p_i \in P$ and suppose that there is no $q^j \in Q(x)$ such that $p_i \in q^j$. Due to the structure of the constructed instance, this implies that there is no path from $u^{\text{org}}(0)$ to v_i^{dest} in $N(x)$ which is a contradiction to the existence of the solution $(x, \mathcal{R})$. Hence, for every $p_i \in P$, there is a $q^j \in Q(x)$ such that $p_i \in q^j$. Furthermore, like in (4.4.3) we observe that for any path $P^j \in N(x)$ from $u^{\text{org}}(0)$ to $(v^j)^{\text{dest}}$ it holds that

$$c^d(x, P^j) = \begin{cases} 1 \text{ if } (\text{tr}^0 - s_0 - \text{Arr}, \text{tr}^j - s_0 - \text{Dep}) \in \mathcal{A}_{\text{fix}}(x), \\ 0 \text{ otherwise.} \end{cases} = \begin{cases} 1 \text{ if } q^j \in Q(x), \\ 0 \text{ otherwise,} \end{cases}$$

and that $c^d(x, P_i) = 1$ for any path $P_i \in N(x)$ from $u^{\text{org}}(0)$ to v_i^{dest}. Hence

$$\begin{aligned} |Q(x)| &= |\{q^j : (\text{tr}^0 - s_0 - \text{Arr}, \text{tr}^j - s_0 - \text{Dep}) \in \mathcal{A}_{\text{fix}}(x)\}| \\ &= |\{(\text{tr}^0 - s_0 - \text{Arr}, \text{tr}^j - s_0 - \text{Dep}) \in \mathcal{A}_{\text{fix}}(x)\}| \\ &= \sum_{j=1}^{n} c^d(x, P^j) \\ &= \sum_{P^j \in \mathcal{R}(x)} c^d(x, P^j) - \sum_{i=1}^{m} c^d(x, P_i) \\ &\leq m + K - m \\ &= K. \end{aligned}$$

So every instance of Set Cover can be transformed in polynomial time to an instance of DMwR with the claimed properties. □

4.4.4 Inapproximability of Delay Management with Routing

To begin with, we remark that if we consider the objective function of minimizing the overall delay, specified in (4.2), unless P = NP, the nonexistence of a polynomial-time approximation algorithm, even for DMwR with only one OD-pair, can directly be concluded from the construction in the proof of Theorem 4.4.1:

Lemma 4.4.24. *There is no polynomial-time algorithm with guaranteed approximation ratio for DMwR with objective function c^d unless $P = NP$, even if we consider only instances of DMwR where*

- *there is only one OD-pair, and*
- *there is only one source delay.*

Proof. Let A be a polynomial-time algorithm for DMwR. Suppose that there is a function f in the input size such that, for any given instance I of DMwR, A returns a solution $(x^A, \mathcal{R}^A)$ with

$$c^d(x^A, \mathcal{R}^A) \leq f(|I|) c^d(x^*, \mathcal{R}^*),$$

where $(x^*, \mathcal{R}^*)$ denotes an optimal solution to I. Given an instance (G, t_1, t_2) of Hamiltonian Path, we construct an instance of DMwR as described in the proof of Theorem 4.4.1, apply A to this instance, and obtain a solution $(x^A, \mathcal{R}^A)$. Let $(x^*, \mathcal{R}^*)$ be an optimal solution to I. Then as described in the proof of Theorem 4.4.1, we apply A to this instance and obtain a solution $(x^A, \mathcal{R}^A)$. Let $(x^*, \mathcal{R}^*)$ be an optimal solution to I. Then

$$c^d(x^A, \mathcal{R}^A) = \begin{cases} 0 & \text{if } c^d(x^*, \mathcal{R}^*) = 0, \\ > 0 & \text{otherwise.} \end{cases}$$

Due to steps (2) and (3) of the proof of Theorem 4.4.1, this implies

$$c^d(x^A, \mathcal{R}^A) = \begin{cases} 0 & \text{if there is a Hamiltonian path in } (G, t_1, t_2), \\ > 0 & \text{otherwise.} \end{cases}$$

Hence, the algorithm decides the Hamiltonian Path problem. □

On the other hand, when choosing the overall travel time to be our objective, we have seen in Lemma 4.4.21 that for one OD-pair Algorithm 1 is a 2-approximation algorithm. Unfortunately, unless $P = NP$, this result cannot be extended to instances with more OD-pairs:

Theorem 4.4.25. *Unless $P = NP$, there is no polynomial-time constant-factor approximation algorithm for DMwR with objective function c, even if we consider only instances of DMwR where*

- *there are only two OD-pairs,*
- *the OD-pairs have the same destination, and*
- *there is only one source delay.*

Proof. Let A be a polynomial-time algorithm for DMwR and for an instance I let $(x^A, \mathcal{R}^A)$ be the solution returned by A and let $(x^*, \mathcal{R}^*)$ be an optimal solution. Suppose that there is a constant $C \geq 1$ such that

4.4 Complexity of Delay Management with Routing

$$\frac{c(x^A, \mathcal{R}^A)}{c(x^*, \mathcal{R}^*)} \le C$$

for all instances of DMwR which fulfill the conditions specified in Theorem 4.4.25.

We show that the existence of such a constant C would lead to a polynomial-time algorithm for the Hamiltonian Path problem. To this end, we modify the network constructed from an instance of Hamiltonian Path with fixed start and end nodes in the proof of Theorem 4.4.1 by adding a station, a train, and an OD-pair and modifying the initial delay and the timetable. Like in the proof of Theorem 4.4.1 and Lemma 4.4.24, in the instance of DMwR we construct here, it will turn out that there is a routing that does not yield any delay if and only if there is a Hamiltonian path from the start node to the end node in the given instance.

Let (G, t_1, t_2) be an instance of Hamiltonian Path with fixed start and end nodes, and let $n \ge 3$ denote the number of nodes in G. We construct an instance $I = (\mathcal{PTN}, \text{TR}, \mathcal{C}, l, \pi, \mathcal{OD}, d)$ of DMwR as follows:

- We set $\tilde{d} := 3C((n-2)(n-3) + 2) + 1$. The edges and edge lengths of the PTN are characterized implicitly by the description of the trains, assuming a driving time function of 1 for all lines.
- The set of stations is

$$S := \{s_u, s_v, s^1, s^{n-1}\} \cup \{s_i : i = 2, \dots, n-1\}$$
$$\cup \{s_i^j : i = 2, \dots, n-1, j = 2, \dots, n-2\} \cup \{s'\}.$$

- The set of trains is TR $:= \{\text{tr}_1, \dots, \text{tr}_n\} \cup \{\text{tr}'\}$.
  - Train tr_1 starts at station s_u at time 0 and arrives at station s^1 at time 1.
  - Train tr_n starts at station s^{n-1} at time $(n-2)(n-3)(\tilde{d}+3) + (2\tilde{d}+6)$ and arrives at station s_v at time $(n-2)(n-3)(\tilde{d}+3) + (2\tilde{d}+7)$.
  - Train tr' starts at station s' at time $(n-2)(n-3)(\tilde{d}+3) + (2\tilde{d}+4)$ and arrives at station s^{n-1} at time $(n-2)(n-3)(\tilde{d}+3) + (2\tilde{d}+5)$.
  - The trains tr_i for $i = 2, \dots, n-1$ start at station s_i at time 0 and go on driving through stations $s^1, s_2^2, s_3^2, \dots, s_{n-1}^2, s_2^3, s_3^3, \dots, s_{n-1}^3, \dots, s_{n-1}^{n-2}, s^{n-1}$ in this order. Note that the trains tr_i do not stop at all of these stations. The stopping pattern and the timetable of the trains $\text{tr}_2, \dots, \text{tr}_{n-1}$ are as follows:
    - For $i = 2, \dots, n-1$, if $\{1, i\} \in G$, tr_i stops at station s^1. The arrival time according to π is $2 + \tilde{d}$, and the departure time is $3 + \tilde{d}$.
    - Train tr_i stops at station s_i^j for every $i = 2, \dots, n-1$ and $j = 2, \dots, n-2$. It arrives there at time $((j-2)(n-2) + (i-2))(\tilde{d}+3) + (4+\tilde{d})$ and departs at time $((j-2)(n-2) + (i-2))(\tilde{d}+3) + (5+\tilde{d})$.
    - Whenever $\{i, i'\} \in G$ for $i, i' \in \{2, \dots, n-1\}$, $\text{tr}_{i'}$ stops at station s_i^j. It arrives there at time $((j-2)(n-2) + (i-2))(\tilde{d}+3) + (5+2\tilde{d})$ and departs at time $((j-2)(n-2) + (i-2))(\tilde{d}+3) + (6+2\tilde{d})$.
    - If $\{i, n\} \in G$, tr_i stops at station s^{n-1}. Its arrival time at this station is $(n-3)(n-2)(\tilde{d}+3) + (4+\tilde{d})$.

- We define the lower bounds on the driving and waiting times such that all corresponding slack times are 0, i.e., $l_{ij} := \pi_j - \pi_i$ for all subsequent arrivals and departures i, j.
- We define

$$\mathcal{C} := \{(\text{tr}_1, \text{tr}_{i'}, s^1) : \{1, i'\} \in G\}$$

$$\cup \bigcup_{j=2}^{n-2} \bigcup_{i=2}^{n-1} \{(\text{tr}_i, \text{tr}_{i'}, s_i^j) : \{i, i'\} \in G\}$$

$$\cup \{(\text{tr}_{i'}, \text{tr}_n, s^{n-1}) : \{i', n\} \in G\}$$

$$\cup \{(\text{tr}', \text{tr}_n, s^{n-1})\}.$$

The lower bounds on the changing times are $l_{(\text{tr}-s-\text{Arr}, \text{tr}'-s-\text{Dep})} := 2$ for all $(\text{tr}, \text{tr}', s) \in \mathcal{C}$. Consequently, there is a slack time of $\tilde{d}$ on each connection.
- A delay of $d_{(\text{tr}_1-s_u-\text{Dep})} := (n-1)\tilde{d}$ occurs in the departure node at station s_u. This is the only delay.
- There are two OD-pairs:
  - $(u, v, \sigma) := (s_u, s_v, 0)$ with $w_{uv\sigma} := 1$, and
  - $(\tilde{u}, v, \tilde{\sigma}) := (s', s_{v'}, (n-2)(n-3)(\tilde{d}+3) + (2\tilde{d}+4))$ with $w_{\tilde{u}v\tilde{\sigma}} := \tilde{d}$.

Let $\mathcal{N}$ be the EAN for instance I. From now on, we regard I as an instance in the EAN, i.e., $I = (\mathcal{N}, l, \pi, \mathcal{OD}, d)$.

An example for this construction is given in Fig. 4.4. Figure 4.4a shows an instance of Hamiltonian Path with fixed start node 1 and fixed end node 5. The EAN constructed for the reduction is shown in Fig. 4.4b. The numbers in the nodes indicate the times the corresponding events are scheduled according to π.

Note that if we remove station s', train tr', and OD-pair $(\tilde{u}, v, \tilde{\sigma})$ and set $\tilde{d} = 1$, we obtain the EAN constructed in the proof of Theorem 4.4.1.

Analogously to step (1) of the proof of Theorem 4.4.1, we see that for a solution $(x, \mathcal{R})$ with $P := P_{uv\sigma} \in \mathcal{R}$,

- $x(\mathcal{R})_{(\text{tr}_1-s_u-\text{Dep})} - \pi_{(\text{tr}_1-s_u-\text{Dep})} = \tilde{d}(n-1)$
- For $(i, j) \in P$

$$x(\mathcal{R})_j - \pi_j$$

$$= \begin{cases} x(\mathcal{R})_i - \pi_i & \text{if } (i,j) \in \mathcal{A}_{\text{drive}} \cup \mathcal{A}_{\text{wait}}, \\ x(\mathcal{R})_i - \pi_i - \tilde{d} & \text{if } (i,j) \in \mathcal{A}_{\text{change}} \text{ and} \\ & P' \cap \{k : \text{tr}(k) = \text{tr}(j), k \leq_{\mathcal{N}} j, \} = \emptyset \\ x(\mathcal{R})_k - \pi_k \geq x(\mathcal{R}')_i - \pi_i & \text{for the direct predecessor } k \text{ of } j \text{ on tr}(j) \\ & \text{otherwise.} \end{cases}$$

4.4 Complexity of Delay Management with Routing

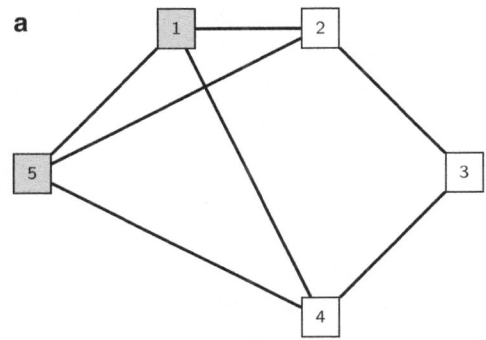

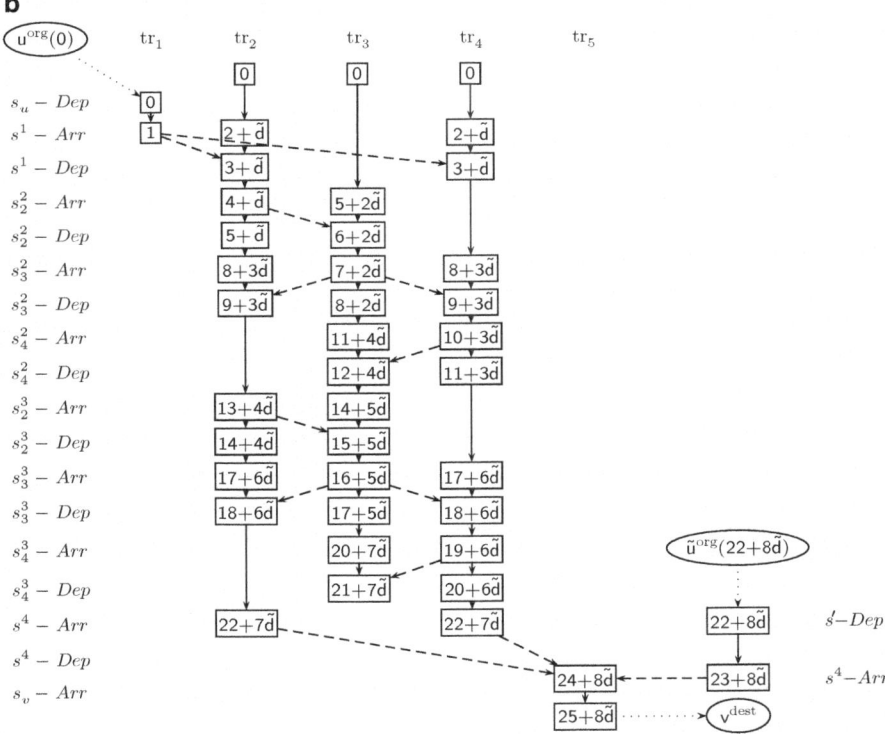

Fig. 4.4 Example for the constructed instance of DMwR in the proof of Theorem 4.4.25. (**a**) Instance of Hamiltonian Path with fixed start node 1 and fixed end node 5. (**b**) EAN and timetable for the instance of DMwR constructed from (**a**) in the proof of Theorem 4.4.25

- That means whenever the path P enters a train that has not been entered before on activity $(i, j) \in \mathcal{A}_{\text{change}}$, the delay in j is reduced by $\tilde{d}$ compared to the delay in i and there is no other possibility to reduce the delay.

- Thus, if the path P enters every train $\text{tr}_1, \ldots, \text{tr}_n$ exactly once,

$$d(x(\mathcal{R}), (\text{tr}_n - s_v - \text{Arr})) = x(\mathcal{R})_{(\text{tr}_n - s_v - \text{Arr})} - \pi_{(\text{tr}_n - s_v - \text{Arr})} = 0,$$

otherwise,

$$d(x(\mathcal{R}), (\text{tr}_n - s_v - \text{Arr})) = x(\mathcal{R})_{(\text{tr}_n - s_v - \text{Arr})} - \pi_{(\text{tr}_n - s_v - \text{Arr})} \geq q\tilde{d}$$

with $q \in \mathbb{N}$.

Since $(\text{tr}_n - s_v - \text{Arr})$ is the last arrival node for both OD-pairs, we conclude analogously to steps (1)–(3) of the proof of Theorem 4.4.1 that there is a solution to the given instance of Hamiltonian Path with fixed start and end nodes if and only if there is a solution $(x, \mathcal{R})$ to the constructed instance of DMwR with $c^d(x, \mathcal{R}) = 0$, or, in terms of the travel time,

$$c(x, \mathcal{R}) = (n-2)(n-3)(\tilde{d}+3) + (2\tilde{d}+7) + 3w_{\tilde{u}\tilde{v}\tilde{\sigma}}.$$

On the other hand, for every solution $(x', \mathcal{R}')$ with $c^d(x', \mathcal{R}') > 0$ and $P' := P_{uv\sigma} \in \mathcal{R}'$, we see that $c^d(x', \mathcal{R}') \geq (w_{\tilde{u}\tilde{v}\tilde{\sigma}} + 1)\tilde{d}$. Hence, in terms of the travel time

$$c(x', \mathcal{R}') \geq (n-2)(n-3)(\tilde{d}+3) + (2\tilde{d}+7) + 3w_{\tilde{u}\tilde{v}\tilde{\sigma}} + (w_{\tilde{u}\tilde{v}\tilde{\sigma}} + 1)\tilde{d}. \quad (4.16)$$

Suppose that there exists a Hamiltonian path from 1 to n in the given instance. In this case for an optimal solution $(x^*, \mathcal{R}^*)$ of the constructed instance I

$$c(x^*, \mathcal{R}^*) = (n-2)(n-3)(\tilde{d}+3) + (2\tilde{d}+7) + 3w_{\tilde{u}\tilde{v}\tilde{\sigma}}$$
$$= ((n-2)(n-3) + 2)\tilde{d} + 3((n-2)(n-3) + 2) + 3w_{\tilde{u}\tilde{v}\tilde{\sigma}} + 1.$$

Then algorithm A has to find an optimal solution to I, because if $(x^A, \mathcal{R}^A)$ is not optimal,

$$c(x^A, \mathcal{R}^A) \geq (n-2)(n-3)(\tilde{d}+3) + (2\tilde{d}+7) + 3w_{\tilde{u}\tilde{v}\tilde{\sigma}} + (w_{\tilde{u}\tilde{v}\tilde{\sigma}} + 1)\tilde{d}$$

due to (4.16), and using $w_{\tilde{u}\tilde{v}\tilde{\sigma}} = \tilde{d} = 3C((n-2)(n-3) + 2) + 1$, we obtain

$$\frac{c(x^A, \mathcal{R}^A)}{c(x^*, \mathcal{R}^*)}$$
$$\geq \frac{(n-2)(n-3)(\tilde{d}+3) + (2\tilde{d}+7) + 3w_{\tilde{u}\tilde{v}\tilde{\sigma}} + (w_{\tilde{u}\tilde{v}\tilde{\sigma}} + 1)\tilde{d}}{(n-2)(n-3)(\tilde{d}+3) + (2\tilde{d}+7) + 3w_{\tilde{u}\tilde{v}\tilde{\sigma}}}$$
$$= 1 + \frac{(w_{\tilde{u}\tilde{v}\tilde{\sigma}} + 1)\tilde{d}}{(n-2)(n-3)(\tilde{d}+3) + (2\tilde{d}+7) + 3w_{\tilde{u}\tilde{v}\tilde{\sigma}}}$$

4.4 Complexity of Delay Management with Routing

$$= 1 + \frac{(3C((n-2)(n-3)+2)+2)\tilde{d}}{(n-2)(n-3)(\tilde{d}+3)+(2\tilde{d}+7)+3w_{\bar{u}\bar{v}\bar{\sigma}}}$$

$$= 1 + \frac{3C((n-2)(n-3)+2)\tilde{d}+2\tilde{d}}{(n-2)(n-3)\tilde{d}+5\tilde{d}+3(n-2)(n-3)+7}$$

$$= 1 + \frac{3C((n-2)(n-3))\tilde{d}+6C\tilde{d}+6C((n-2)(n-3)+2)+2}{(n-2)(n-3)\tilde{d}+5\tilde{d}+3(n-2)(n-3)+7}$$

$$= 1 + \frac{3C((n-2)(n-3))\tilde{d}+6C\tilde{d}+6C(n-2)(n-3)+12C+2}{(n-2)(n-3)\tilde{d}+5\tilde{d}+3(n-2)(n-3)+7}$$

$$\geq 1 + \frac{C((n-2)(n-3))\tilde{d}+5C\tilde{d}+3C(n-2)(n-3)+7C}{(n-2)(n-3)\tilde{d}+5\tilde{d}+3(n-2)(n-3)+7}$$

$$= 1 + C > C$$

Hence, if (and only if) there is a Hamiltonian path from t_1 to t_2 in G, algorithm A returns a solution $(x^*, \mathcal{R}^*)$ with delay $c^d(x^*, \mathcal{R}^*) = 0$. Thus, A solves the Hamiltonian Path problem, which would imply that P = NP. □

However, if the delay is bounded with regard to the nominal travel time of the OD-pairs, an easy approximation algorithm is to maintain all connections. This approach is known in the literature under the name of *all-wait strategy* (see, e.g., [Bau10]).

Lemma 4.4.26. *Let $I = (\mathcal{N}, l, \pi, \mathcal{OD}, d)$ be an instance of DMwR. For every $(u_k, v_k, \sigma_k) \in \mathcal{OD}$ denote by $\hat{P}_k$, a path from $u_k^{org}(\sigma_k)$ to v_k^{dest} with last arrival node $\hat{j}_k$ that minimizes the nominal travel time $SP_k = \pi_{\hat{j}_k} - \sigma_k$. Set $M := \frac{\max_{i \in \mathcal{E}_{op}} d_i}{\min_{(u_k, v_k, \sigma_k) \in \mathcal{OD}} (\pi_{\hat{j}_k} - \sigma_k)}$. Then*

$$\frac{c(x(\mathcal{A}_{change}), \mathcal{R}(x(\mathcal{A}_{change})))}{c(x^*, \mathcal{R}^*)} \leq M + 1,$$

for an optimal solution $(x^*, \mathcal{R}^*)$ to I.

Proof. Denote $\hat{R} := \{\hat{P}_k : (u_k, v_k, \sigma_k) \in \mathcal{OD}\}$. Since $\hat{R}$ is feasible for $x(\mathcal{A}_{change})$,

$$c(x(\mathcal{A}_{change}), \mathcal{R}(x(\mathcal{A}_{change}))) \leq c(x(\mathcal{A}_{change}), \hat{R}).$$

Due to Corollary 4.3.9

$$c(x(\mathcal{A}_{change}), \hat{R}) \leq \sum_{(u_k, v_k, \sigma_k) \in \mathcal{OD}} w_k (\pi_{\hat{j}_k} + \max_{i \in \mathcal{E}_{op}} d_i - \sigma_k).$$

Hence, since $c(x^*, \mathcal{R}^*) \geq \sum_{(u_k, v_k, \sigma_k) \in \mathcal{OD}} w_k (\pi_{\hat{j}_k} - \sigma_k)$,

$$\frac{c(x(\mathcal{A}_{\text{change}}), \mathcal{R}(x(\mathcal{A}_{\text{change}})))}{c(x^*, \mathcal{R}^*)}$$

$$\leq \frac{c(x(\mathcal{A}_{\text{change}}), \hat{R})}{c(x^*, \mathcal{R}^*)}$$

$$\leq \frac{\sum_{(u_k, v_k, \sigma_k) \in \mathcal{OD}} w_k (\pi_{\hat{j}_k} + \max_{i \in \mathcal{E}_{\text{op}}} d_i - \sigma_k)}{\sum_{(u_k, v_k, \sigma_k) \in \mathcal{OD}} w_k (\pi_{\hat{j}_k} - \sigma_k)}$$

$$\leq \frac{\sum_{(u_k, v_k, \sigma_k) \in \mathcal{OD}} w_k (\pi_{\hat{j}_k} - \sigma_k + M \cdot \min_{(u_{k'}, v_{k'}, \sigma_{k'}) \in \mathcal{OD}} \{\pi_{\hat{j}_{k'}} - \sigma_{k'}\})}{\sum_{(u_k, v_k, \sigma_k) \in \mathcal{OD}} w_k (\pi_{\hat{j}_k} - \sigma_k)}$$

$$\leq \frac{\sum_{(u_k, v_k, \sigma_k) \in \mathcal{OD}} w_k (M+1)(\pi_{\hat{j}_k} - \sigma_k)}{\sum_{(u_k, v_k, \sigma_k) \in \mathcal{OD}} w_k (\pi_{\hat{j}_k} - \sigma_k)}$$

$$\leq \frac{(M+1) \sum_{(u_k, v_k, \sigma_k) \in \mathcal{OD}} w_k (\pi_{\hat{j}_k} - \sigma_k)}{\sum_{(u_k, v_k, \sigma_k) \in \mathcal{OD}} w_k (\pi_{\hat{j}_k} - \sigma_k)}$$

$$= M + 1.$$

□

4.5 Integer Programming Formulation

In this section we present an integer programming formulation for DMwR from [DHSS09, DHSS12]. The model is based on an IP formulation for delay management with constant weights introduced in [Sch06].

Like for cLPwR and for the first IP formulation for TTwR, we use a passenger flow-based formulation for modeling DMwR in an IP.

Let $I = (\mathcal{N}, l, \pi, \mathcal{OD}, d)$ be an instance of DMwR. The first type of variables decides whether a changing activity is maintained or not. That is, for every changing activity $a \in \mathcal{A}_{\text{change}}$, there is a binary variable z_a, which is defined as

$$z_a = \begin{cases} 1 & \text{if } a \text{ is maintained,} \\ 0 & \text{otherwise.} \end{cases}$$

The task of the next set of variables is the calculation of the disposition timetable. For every event $e \in \mathcal{E}_{\text{op}}$, we define the variable x_e as

$x_e :=$ time that event e takes place in the disposition timetable.

These variables suffice to model delay management with constant weights (see, e.g., [Sch06]).

4.5 Integer Programming Formulation

However, since in DMwR the routing is not part of the input but of the solution that has to be found, we need additional variables to model the passenger flows:

$$q_a^k = \begin{cases} 1 & \text{if activity } a \text{ is used by OD-pair } (u_k, v_k, \sigma_k) \\ 0 & \text{otherwise.} \end{cases}$$

Furthermore, for every OD-pair $(u_k, v_k, \sigma_k) \in \mathcal{OD}$, we introduce variables

$$t^k := \text{arrival time of OD-pair } (u_k, v_k, \sigma_k).$$

We obtain the following IP:

$$\min \sum_{(u_k, v_k, \sigma_k) \in \mathcal{OD}} w_k (t^k - \sigma_k) \tag{4.17}$$

$$\text{s.t. } x_e \geq \pi_e + d_e \qquad \forall e \in \mathcal{E}_{\text{op}}, \tag{4.18}$$

$$x_e \geq x_{e'} + l_a \qquad \forall a = (e', e) \in \mathcal{A}_{\text{drive}} \cup \mathcal{A}_{\text{wait}}, \tag{4.19}$$

$$x_e \geq x_{e'} + l_a - M_1(1 - z_a) \qquad \forall a = (e', e) \in \mathcal{A}_{\text{change}}, \tag{4.20}$$

$$\sum_{a \in \delta^-(u_k^{\text{org}}(\sigma_k))} q_a^k = 1 \qquad \forall u_k^{\text{org}}(\sigma_k) \in \mathcal{E}'_{\text{org}}, \tag{4.21}$$

$$\sum_{a \in \delta^-(e)} q_a^k = \sum_{a \in \delta^+(e)} q_a^k \qquad \forall (u_k, v_k, \sigma_k) \in \mathcal{OD}, e \in \mathcal{E}_{\text{op}}, \tag{4.22}$$

$$\sum_{a \in \delta^+(v_k^{\text{dest}})} q_a^k = 1 \qquad \forall v_k^{\text{dest}} \in \mathcal{E}_{\text{dest}}, \tag{4.23}$$

$$q_a^k \leq z_a \qquad \forall (u_k, v_k, \sigma_k) \in \mathcal{OD}, a \in \mathcal{A}_{\text{change}}, \tag{4.24}$$

$$t^k \geq x_e - M_2^k(1 - q_a^k) \qquad \forall v_k^{\text{dest}} \in \mathcal{E}_{\text{dest}}, a = (e, v_k^{\text{dest}}) \in \delta^+(v_k^{\text{dest}}), \tag{4.25}$$

$$z_a \in \{0, 1\} \qquad \forall a \in \mathcal{A}_{\text{change}}, \tag{4.26}$$

$$q_a^k \in \{0, 1\} \qquad \forall (u_k, v_k, \sigma_k) \in \mathcal{OD}, a \in \mathcal{A}, \tag{4.27}$$

$$x_e \in \mathbb{N} \qquad \forall e \in \mathcal{E}_{\text{op}}, \tag{4.28}$$

$$t^k \in \mathbb{N} \qquad \forall (u_k, v_k, \sigma_k) \in \mathcal{OD}. \tag{4.29}$$

Before proving correctness of the IP formulation (4.17)–(4.29) in Lemmas 4.5.1 and 4.5.2, we explain informally the intention of the constraints: The objective function (4.17) minimizes the overall arrival time of all passengers. Constraints (4.18) imply that events cannot take place earlier than in the original timetable and that source delays are taken into account. To make sure that delays are propagated through the network correctly, constraints (4.19) transfer the delay from the start of activity a to its end. For maintained changing activities, that is changing activities for which $z_a = 1$, constraints (4.20) transfer delays from the feeder train to the connecting train. The value of M_1 should be chosen large enough for these constraints to be correct. In [Sch06] it has been shown that $M_1 = \max_{e \in \mathcal{E}} d_e$ is large enough. Constraints (4.18)–(4.20) are also used for modeling delay management with constant weights.

The flow constraints (4.21)–(4.23) ensure that every OD-pair leaves its origin, travels until its destination, and stops traveling there. Constraint (4.24) ensures that only maintained changing activities are used for traveling. Finally, constraint (4.25) defines the arrival time for OD-pair (u_k, v_k, σ_k) where M_2^k is a large number.

For $(u_k, v_k, \sigma_k) \in \mathcal{OD}$, denote by $\mathcal{E}_{\text{arr}}^k := \{e \in \mathcal{E}_{\text{arr}} : (e, v_k^{\text{dest}}) \in \mathcal{A}_{\text{dest}}\}$ the set of arrival events at station v_k like in Definition 3.6.1.

Lemma 4.5.1. *Let $I = (\mathcal{N}, l, \pi, \mathcal{OD}, d)$ be an instance of DMwR. Then for any $M_1 \in \mathbb{N}_0$ and $\{M_2^k : (u_k, v_k, \sigma_k) \in \mathcal{OD}\} \subset \mathbb{N}_0^{|\mathcal{OD}|}$, it holds that a feasible solution (z, x, q, t) to the IP (4.17)–(4.29) defines a feasible solution $(x', \mathcal{R}')$ to instance I of DMwR and*

$$c(x', \mathcal{R}') \leq \sum_{(u_k, v_k, \sigma_k) \in \mathcal{OD}} w_k(t^k - \sigma_k).$$

Proof. Let $I = (\mathcal{N}, l, \pi, \mathcal{OD}, d)$ be an instance of DMwR. We consider the corresponding IP formulation (4.17)–(4.29). Let (z, x, q, t) be feasible for the IP (4.17)–(4.29), i.e., (z, x, q, t) fulfills constraints (4.18)–(4.29). We set $x'_e := x_e$ $\forall e \in \mathcal{E}_{\text{op}}$. Then x' defines a disposition timetable due to (4.18) and (4.19). Because of (4.20), $z_a = 0$ for all $a \in \mathcal{A}_{\text{change}} \setminus \mathcal{A}_{\text{fix}}(x')$. Hence,

$$q_a^k = 0 \quad \forall a \in \mathcal{A}_{\text{change}} \setminus \mathcal{A}_{\text{fix}}(x') \quad (4.30)$$

due to (4.24). For every $(u_k, v_k, \sigma_k) \in \mathcal{OD}$, the set of variables $\{q_a^k : a \in \mathcal{A}\}$ defines a path P_k from $u_k^{\text{org}}(\sigma_k)$ to v_k^{dest} in $\mathcal{N}$ due to (4.21)–(4.23). Due to (4.30), for every k P_k is a path in $\mathcal{N}(x')$ and hence feasible for x'. Thus, $(x', \{P_k : (u_k, v_k, \sigma_k) \in \mathcal{OD}\})$ is a feasible solution to I. Furthermore, from (4.25) it follows that for every $(u_k, v_k, \sigma_k) \in \mathcal{OD}$ we have $t^k \geq x_{e_k} = x'_{e_k}$ for the last arrival node e_k on P_k and we have

$$c(x', \mathcal{R}') = \sum_{(u_k, v_k, \sigma_k) \in \mathcal{OD}} w_k c(x', P_k)$$

$$= \sum_{(u_k, v_k, \sigma_k) \in \mathcal{OD}} w_k (x'_{e_k} - \sigma_k)$$

$$\leq \sum_{(u_k, v_k, \sigma_k) \in \mathcal{OD}} w_k (t^k - \sigma_k).$$

□

4.5 Integer Programming Formulation

Lemma 4.5.2. *Let* $I = (\mathcal{N}, l, \pi, \mathcal{OD}, d)$ *be an instance of DMwR and let the numbers* M_1 *and* M_2^k *for every* (u_k, v_k, σ_k) *be chosen in such a way that*

$$M_1 \geq \max_{e \in \mathcal{E}_{op}} d_e \quad \text{and} \quad M_2^k \geq \max_{e \in \mathcal{E}_{op}} d_e + \max_{(i,j) \in \mathcal{E}_{arr}^k \times \mathcal{E}_{arr}^k} (\pi_j - \pi_i).$$

Then a solution $(x', \mathcal{R}')$*, defined by an optimal solution* (z, x, q, t) *to the IP (4.17)–(4.29) is an optimal solution to DMwR for instance* I*, and*

$$c(x', \mathcal{R}') = \sum_{(u_k, v_k, \sigma_k) \in \mathcal{OD}} w_k (t^k - \sigma_k).$$

Proof. Let $I = (\mathcal{N}, l, \pi, \mathcal{OD}, d)$ be an instance of DMwR. We consider the corresponding IP formulation (4.17)–(4.29).

1. Let $(x', \mathcal{R}')$ be a feasible solution to I. With $\mathcal{R}' := \{P_k : (u_k, v_k, \sigma_k) \in \mathcal{OD}\}$, we set

$$x_e := x'_e \qquad \forall e \in \mathcal{E}_{op},$$

$$q_a^k := \begin{cases} 1 & \text{if } a \in P_k, \\ 0 & \text{otherwise} \end{cases} \qquad \forall a \in \mathcal{A}, (u_k, v_k, \sigma_k) \in \mathcal{OD},$$

$$z_a := \begin{cases} 1 & \text{if } a \in \mathcal{A}_{\text{fix}}(x'), \\ 0 & \text{otherwise}, \end{cases} \qquad \forall a \in \mathcal{A}_{\text{change}},$$

$$t^k := c(x', P_k) + \sigma_k \qquad \forall (u_k, v_k, \sigma_k) \in \mathcal{OD}.$$

We show that constraints (4.18)–(4.29) are fulfilled for (x, q, z, t): Constraints (4.18) and (4.19) hold because x' is a disposition timetable. For $a \in \mathcal{A}_{\text{fix}}(x')$, (4.20) holds due to the definition of $\mathcal{A}_{\text{fix}}$ (4.1). For $a = (e', e) \in \mathcal{A}_{\text{change}} \setminus \mathcal{A}_{\text{fix}}(x')$, it follows from Corollary 4.3.9 that

$$x_e \geq \pi_e \geq \pi_{e'} + l_a \geq x_{e'} + l_a - \max_{i \in \mathcal{E}_{op}} d_e \geq x_{e'} + l_a - M_1(1 - z_a).$$

Constraints (4.21)–(4.23) and (4.24) are fulfilled due to the definition of q and z. For every k, for the last arrival node e_k on P_k, (4.25) is trivially fulfilled. For $e \in \mathcal{E}_{\text{arr}}^k \setminus \{e_k\}$, using Corollary 4.3.9, we obtain

$$t^k = x_{e_k} \geq \pi_{e_k} = \pi_e - (\pi_e - \pi_{e_k}) \geq x_e - \max_{i \in \mathcal{E}_{op}} d_i - (\pi_e - \pi_{e_k})$$

$$\geq x_e - M_2^k = x_e - M_2^k(1 - q_{ev_k^{\text{dest}}}^k).$$

Furthermore, the definition of t^k implies that

$$\sum_{(u_k,v_k,\sigma_k)\in \mathcal{OD}} w_k(t^k - \sigma_k) = c(x', \mathcal{R}').$$

2. Let (z, x, q, t) be an optimal solution to (4.17)–(4.29). Then due to Lemma 4.5.1, there is a corresponding feasible solution $(x', \mathcal{R}')$ to I with

$$c(x', \mathcal{R}') \leq \sum_{(u_k,v_k,\sigma_k)\in \mathcal{OD}} w_k(t^k - \sigma_k).$$

Suppose that for an optimal solution $(x^*, \mathcal{R}^*)$ to DMwR it holds that $c(x^*, \mathcal{R}^*) < \sum_{(u_k,v_k,\sigma_k)\in \mathcal{OD}} w_k(t^k - \sigma_k)$. Then according to (4.5.2), there is a corresponding feasible solution (z^*, x^*, q^*, t^*) to (4.17)–(4.29) with

$$\sum_{(u_k,v_k,\sigma_k)\in \mathcal{OD}} w_k((t^*)^k - \sigma_k) = c(x^*, \mathcal{R}^*) < \sum_{(u_k,v_k,\sigma_k)\in \mathcal{OD}} w_k(t^k - \sigma_k).$$

This contradicts the optimality of (z, x, q, t).

□

Given an instance $I = (\mathcal{N}, l, \pi, \mathcal{OD}, d)$ of DMwR, formulation (4.17)–(4.29) can be tightened using the valid inequalities provided by Lemma 4.4.22, e.g., we add

$$t^k \geq \text{lb}_k + \sigma_k \quad \forall (u_k, v_k, \sigma_k) \in \mathcal{OD},$$

where lb_k denotes the output of Algorithm 1 for instance $I = (\mathcal{N}, l, \pi, \{(u_k, v_k, \sigma_k)\}, d)$.

As we have seen in Lemma 4.2.7, DMwR can equivalently be formulated using the overall delay instead of the overall travel time as an objective. Hence, we could replace (4.17) with the equivalent objective function

$$\sum_{(u_k,v_k,\sigma_k)\in \mathcal{OD}} w_k(t^k - SP_k). \tag{4.31}$$

Using (4.31) instead of the overall travel time, the proposed integer programming formulation of DMwR specified by the modified objective (4.31) and the constraints (4.18)–(4.29) has been tested in [DHSS09, DHSS12] on several instances based on data from Dutch Railways and compared to an IP formulation for delay management with constant weights.

Chapter 5
An Iterative Solution Approach for General Network Problems with Routing

In this chapter we present an iterative approach for solving network problems with routing which alternatingly calculates routings and decides about the network structure.

To this end, in Sect. 5.1 we identify structural similarities between the problems considered in the preceding sections. We classify the problems in problem classes in order to find out to which extent they are tractable by our iterative approach which is presented and analyzed in Sect. 5.2 for problems from the defined problem classes.

In Sect. 5.3 we define the *price of sequentiality*, an indicator for the quality of the solutions found by the iterative approach which we analyze for the problems TTwR and *arc speed-up* in Sects. 5.4 and 5.5.

5.1 Classification of Network Problems with Routing

Consider the definitions of uLPwR, scLPwR, and cLPwR in the CGN (Definitions 2.2.5) and the definitions of TTwR and DMwR in the EAN (Definitions 3.2.6 and 4.2.5). We notice that common components in the input of the problems are

- a directed graph $N = (V, A)$, and
- the specification of a subset of $V \times V$ by means of the OD-pairs.

We call this representation of the OD-pairs *node-OD-pairs*.

Definition 5.1.1. Given a network $N = (V, A)$, a set of *node-OD-pairs* in N is a subset of the Cartesian product of the node set of N, i.e., $\mathcal{OD}_{\text{node}} \subset V \times V$. For every node-OD-pair (u, v), the *number of passengers* or *passenger weight* w_{uv} is specified.

We restate the definition of paths and routings in terms of the node-OD-pairs.

Notation 5.1.2. Let $N = (V, A)$ be an undirected graph and $\mathcal{OD}_{\text{node}}$ a set of node-OD-pairs in N.

- A path for node-OD-pair $(u, v) \in \mathcal{OD}_{\text{node}}$ *is a path* P_{uv} *from u to v in N*.
- *A routing for node-OD-pair* $(u, v) \in \mathcal{OD}_{\text{node}}$ *with passenger weight* w_{uv} *is a collection of paths* $\mathcal{R}_{uv}$ *for OD-pair* (u, v), *i.e.*,

$$\mathcal{R}_{uv} = \{P : P \text{ path for } (u, v)\}.$$

Every path P is assigned a number of passengers w_{uv}^P, *such that* $\sum_{P \in \mathcal{R}_{uv}} w_{uv}^P = w_{uv}$.
- *A routing for the set of node-OD-pairs* $\mathcal{OD}_{\text{node}}$ *is defined as*

$$\mathcal{R} = \bigcup_{(u,v) \in \mathcal{OD}_{\text{node}}} \mathcal{R}_{uv}.$$

We capture the identified similarities of the problems uLPwR, scLPwR, cLPwR, TTwR, and DMwR in the definition of a joint problem class $\mathcal{Q}$.

Definition 5.1.3. *A minimization problem* Q *with objective function* c *belongs to the class* $\mathcal{Q}$ *of network problems with routing if it has the following properties:*

1. The input of an instance of Q specifies a (directed or undirected) graph $N = (V, A)$ and a set of node-OD-pairs $\mathcal{OD}_{\text{node}} \subset V \times V$.
2. A solution $(\mathcal{S}, \mathcal{R})$ to an instance of Q consists of a routing $\mathcal{R}$ as defined in Notation 5.1.2 and a network component $\mathcal{S}$ which can be interpreted as an assignment of length and capacity to the arcs

$$f_{\mathcal{S}} : A \to \mathbb{Q} \times \mathbb{Z}, \quad a \mapsto (\text{length}(a), \text{capacity}(a)).$$

3. There is $lb_Q \in \mathbb{Z}$ such that for every solution $(\mathcal{S}, \mathcal{R})$ to Q, $c(\mathcal{S}, \mathcal{R}) \geq lb_Q$.

Indeed, all problems considered in this text so far fit into this description.

Lemma 5.1.4. *uLPwR, scLPwR, cLPwR, TTwR, and DMwR can be regarded as problems belonging to* $\mathcal{Q}$.

Proof. • For an instance $I = (N^{\text{CGN}}, \mathcal{L}, \mathcal{OD}, B)$ of uLPwR, we set $N := N^{\text{CGN}}$ for the CGN N^{CGN} and $\mathcal{OD}_{\text{node}} := \{(u^{\text{org}}, v^{\text{org}}) : (u, v) \in \mathcal{OD}\}$. A solution $(\mathcal{L}', \mathcal{R})$ to I consists of a routing $\mathcal{R}$ and a line concept $\mathcal{L}'$ that defines which part of the network is built, i.e., can be used by the passengers, and which part is deleted. Hence, we can regard $\mathcal{L}'$ as an assignment

$$f_{\mathcal{L}'}((i, j)) := \begin{cases} (c_{ij}, \sum_{(u,v) \in \mathcal{OD}} w_{uv}) & \text{if } (i, j) \in A_{\text{drive}}, l((i, j)) \in \mathcal{L}', \\ (c_{ij}, \sum_{(u,v) \in \mathcal{OD}} w_{uv}) & \text{if } (i, j) \in A_{\text{trans}}, l(i), l(j) \in \mathcal{L}', \\ (c_{ij}, \sum_{(u,v) \in \mathcal{OD}} w_{uv}) & \text{if } (i, j) \in A_{\text{org}}, l(j) \in \mathcal{L}', \\ (c_{ij}, \sum_{(u,v) \in \mathcal{OD}} w_{uv}) & \text{if } (i, j) \in A_{\text{dest}}, l(i) \in \mathcal{L}', \\ (c_{ij}, 0) & \text{otherwise.} \end{cases}$$

Furthermore, $c(\mathcal{L}', \mathcal{R}) \geq 0$ for all solutions $(\mathcal{L}', \mathcal{R})$ to I.

5.1 Classification of Network Problems with Routing 169

- Similarly, for an instance $I = (N^{\text{CGN}}, \mathcal{L}, \mathcal{OD}, B, \text{Cap})$ of scLPwR or cLPwR, $N := N^{\text{CGN}}$ and $\mathcal{OD}_{\text{node}} := \{(u^{\text{org}}, v^{\text{org}}) : (u,v) \in \mathcal{OD}\}$ are defined by the input. A solution $(\mathcal{L}', \mathcal{R})$ to I consists of a routing $\mathcal{R}$ and a line concept $\mathcal{L}'$ with frequencies $f_l > 0$ for $l \in \mathcal{L}'$. $\mathcal{L}'$ defines which part of the network is built and the capacity of the network by the specification of the frequencies. We could regard $\mathcal{L}'$ as a function

$$f_{\mathcal{L}'}((i,j)) := \begin{cases} (c_{ij}, f_l \cdot \text{Cap}) & \text{if } (i,j) \in A_{\text{drive}}, l((i,j)) \in \mathcal{L}', \\ (c_{ij}, \sum_{(u,v) \in \mathcal{OD}} w_{uv}) & \text{if } (i,j) \in A_{\text{trans}}, l(i), l(j) \in \mathcal{L}', \\ (c_{ij}, \sum_{(u,v) \in \mathcal{OD}} w_{uv}) & \text{if } (i,j) \in A_{\text{org}}, l(j) \in \mathcal{L}', \\ (c_{ij}, \sum_{(u,v) \in \mathcal{OD}} w_{uv}) & \text{if } (i,j) \in A_{\text{dest}}, l(i) \in \mathcal{L}', \\ (c_{ij}, 0) & \text{otherwise.} \end{cases}$$

where f_l denotes the frequency of line $l \in \mathcal{L}'$. Again it holds that $c(\mathcal{L}', \mathcal{R}) \geq 0$ for all solutions $(\mathcal{L}', \mathcal{R})$ to I.

- For an instance $(\mathcal{N}, l, u, \mathcal{OD})$ of TTwR, we set $N := \mathcal{N}$ and $\mathcal{OD}_{\text{node}} := \{(u^{\text{org}}, v^{\text{org}}) : (u,v) \in \mathcal{OD}\}$. A solution $(\pi, \mathcal{R})$ to I consists of a routing $\mathcal{R}$ and a timetable π which defines the arc lengths of the network $N(\pi)$. Hence, we can understand π as an assignment of $L_{ij} = \pi_j - \pi_i$ to the arcs:

$$f_\pi((i,j)) := \begin{cases} (\pi_j - \pi_i, \sum_{(u,v) \in \mathcal{OD}} w_{uv}) & \text{if } (i,j) \in \mathcal{A}_{\text{op}}, \\ (0, \sum_{(u,v) \in \mathcal{OD}} w_{uv}) & \text{otherwise.} \end{cases}$$

Furthermore, $c(\pi, \mathcal{R}) \geq 0$ for all solutions $(\pi, \mathcal{R})$ to I.

- For an instance $(\mathcal{N}, l, \pi, \mathcal{OD}, d)$ of DMwR, we set $\mathcal{OD}_{\text{node}} := \{(u^{\text{org}}(\sigma), v^{\text{org}}) : (u,v,\sigma) \in \mathcal{OD}\}$ and $N := \mathcal{N}$. A solution $(x, \mathcal{R})$ to I consists of a routing $\mathcal{R}$ and a disposition timetable x which defines the arc lengths of the network $\mathcal{N}(x)$ and the existence of changing activities; hence, we identify x with the following function

$$f_x((i,j)) := \begin{cases} (x_j - x_i, \sum_{(u,v) \in \mathcal{OD}} w_{uv}) & \text{if } (i,j) \in \mathcal{A}_{\text{drive}} \cup \mathcal{A}_{\text{wait}} \cup \mathcal{A}_{\text{fix}}(x), \\ (x_j - x_i, 0) & \text{if } (i,j) \in \mathcal{A}_{\text{change}} \setminus \mathcal{A}_{\text{fix}}(x), \\ (x_j - \sigma, \sum_{(u,v) \in \mathcal{OD}} w_{uv}) & \text{if } (i,j) \in \mathcal{A}'_{\text{org}}, \\ (0, \sum_{(u,v) \in \mathcal{OD}} w_{uv}) & \text{if } (i,j) \in \mathcal{A}_{\text{dest}}. \end{cases}$$

Furthermore, $c(x, \mathcal{R}) \geq 0$ for all solutions $(x, \mathcal{R})$ to I. □

The goal of this section is to identify for which of the problems considered in this book it is promising to determine a solution iteratively, i.e., to iterate the calculation of $\mathcal{R}$ and $\mathcal{S}$ while fixing $\mathcal{S}$ or $\mathcal{R}$, respectively.

Obviously, this approach only makes sense if for a given routing we can determine a feasible network component and vice versa:

Definition 5.1.5. Let $Q \in \mathcal{Q}$ be a network problem with routing and I be an instance of Q.

- Let $\mathcal{R}$ be a routing for I. We define the set of *network components feasible for $\mathcal{R}$*

$$\text{FEAS}(\mathcal{R}) := \{\mathcal{S} : (\mathcal{S}, \mathcal{R}) \text{ is feasible}\}.$$

- Let $\mathcal{S}$ be a network component for I. We define the set of *routings feasible for $\mathcal{S}$*

$$\text{FEAS}(\mathcal{S}) := \{\mathcal{R} : (\mathcal{S}, \mathcal{R}) \text{ is feasible}\}.$$

This notion of feasibility for a routing or a network component for uLPwR and scLPwR is captured in Definition 2.2.4. To obtain a feasible solution $(\mathcal{L}', \mathcal{R})$ to cLPwR, additionally to Definition 2.2.4, we need that $\mathcal{R}$ is a shortest-path routing in $N(\mathcal{L}')$. For DMwR see Definition 4.2.3. In TTwR, every feasible timetable is feasible for every routing and vice versa.

In order to improve the objective values during an iterative procedure, we aim at finding the best possible solution in every step. To define problem classes for which this approach is possible, we introduce the following notation.

Definition 5.1.6. Let $Q \in \mathcal{Q}$ be a network problem with routing and I be an instance of Q.

- Let $\mathcal{R}$ be a routing for I such that $\text{FEAS}(\mathcal{R}) \neq \emptyset$. We define

$$\text{OPT}(\mathcal{R}) := \{\mathcal{S}(\mathcal{R}) \in \text{FEAS}(\mathcal{R}) : c(\mathcal{S}(\mathcal{R}), \mathcal{R}) \leq c(\mathcal{S}', \mathcal{R}) \forall \mathcal{S}' \in \text{FEAS}(\mathcal{R})\}.$$

- Let $\mathcal{S}$ be a network component for I such that $\text{FEAS}(\mathcal{S}) \neq \emptyset$. We define

$$\text{OPT}(\mathcal{S}) := \{\mathcal{R}(\mathcal{S}) \in \text{FEAS}(\mathcal{S}) : c(\mathcal{S}, \mathcal{R}(\mathcal{S})) \leq c(\mathcal{S}, \mathcal{R}') \forall \mathcal{R}' \in \text{FEAS}(\mathcal{S})\}.$$

Note that the definition of $\text{OPT}(\mathcal{S})$ corresponds to the definitions of $\text{OPT}(\mathcal{L}')$ for a given line concept $\mathcal{L}'$ in uLPwR (see Definition 2.3.3), $\text{OPT}(\pi)$ for a given timetable in TTwR, and $\text{OPT}(x)$ for a given disposition timetable x in DMwR (see Definition 4.3.1).

The definition of $\text{OPT}(\mathcal{R})$ corresponds to the definition of $\text{OPT}(\mathcal{R})$ for a given routing in TTwR. For an instance I of uLPwR with a given routing $\mathcal{R}$, we have $\mathcal{L}(\mathcal{R}) \in \text{OPT}(\mathcal{R})$ in the sense of Definition 5.1.6 (see Definition 2.3.1), and for an instance I of DMwR with a given routing $\mathcal{R}$, we have $x(\mathcal{R}) \in \text{OPT}(\mathcal{R})$ in the sense of Definition 5.1.6.

Definition 5.1.7. Let $Q \in \mathcal{Q}$. We define the subclass $\mathcal{Q}^{\text{opt}} \subset \mathcal{Q}$ as follows: $Q \in \mathcal{Q}^{\text{opt}}$ if the following conditions hold:

1. Let $\mathcal{R}$ be a routing such that $\text{FEAS}(\mathcal{R}) \neq \emptyset$. Then $\text{OPT}(\mathcal{R}) \neq \emptyset$.
2. Let $\mathcal{S}$ be a network component such that $\text{FEAS}(\mathcal{S}) \neq \emptyset$. Then $\text{OPT}(\mathcal{S}) \neq \emptyset$.

Lemma 5.1.8. *uLPwR, scLPwR, cLPwR, TTwR, and DMwR are in $\mathcal{Q}^{opt}$.*

5.1 Classification of Network Problems with Routing

Proof. For every instance I of a problem $Q \in \{uLPwR, scLPwR, cLPwR, TTwR, DMwR\}$, there is only a finite number of solutions $(\mathcal{S}, \mathcal{R})$. Hence, for any routing $\mathcal{R}$ for I, FEAS$(\mathcal{R})$ is a finite set, and FEAS$(\mathcal{R}) \neq \emptyset$ implies OPT$(\mathcal{R}) = \text{argmin}_{\mathcal{S} \in \text{FEAS}(\mathcal{R})} c(\mathcal{S}, \mathcal{R}) \neq \emptyset$. Analogously, for any network component $\mathcal{S}$ for I, FEAS$(\mathcal{S})$ is a finite set and FEAS$(\mathcal{S}) \neq \emptyset$ implies OPT$(\mathcal{S}) = \text{argmin}_{\mathcal{R} \in \text{FEAS}(\mathcal{S})} c(\mathcal{S}, \mathcal{R}) \neq \emptyset$. □

Definition 5.1.9. Let $Q \in \mathcal{Q}$. We define the subclass $\mathcal{Q}_{\text{pol}}^{\text{opt}} \subset \mathcal{Q}$ as follows: $Q \in \mathcal{Q}_{\text{pol}}^{\text{opt}}$ if $Q \in \mathcal{Q}^{\text{opt}}$ and the following conditions hold:

1. Let $\mathcal{R}$ be a routing such that FEAS$(\mathcal{R}) \neq \emptyset$. Then we can find a network component $\mathcal{S}(\mathcal{R}) \in \text{OPT}(\mathcal{R})$ in polynomial time.
2. Let $\mathcal{S}$ be a network component such that FEAS$(\mathcal{S}) \neq \emptyset$. Then we can find a routing $\mathcal{R}(\mathcal{S}) \in \text{OPT}(\mathcal{S})$ in polynomial time.

Lemma 5.1.10. *uLPwR, TTwR, and DMwR are in $\mathcal{Q}_{\text{pol}}^{\text{opt}}$.*

Proof. • Let $(N^{\text{CGN}}, \mathcal{L}, \mathcal{OD}, B)$ be an instance of uLPwR and let $\mathcal{R}$ be a routing for $\mathcal{OD}$ (or the corresponding set of node-OD-pairs $\mathcal{OD}_{\text{node}}$, respectively). We find $\mathcal{L}(\mathcal{R})$ as defined in Definition 2.3.1 in polynomial time. According to Lemma 2.3.2, $\mathcal{L}'$ is feasible for $\mathcal{R}$ if there exists a feasible line concept for $\mathcal{R}$. Furthermore, $c(\mathcal{L}', \mathcal{R})$ does not depend on the line concept $\mathcal{L}'$, as long as it is feasible for $\mathcal{R}$. Hence, $\mathcal{L}(\mathcal{R}) \in \text{OPT}(\mathcal{R})$.
Now let $\mathcal{L}'$ be a line concept. A shortest-path routing $\mathcal{R}(\mathcal{L}')$ in $N(\mathcal{L}')$ can be found in polynomial time.

• Let $(\mathcal{N}, l, u, \mathcal{OD})$ be an instance of TTwR. According to Definition 3.3.4, Lemma 3.3.3, and Corollary 3.3.5, given a routing $\mathcal{R}$ a timetable $\pi(\mathcal{R}) \in \text{OPT}(\mathcal{R})$ can be found in polynomial time. On the other hand, Definition 3.3.6 and Lemma 3.3.7 imply that given a timetable π a routing $\mathcal{R}(\pi) \in \text{OPT}(\pi)$ can be found in polynomial time.

• Let $(\mathcal{N}, l, \pi, \mathcal{OD}, d)$ be an instance of DMwR. According to Definition 4.3.10, Definition 4.3.4, and Corollary 4.3.11, given a routing $\mathcal{R}$ the disposition timetable $x(\mathcal{R})$ can be found in polynomial time. Furthermore, for a disposition timetable x, Lemma 4.3.2 ensures that a routing $\mathcal{R}(x) \in \text{OPT}(x)$ as defined in Definition 4.3.1 can be found in polynomial time. □

Even if a problem Q is not in $\mathcal{Q}_{\text{pol}}^{\text{opt}}$, if we have polynomial-time heuristics which find network components for given routings and routings for given network components with good objective values, an iterative approach could be successfully used to improve the solution quality of an initial solution. The class of problems for which this could be possible is described in the following definition:

Definition 5.1.11. Let $Q \in \mathcal{Q}$. We define the subclass $\mathcal{Q}_{\text{pol}}^{\text{feas}} \subset \mathcal{Q}$ as follows: $Q \in \mathcal{Q}_{\text{pol}}^{\text{feas}}$ if the following conditions hold:

1. Let $\mathcal{R}$ be a routing such that FEAS($\mathcal{R}$) $\neq \emptyset$. Then we can find a network component $\mathcal{S} \in$ FEAS($\mathcal{R}$) in polynomial time.
2. Let $\mathcal{S}$ be a network component such that FEAS($\mathcal{S}$) $\neq \emptyset$. Then we can find a routing $\mathcal{R} \in$ FEAS($\mathcal{S}$) in polynomial time.

However, none of the problems considered in this book is in $\mathcal{Q}_{pol}^{feas} \setminus \mathcal{Q}_{pol}^{opt}$:

Lemma 5.1.12. *scLPwR and cLPwR are not in $\mathcal{Q}_{pol}^{feas}$.*

Proof. This follows from Theorem 2.3.8. □

In order to obtain a polynomial-time iterative algorithm, we do not only need the property that the iterations can be performed in polynomial time, but we also need to find an initial solution in polynomial time.

Definition 5.1.13. We define the subclasses $\mathcal{Q}_{rout}^{trac}, \mathcal{Q}_{comp}^{trac} \subset \mathcal{Q}_{pol}^{opt}$ as follows: Let $Q \in \mathcal{Q}_{pol}^{opt}$ be a problem for which a feasible solution ($\mathcal{S}, \mathcal{R}$) exists.

- $Q \in \mathcal{Q}_{rout}^{trac}$ if a routing $\mathcal{R}$ with FEAS($\mathcal{R}$) $\neq \emptyset$ can be found in polynomial time.
- $Q \in \mathcal{Q}_{comp}^{trac}$ if a network component $\mathcal{S}$ with FEAS($\mathcal{S}$) $\neq \emptyset$ can be found in polynomial time.

Lemma 5.1.14. *Let Q be a problem in $\mathcal{Q}_{pol}^{opt}$.*

$$Q \in \mathcal{Q}_{rout}^{trac} \Leftrightarrow Q \in \mathcal{Q}_{comp}^{trac}$$

⇔ *For every instance of Q we can find a feasible solution in polynomial time.*

Proof. Let I be an instance of Q.

- If $Q \in \mathcal{Q}_{rout}^{trac}$ we find a routing $\mathcal{R}^0$ with FEAS($\mathcal{R}^0$) $\neq \emptyset$ in polynomial time. Since $Q \in \mathcal{Q}_{pol}^{opt}$ we find a network component $\mathcal{S}^0 \in$ OPT($\mathcal{R}^0$) in polynomial time. Then $\mathcal{R}^0 \in$ FEAS($\mathcal{S}^0$), hence FEAS($\mathcal{S}^0$) $\neq \emptyset$ which implies $Q \in \mathcal{Q}_{comp}^{trac}$.
- If $Q \in \mathcal{Q}_{comp}^{trac}$ we find a network component $\mathcal{S}^0$ with FEAS($\mathcal{S}^0$) $\neq \emptyset$ in polynomial time. Since $Q \in \mathcal{Q}_{pol}^{opt}$ we find a solution routing $\mathcal{R}(\mathcal{S}^0) \in$ OPT($\mathcal{S}^0$) in polynomial time. Then ($\mathcal{S}^0$, OPT($\mathcal{S}^0$)) is a feasible solution to I.
- This is obvious, since a feasible solution ($\mathcal{S}, \mathcal{R}$) contains a routing $\mathcal{R}$ with FEAS($\mathcal{R}$) $\supset \{\mathcal{S}\}$. □

Definition 5.1.15. We hence define $\mathcal{Q}^{trac} := \mathcal{Q}_{rout}^{trac} = \mathcal{Q}_{comp}^{trac}$.

Lemma 5.1.16. *The restriction of uLPwR to linear PTNs, TTwR, and DMwR are in $\mathcal{Q}^{trac}$.*

Proof. • Let ($N, \mathcal{L}, \mathcal{OD}, B$) be an instance of uLPwR. A feasible line plan can be found in polynomial time as described in the proof of Theorem 2.5.9 if it exists. Hence, the problem uLPwR restricted on linear PTNs is contained in $\mathcal{Q}_{comp}^{trac}$.

- Let $(\mathcal{N}, l, u, \mathcal{OD})$ be an instance of TTwR. A routing $\mathcal{R}$ in $\mathcal{N}$ (e.g., a shortest-path routing taking lower bounds as arc lengths) can be found in polynomial time. FEAS$(\mathcal{R}) = \{\pi : \pi$ is a feasible timetable $\}$. Hence, TTwR is contained in $\mathcal{Q}_{\text{rout}}^{\text{trac}}$.
- Let $(\mathcal{N}, l, \pi, \mathcal{OD}, d)$ be an instance of DMwR. A routing $\mathcal{R}$ in $\mathcal{N}$ (e.g., a shortest-path routing taking lower bounds as arc lengths) can be found in polynomial time. Furthermore, $x(\mathcal{R}) \in$ FEAS$(\mathcal{R})$ (see Definition 4.3.10). Hence, DMwR is contained in $\mathcal{Q}_{\text{rout}}^{\text{trac}}$.

□

Lemma 5.1.17. *uLPwR is not in $\mathcal{Q}^{trac}$.*

Proof. This follows from Theorem 2.5.1. □

The results of this section are summarized in the following table.

	$\mathcal{Q}^{\text{opt}}$	Find $\mathcal{R}(\mathcal{S}) \in$ OPT$(\mathcal{S})$ in polynomial time	Find $\mathcal{S}(\mathcal{R}) \in$ OPT$(\mathcal{R})$ in polynomial time	$\mathcal{Q}_{\text{pol}}^{\text{opt}}$	$\mathcal{Q}^{\text{trac}}$
uLPwR in linear PTNs	x	x	x	x	x
uLPwR	x	x	x	x	−
scLPwR	x	−	x	−	−
cLPwR	x	−	x	−	−
TTwR	x	x	x	x	x
DMwR	x	x	x	x	x

5.2 An Iterative Heuristic

We propose the heuristic presented in Algorithm 2 to solve problems in problem class $\mathcal{Q}^{\text{opt}}$.

Due to Lemma 5.1.14 we can assume without loss of generality that we start with a given routing $\mathcal{R}^0$. However, we could equivalently formulate an iterative heuristic that starts with a feasible network component $\mathcal{S}$.

Note that, in general, the sets OPT$(\mathcal{S}^i)$ and OPT$(\mathcal{R}^i)$ contain more than one element and that the performance of the heuristic certainly depends on the choices made in steps 1, 2, 4, and 5.

However, we assume that the choices of the algorithm are deterministic and only depend on the network component $\mathcal{S}^{i-1}$ in step 4 and on the routing $\mathcal{R}^0$ or $\mathcal{R}^i$, respectively, in steps 2 and 5. Hence, if $\mathcal{S}^i = \mathcal{S}^{i-1}$ for an $i \in \{1, \ldots, s\}$, Algorithm 2 would find the same solution $(\mathcal{S}^j, \mathcal{R}^j) = (\mathcal{S}^i, \mathcal{R}^i)$ for every $j = i, \ldots, s$. This justifies the stopping criterion in steps 6–8.

Algorithm 2 Iterative heuristic

Require: Instance I of problem $Q \in \mathcal{Q}^{\mathrm{opt}}$. Maximal number of iterations $s \in \mathbb{N} \cup \{\infty\}$.
Ensure: Solution $(\mathcal{S}, \mathcal{R})$ to I (if it exists)
1: Find an initial routing $\mathcal{R}^0$ with $\mathrm{OPT}(\mathcal{R}^0) \neq \emptyset$
2: Choose $\mathcal{S}^0 \in \mathrm{OPT}(\mathcal{R}^0)$
3: **for** $i = 1, \ldots, s$ **do**
4:     Choose $\mathcal{R}^i \in \mathrm{OPT}(\mathcal{S}^{i-1})$
5:     Choose $\mathcal{S}^i \in \mathrm{OPT}(\mathcal{R}^i)$
6:     **if** $\mathcal{S}^i = \mathcal{S}^{i-1}$ **then**
7:         Stop. **return** $(\mathcal{S}^i, \mathcal{R}^i)$
8:     **end if**
9: **end for**
10: Stop. **return** $(\mathcal{S}^s, \mathcal{R}^s)$

Lemma 5.2.1. *For $Q \in \mathcal{Q}^{opt}$, steps 2, 4, and 5 of Algorithm 2 are well-defined, i.e.,*

$$OPT(\mathcal{S}^{i-1}) \neq \emptyset \text{ and } OPT(\mathcal{R}^i) \neq \emptyset \text{ for } i = 1, \ldots, s.$$

Proof. This lemma is shown inductively:

- $\mathcal{R}^0$ is chosen such that $\mathrm{OPT}(\mathcal{R}^0) \neq \emptyset$ in step 1 of Algorithm 2.
- For $i = 1, \ldots, s + 1$ suppose that $\mathrm{OPT}(\mathcal{R}^{i-1}) \neq \emptyset$. Then in step 2 of iteration $i-1$, $\mathcal{S}^{i-1}$ is chosen from $\mathrm{OPT}(\mathcal{R}^{i-1}) \neq \emptyset$. Hence, $(\mathcal{S}^{i-1}, \mathcal{R}^{i-1})$ is a feasible solution to I. According to (1.) in Definition 5.1.7, $\mathrm{OPT}(\mathcal{S}^{i-1}) \neq \emptyset$ follows.
- For $i = 1, \ldots, s$ suppose that $\mathrm{OPT}(\mathcal{S}^{i-1}) \neq \emptyset$. $\mathcal{R}^i$ is chosen from $\mathrm{OPT}(\mathcal{S}^{i-1})$ for $i = 1, \ldots, s$. Hence, $(\mathcal{S}^{i-1}, \mathcal{R}^i)$ is a feasible solution to I. According to (2.) in Definition 5.1.7, $\mathrm{OPT}(\mathcal{R}^i) \neq \emptyset$ follows. □

Lemma 5.2.2. *Let I be an instance of a problem $Q \in \mathcal{Q}^{trac}$. For the solutions $(\mathcal{S}^i, \mathcal{R}^i)$ calculated in Algorithm 2, it holds that*

$$c(\mathcal{S}^i, \mathcal{R}^i) \leq c(\mathcal{S}^{i-1}, \mathcal{R}^{i-1}) \text{ for } i = 1, \ldots, s.$$

In particular, Algorithm 2 converges to a solution $(\mathcal{S}^\infty, \mathcal{R}^\infty)$ for $s \to \infty$.

Proof. Since $\mathcal{S}^i \in \mathrm{OPT}(\mathcal{R}^i)$ it follows that

$$c(\mathcal{S}^i, \mathcal{R}^i) \leq c(\mathcal{S}^{i-1}, \mathcal{R}^i) \text{ for all } i = 1, \ldots, s,$$

and since $\mathcal{R}^i \in \mathrm{OPT}(\mathcal{S}^{i-1})$ it follows that

$$c(\mathcal{S}^{i-1}, \mathcal{R}^i) \leq c(\mathcal{S}^{i-1}, \mathcal{R}^{i-1}) \text{ for all } i = 1, \ldots, s$$

which implies the lemma. Since c is bounded from below (see Definition 5.1.3), we obtain convergence. □

However, $(\mathcal{S}^\infty, \mathcal{R}^\infty)$ does not need to be optimal as we see in Sects. 5.5.

We now regard the running time of Algorithm 2.

Lemma 5.2.3. *If $Q \in \mathcal{Q}_{pol}^{opt}$, steps 2–9 of Algorithm 2 need polynomial time. In particular, every iteration of Algorithm 2 is in polynomial time.*

If also the first routing can be found in polynomial time, Algorithm 2 has polynomial running time for a fixed maximal number of iterations s.

Lemma 5.2.4. *If $Q \in \mathcal{Q}^{trac}$, for a fixed maximal number of iterations s, Algorithm 2 runs in polynomial time.*

Note that for a problem $Q \in \mathcal{Q}$, given heuristics for finding a network component for a fixed routing in a network problem and vice versa, we could define an alternative iterative algorithm, substituting OPT($\mathcal{R}^i$) with FEAS($\mathcal{R}^i$) and OPT($\mathcal{S}^i$) with FEAS($\mathcal{S}^i$). However, Lemma 5.2.2 does not hold in this case. Therefore, and since none of the problems uLPwR, cLPwR, scLPwR, TTwR, and DMwR is contained in $\mathcal{Q}_{pol}^{feas} \setminus \mathcal{Q}_{pol}^{opt}$, we do not detail the alternative algorithm in this book.

5.3 The Price of Sequentiality

In the following let Q be a problem in $\mathcal{Q}^{opt}$ and let I be an instance of Q specifying a network N and a set of node-OD-pairs $\mathcal{OD}_{node}$. Let $(\mathcal{S}^*, \mathcal{R}^*)$ be an optimal solution to I.

Naturally, the outcome of Algorithm 2 depends on the initial routing chosen in step 1 as well as on the on the choices made in steps 2 and 4 and 5 of every iteration of Algorithm 2.

Notation 5.3.1. *Denote by $S^0(I)$ the set of all solutions $(\mathcal{S}^0, \mathcal{R}^0)$ that may be returned at the end of steps 1 and 2 of Algorithm 2 and by $S^i(I)$ for $i = 1, \ldots, s$ the set of solutions $(\mathcal{S}^i, \mathcal{R}^i)$ that may be returned at the end of iteration i of steps 4 and 5 of Algorithm 2.*

We now define the price of sequentiality which specifies the worst-case relative error of Algorithm 2 after each iteration, analogously to the ratio often used in the analysis of approximation algorithms.

Definition 5.3.2. Let I be an instance of a problem $Q \in \mathcal{Q}^{opt}$.

- For $i = 0, \ldots, s$ the *price of sequentiality for I in iteration i* is defined as

$$PoS^i(I) = \max_{(\mathcal{S}^i, \mathcal{R}^i) \in S^i(I)} \frac{c(\mathcal{S}^i, \mathcal{R}^i)}{c(\mathcal{S}^*, \mathcal{R}^*)}.$$

- The *price of sequentiality for I* is defined as

$$PoS(I) = \lim_{i \to \infty} PoS^i(I),$$

(for $s := \infty$ and ignoring the stopping criterion defined in steps 6–8 of Algorithm 2).

Note that if for an instance I of a problem $Q \in \mathcal{Q}^{\text{opt}}$, it holds that $|\text{OPT}(\mathcal{R})| = |\text{OPT}(\mathcal{S})| = 1$ for all considered routings $\mathcal{R}$ and network components $\mathcal{S}$, $PoS^i(I)$ depends only on the initial solution.

Since Algorithm 2 decreases in every step and converges (see Lemma 5.2.2), we see the following lemma.

Lemma 5.3.3. *Let $Q \in \mathcal{Q}^{opt}$ and I be an instance of Q. Then*

- *For every $i = 1, \ldots, s$, $PoS^i(I) \leq PoS^{i-1}(I)$, and*
- $PoS(I) = \inf_{i \in \mathbb{N}} PoS^i(I)$.

We are not only interested in the outcome of Algorithm 2 for a specific instance I of a problem Q, but in bounds that more generally hold for whole classes of instances of Q. Therefore, we generalize the definition of the price of sequentiality to classes of instances $\mathcal{I}$.

Definition 5.3.4. *Let Q be a problem and $\mathcal{I}$ be a class of instances of Q.*

- *For $i = 0, \ldots, s$ the price of sequentiality for $\mathcal{I}$ in iteration i is defined as*

$$PoS^i(\mathcal{I}) = \sup_{I \in \mathcal{I}} PoS^i(I).$$

- *We define the price of sequentiality for $\mathcal{I}$ as*

$$PoS(\mathcal{I}) = \sup_{I \in \mathcal{I}} PoS(I).$$

Naturally, the statement of Lemma 5.3.3 also holds for classes of instances:

Corollary 5.3.5. *Let $Q \in \mathcal{Q}^{opt}$ and I be an instance of Q. Then*

- *For every $i = 1, \ldots, s$, $PoS^i(\mathcal{I}) \leq PoS^{i-1}(\mathcal{I})$, and*
- $PoS(\mathcal{I}) = \inf_{i \in \mathbb{N}} PoS^i(\mathcal{I})$.

5.4 Analysis of the Iterative Heuristic for Timetabling with Routing

In this section we illustrate the theoretical concepts of the iterative heuristic and the price of sequentiality using the problem TTwR.

We have seen in Lemma 5.1.16 that TTwR is in $\mathcal{Q}^{\text{trac}}$. We restate Algorithm 2 for application for TTwR in Algorithm 3. As described in steps 3 and 7 of

5.4 Analysis of the Iterative Heuristic for Timetabling with Routing

Algorithm 3 Iterative heuristic for TTwR

Require: Instance $I = (\mathcal{N}, l, u, \mathcal{OD})$ of TTwR. Maximal number of iterations s.
Ensure: Solution $(\pi, \mathcal{R})$ to I.
1: Find a routing $\mathcal{R}^0$ for $\mathcal{OD}$ in $\mathcal{N}$.
2: Calculate activity weights $w_a^0 := \sum_{P_{(u,v)} \in \mathcal{R}^0, P_{uv} \ni a} w_{uv}$.
3: Find a timetable π^0 by solving sTT for the activity weights w_a^0 using the linear program (3.5)–(3.8)
4: **for** $i = 1, \ldots, s$ **do**
5:     Find a shortest-path routing $\mathcal{R}^i$ in $\mathcal{N}(\pi^{i-1})$.
6:     Calculate activity weights $w_a^i := \sum_{P_{(u,v)} \in \mathcal{R}^i, P_{uv} \ni a} w_{uv}$.
7:     Find a timetable π^i by solving sTT for the activity weights w_a^i using the linear program (3.5)–(3.8).
8:     **if** $\pi^i = \pi^{i-1}$ **then**
9:         Stop. **return** $(\pi^i, \mathcal{R}^i)$.
10:     **end if**
11: **end for**
12: Stop. **return** $(\pi^s, \mathcal{R}^s)$.

Algorithm 3, for the calculation of a timetable $\pi(\mathcal{R}) \in \text{OPT}(\mathcal{R})$, we can use the linear programming approach described in Sect. 3.3. Certainly, any other algorithm for finding a $\pi(\mathcal{R}) \in \text{OPT}(\mathcal{R})$ could be used instead.

For the calculation of an initial routing $\mathcal{R}^0$ in step 1 of Algorithm 3, we could, e.g., use one of the following strategies.

1. LOW: We calculate shortest paths in the EAN $\mathcal{N}$ with respect to arc lengths $c_a := l_a$ for $a \in \mathcal{A}_{\text{op}}$ and $c_a := 0$ for $a \in \mathcal{A} \setminus \mathcal{A}_{\text{op}}$.
2. UPP: We calculate shortest paths in the EAN $\mathcal{N}$ with respect to arc lengths $c_a := u_a$ for $a \in \mathcal{A}_{\text{op}}$ and $c_a := 0$ for $a \in \mathcal{A} \setminus \mathcal{A}_{\text{op}}$.
3. SUM: We calculate shortest paths in the EAN $\mathcal{N}$ with respect to arc lengths $c_a := \frac{l_a + u_a}{2}$ for $a \in \mathcal{A}_{\text{op}}$ and $c_a := 0$ for $a \in \mathcal{A} \setminus \mathcal{A}_{\text{op}}$.

However, *any* method, providing a routing or a timetable, could be used to create an initial solution.

We now give a small example which illustrates the approach of Algorithm 3 as well as it proves an infinite price of sequentiality for TTwR.

Consider the instance $I_x = (\mathcal{N}, l, u, \mathcal{OD})$ of TTwR depicted in Fig. 5.1a with $w_{uv} := 1$ and $x \in \mathbb{N}$. Suppose that the train-route P following tr_2 is chosen for the first routing $\mathcal{R}^0$ (the nodes of the corresponding path are marked in gray) in step 1 of Algorithm 3. This choice of a first routing is possible for all strategies LOW, UPP, and SUM.

In Fig. 5.1b a timetable $\pi(\mathcal{R}^0) \in \text{OPT}(\mathcal{R}^0)$ is shown. Suppose that this timetable $\pi^0 := \pi(\mathcal{R}^0)$ is chosen in step 3 of Algorithm 3. Then $\mathcal{R}^1 := \{P\}$ is a shortest-path routing in $N(\pi^0)$ and might hence be chosen in the first iteration of step 5. Then in step 7 we find $\pi^1 = \pi^0$ and the algorithm stops with the solution $(\pi^1, \mathcal{R}^1)$ $= (\pi^0, \mathcal{R}^0)$ and objective value $c(\pi^1, \mathcal{R}^1) = x$.

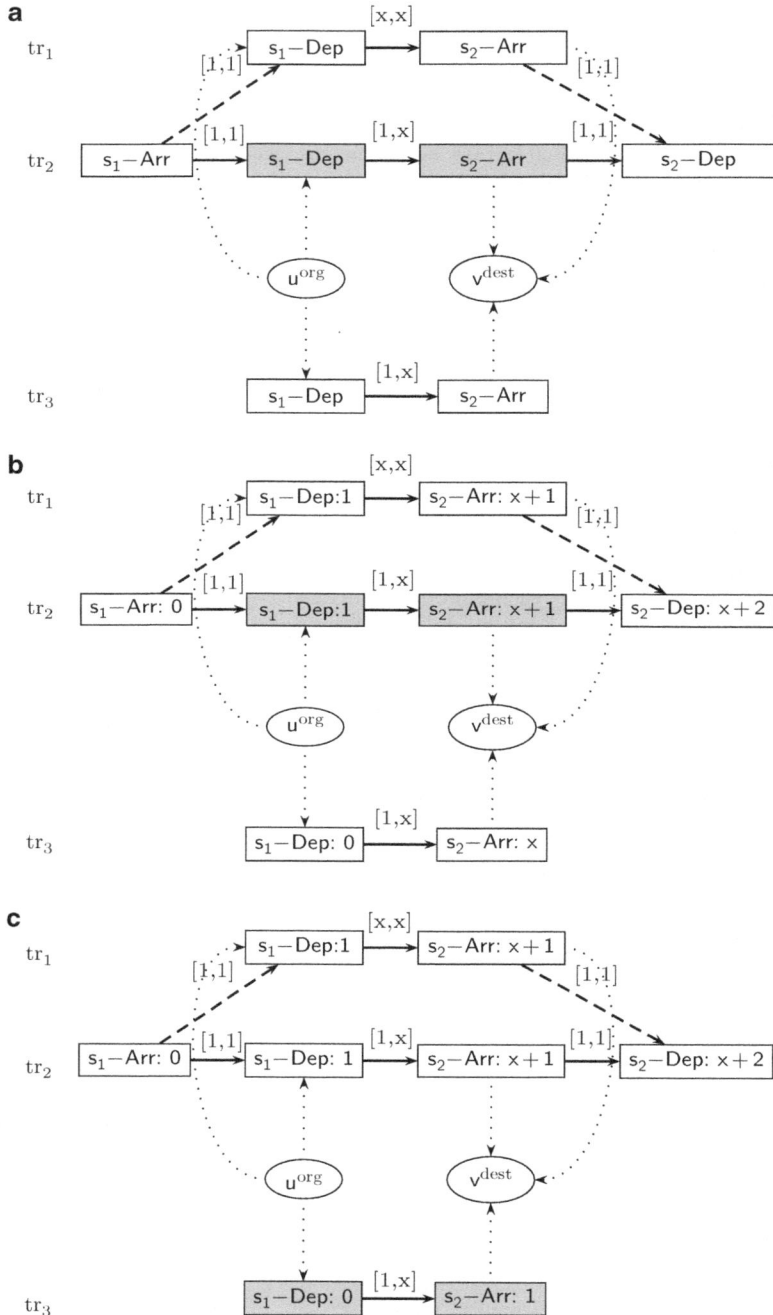

Fig. 5.1 Instance I of TTwR with $PoS(I) = \max_{a \in \mathcal{A}_{\mathrm{op}}} \frac{u_a}{l_a}$. (**a**) EAN with bounds (**b**) Timetable and routing calculated by Algorithm 3 (**c**) Optimal solution

5.4 Analysis of the Iterative Heuristic for Timetabling with Routing

The optimal solution $(\pi^*, \mathcal{R}^*)$ in this situation is depicted in Fig. 5.1c (again the gray nodes indicate the path); it has objective value $c(\pi^*, \mathcal{R}^*) = 1$.

Hence, for the constructed instance I_x, we have $PoS(I_x) = \frac{x}{1}$. Let $\mathcal{I}_{\text{TTwR}}$ denote all instances of TTwR. We obtain

$$PoS(\mathcal{I}_{\text{TTwR}}) \geq \sup_{x \in \mathbb{N}} PoS(I_x) = \infty.$$

Lemma 5.4.1. *Let $\mathcal{I}_{TTwR}$ denote all instances of TTwR. Then*

$$PoS(\mathcal{I}_{TTwR}) = \infty.$$

However, the result of Theorem 3.4.4 provides us with bounds on the price of sequentiality in case of bounded upper and lower bounds for the activity lengths. These bounds are reached in the example shown in Fig. 5.1, since $PoS(I_x) = \frac{x}{1} = \max_{a \in \mathcal{A}_{op}} \frac{u_a}{l_a}$ for the instance I_x considered there. We obtain the following lemma.

Lemma 5.4.2. 1. *Let $\mathcal{I}(r^{\max})$ denote the class of instances of TTwR with $\max_{a \in \mathcal{A}_{op}} \frac{u_a}{l_a} \leq r^{\max}$. Then*

$$PoS(\mathcal{I}(r^{\max})) = PoS^0(\mathcal{I}(r^{\max})) = r^{\max}.$$

2. *Let $\mathcal{I}(R^{\max})$ denote the class of instances of TTwR with $\frac{\sum_{(u_k, v_k) \in \mathcal{OD}} w_k l_k^{upp}}{\sum_{(u_k, v_k) \in \mathcal{OD}} w_k l_k^{low}} \leq R^{\max}$ as defined in Notation 3.4.3. Then*

$$PoS(\mathcal{I}(R^{\max})) = PoS^0(\mathcal{I}(R^{\max})) = R^{\max}.$$

The lower bound $l_{(\text{tr}_2 - s_1 - \text{Dep}, \text{tr}_2 - s_2 - \text{Arr})} = 1$ on $(\text{tr}_2 - s_1 - \text{Dep}, \text{tr}_2 - s_2 - \text{Arr})$ in the example given in Fig. 5.1 may be called *unrealistic*, since there is no feasible timetable π such that $\pi_{(\text{tr}_2 - s_2 - \text{Arr})} - \pi_{(\text{tr}_2 - s_1 - \text{Dep})} = 1$.

However, the price of sequentiality stays unbounded for general instances of TTwR, even if we require that the instances must have *realistic* time bounds, in the sense that for every $(i, j) \in \mathcal{A}$, and every $x_{ij} \in [l_{ij}, u_{ij}]$ there exists a feasible timetable π such that $\pi_j - \pi_i = x_{ij}$. An example is given in Fig. 5.2.

Consider the instance $I = (\mathcal{N}, l, u, \mathcal{OD})$ of TTwR depicted in Fig. 5.2a with $w_{uv} := 1$ and $x \geq 3 \in \mathbb{N}$. Note that the time bounds in this instance are *realistic*.

We have four different paths for OD-pair (u, v). In this example, for the first routing $\mathcal{R}^0$, the path P following tr_2 (first and last even are marked in gray) is chosen for all choices of initial routings specified in (1.)–(3.) in step 1 of Algorithm 3.

In Fig. 5.2a timetable $\pi(\mathcal{R}^0) \in \text{OPT}(\mathcal{R}^0)$ is shown. Suppose that this timetable $\pi^0 := \pi(\mathcal{R}^0)$ is chosen in step 3 of Algorithm 3. Then $\mathcal{R}^1 := \{P\}$ is a shortest-path routing in $N(\pi^0)$ and might hence be chosen in the first iteration of step 5. In step 7 we find $\pi^1 = \pi^0$ and the algorithm stops with the solution $(\pi^1, \mathcal{R}^1) = (\pi^0, \mathcal{R}^0)$ depicted in Fig. 5.2b and objective value $c(\pi^1, \mathcal{R}^1) = x + 2$.

180 5 An Iterative Solution Approach for General Network Problems with Routing

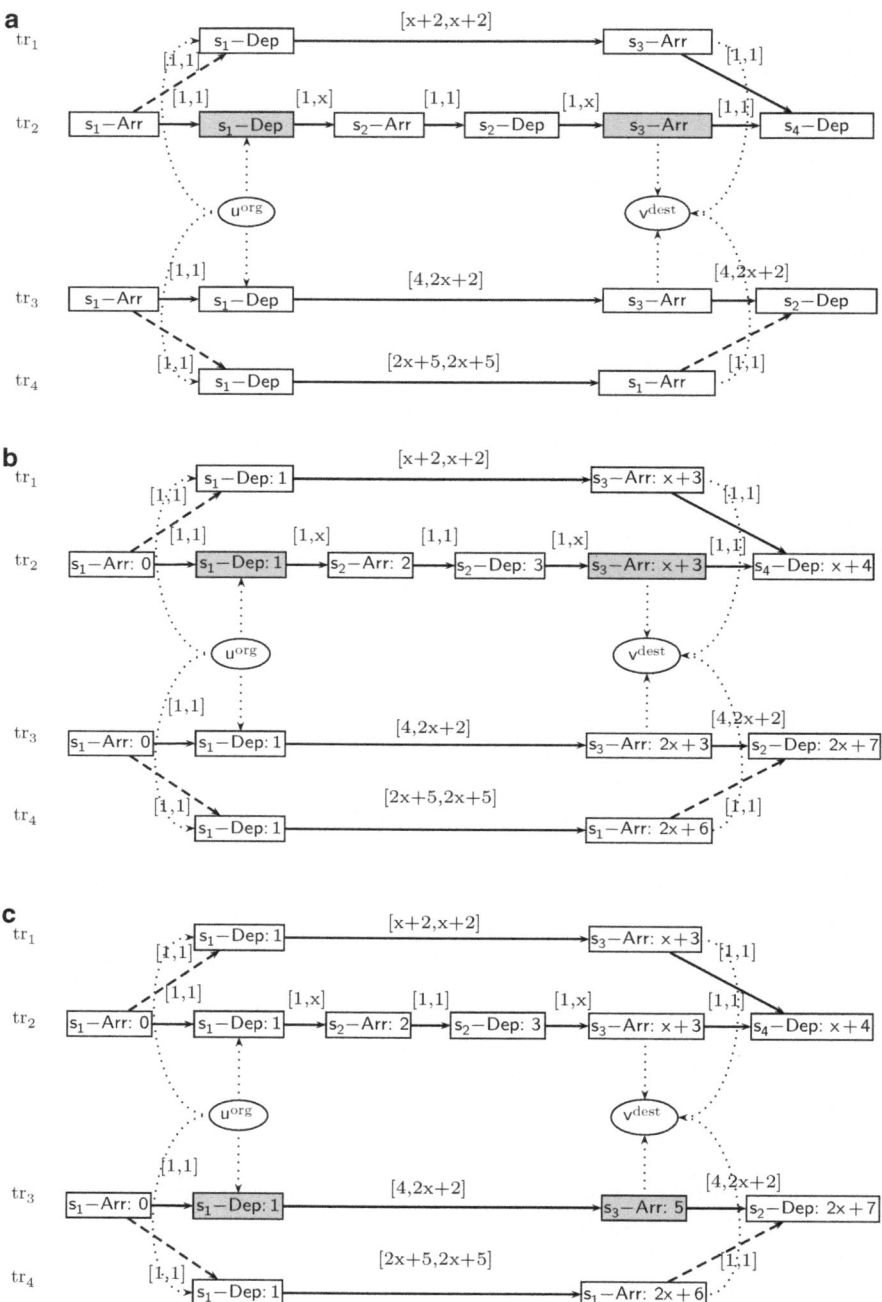

Fig. 5.2 Example for unbounded price of sequentiality, even for realistic instances. (**a**) EAN with bounds (**b**) Timetable and routing calculated by Algorithm 3 (**c**) Optimal solution

On the other hand, the optimal solution $(\pi^*, \mathcal{R}^*)$ in this situation is depicted in Fig. 5.2c (again the gray nodes indicate first and last event of the path), it has objective value $c(\pi^*, \mathcal{R}^*) = 4$.

Hence, for the constructed instance I_x, we have $PoS(I_x) = \frac{x+2}{4}$. If we construct a sequence of instances I_x with $x \to \infty$ we obtain $PoS(\mathcal{I}_{TTwR}) = \infty$ with $\mathcal{I}_{TTwR}$ being the class of all possible instances for TTwR.

A numerical investigation of the iterative heuristic for TTwR and a comparison to solutions obtained by solving the IP formulations presented in Sect. 3.7 can be found in [Anh12]. For the instances tested there, the iterative heuristic provides close-to-optimal results and outperforms the IP formulations with respect to running times.

5.5 Analysis of the Iterative Heuristic for Arc Speed-Up

Similarly to the preceding section, in this section we analyze the iterative heuristic using the price of sequentiality as a quality measure. Since TTwR and the other problems considered in Chaps. 2–4 are already quite complicated, in this section we analyze the structure of the price of sequentiality using the *arc speed-up* problem as an example. Arc speed-up is a simple network problem, which is defined in Sect. 5.5.1. In Sect. 5.5.2 we show that arc speed-up is in $\mathcal{Q}^{\text{trac}}$ and detail how the iterative heuristic given in Algorithm 2 can be applied to solve arc speed-up heuristically. In Sect. 5.5.3 we state an exact algorithm to solve the problem. These two approaches are compared in Sect. 5.5.4 by means of the price of sequentiality.

5.5.1 Definition of the Problem Arc Speed-Up

Let $N = (V, A)$ be a directed network with arc lengths $L_a \in \mathbb{N}$ and let $OD \subset V \times V$ be a set of OD-pairs in N. Like in every network problem with routing, also in the problem *arc speed-up* defined in this section, our task is to make a decision on the network component and find a routing such that the overall travel time is minimized.

In arc speed-up we do not have any capacity restrictions, hence, like in uLPwR, TTwR, and DMwR, we assume that all passengers of an OD-pair use the same path and define a *routing* in N as a collection of paths $\mathcal{R} := \{P_{uv} : (u, v) \in \mathcal{OD}\}$ where P_{uv} is a path from u to v.

Besides finding a routing, the task of arc speed-up is to choose an *arc speed-up* $a \in A$ whose length is decreased by a multiplication with a *speed-up factor* $\alpha \in [0, 1)$.

Given a speed-up factor α, for $a' \in A$, we denote by $N(a')$ the network N with arc lengths

$$L_a^{a'} := \begin{cases} \alpha \cdot L_a & \text{if } a = a' \\ L_a & \text{otherwise.} \end{cases}$$

We call $N(a')$ the *speed-up network corresponding to a'*.

Let $a' \in A$ be given and let $\mathcal{R}$ be a routing. Then the *travel time along $P_{uv} \in \mathcal{R}$ in $N(a')$* is

$$c(a', P_{uv}) := \sum_{a \in P_{uv}} L_a^{a'}.$$

The overall *travel time* in $N(a')$ is

$$c(a', \mathcal{R}) := \sum_{P_{uv} \in \mathcal{R}} w_{uv} c(a', P_{uv}).$$

Note that

$$c(a', P_{uv}) = \sum_{a \in P_{uv}} L_a^{a'} = \begin{cases} \sum_{a \in P_{uv}} L_a & \text{if } a' \notin P_{uv} \\ \sum_{a \in P_{uv}} L_a - (1-\alpha) L_{a'} & \text{otherwise.} \end{cases} \quad (5.1)$$

To prove the statements of this section, we make use of reformulations of the objective function. To this end, we define *arc weights*:

Definition 5.5.1. Let $N = (V, A)$ be a directed network, $\mathcal{OD}$ a set of OD-pairs and $\mathcal{R}$ a routing for $\mathcal{OD}$. We denote by $w(\mathcal{R})_a := \sum_{P_{uv} \in \mathcal{R}, P_{uv} \ni a} w_{uv}$ the *arc weight* of arc a in routing $\mathcal{R}$.

Then, like in (5.1), $c(a', \mathcal{R})$ can be rewritten as

$$\begin{aligned} c(a', \mathcal{R}) &= \sum_{(u,v) \in \mathcal{OD}} w_{uv} \sum_{a \in P(u,v)} L_a^{a'} = \sum_{a \in A} w(\mathcal{R})_a L_a^{a'} \\ &= \sum_{a \in A} w(\mathcal{R})_a L_a - (1-\alpha) w(\mathcal{R})_{a'} L_{a'} \\ &= \sum_{(u,v) \in \mathcal{OD}} w_{uv} \sum_{a \in P(u,v)} L_a - (1-\alpha) w(\mathcal{R})_{a'} L_{a'}. \end{aligned}$$
(5.2)

We now define the problem *arc speed-up*.

Definition 5.5.2. An instance $(N, \mathcal{OD}, \alpha)$ of *arc speed-up (ASU)* consists of a directed network $N = (V, A)$ with edge lengths L_a, a set of OD-pairs $OD \subset V \times V$ and a speed-up factor $\alpha \in [0, 1)$. The task is to choose a *arc speed-up $a \in A$* and a routing $\mathcal{R}$, such that the overall travel time $c(a, \mathcal{R})$ in $N(a)$ is minimized.

Lemma 5.5.3. *ASU is in $\mathcal{Q}$.*

Proof. An instance $(N, \mathcal{OD}, \alpha)$ of ASU contains a network N and a set of OD-pairs $\mathcal{OD}$ as required in Definition 5.1.3. A solution $(a', \mathcal{R})$ specifies a routing $\mathcal{R}$ and an arc a' which can be understood as an assignment $f_{a'} : A \to \mathbb{Q} \times \mathbb{Z}$ to the arcs, namely

$$f_{a'}(a) := \left(L_a^{a'}, \sum_{(u,v) \in \mathcal{OD}} w_{uv} \right).$$

Furthermore, $c(a', \mathcal{R}) \geq 0$ for every solution $(a', \mathcal{R})$ to an instance I of ASU. Hence, ASU is in $\mathcal{Q}$. □

5.5.2 The Iterative Heuristic for Arc Speed-Up

In this section we describe how Algorithm 2 can be applied to ASU. To this end, we explain how to find solutions to ASU when part of the solution, namely, the routing or the arc speed-up is fixed.

Let $I = (N, \mathcal{OD}, \alpha)$ be an instance of ASU with $N = (V, A)$ and let $n := |V|$ denote the number of nodes in N.

First we note that given an arc speed-up $a' \in A$, we can find a shortest-path routing $\mathcal{R}(a')$ in $N(a')$ using, e.g., the Floyd-Warshall algorithm, in $O(n^3)$, where n is the number of nodes in the considered network.

Lemma 5.5.4. *Let $I = (N, \mathcal{OD}, \alpha)$ be an instance of ASU with $\mathcal{OD} = \{(u_k, v_k) : k \in K\}$, $N = (V, A)$ and $a' \in A$. Let n denote the number of nodes in N. Then $OPT(a')$ (as it is defined in Definition 5.1.6) is the set of shortest-path routings in $N(a')$, i.e.,*

$$OPT(a') = \{\{P_1, P_2, \ldots, P_{|K|}\} : P_k \text{ is a shortest path from } u_k \text{ to } v_k \text{ in } N(a')\}$$

and a routing $\mathcal{R}(a') \in OPT(a')$ can be found in $O(n^3)$.

Given a routing $\mathcal{R}$, we can also easily find a corresponding arc speed-up $a(\mathcal{R})$:

Lemma 5.5.5. *Let $I = (N, \mathcal{OD}, \alpha)$ be an instance of ASU with $N = (V, A)$ and $\mathcal{R}$ be a routing. Let n denote the number of nodes in N. Then $OPT(\mathcal{R})$ (as it is defined in Definition 5.1.6) is given by*

$$OPT(\mathcal{R}) := \underset{a \in A}{\operatorname{argmax}} \{w(\mathcal{R})_a L_a\}.$$

$OPT(\mathcal{R})$ can be found in $O(n^3)$.

Proof. Given a routing $\mathcal{R}$ and an arc $a' \in A$ we have

$$c(a', \mathcal{R}) = \sum_{(u,v) \in \mathcal{OD}} w_{uv} \sum_{a \in P(u,v)} L_a - (1-\alpha) w(\mathcal{R})_{a'} L_{a'}.$$

due to (5.2). Hence, OPT($\mathcal{R}$) = $\operatorname{argmax}_{a \in A} \{w(\mathcal{R})_a L_a\}$.

In order to find this set we calculate and compare $w(\mathcal{R})_{a'} L_{a'}$ for every $a' \in A$. Since every path $P \in \mathcal{R}$ contains at most $n-1$ arcs, we calculate $\{w(\mathcal{R})_a L_a : a \in A \text{ s.t. } w(\mathcal{R})_a L_a \neq 0\}$ in time $O(n \cdot |\mathcal{OD}|)$ which is in $O(n^3)$ since $|\mathcal{OD}| \in O(n^2)$. □

Hence, ASU is in $Q_{\text{pol}}^{\text{opt}}$. Since any combination of arc $a' \in A$ and routing $\mathcal{R}$ is a feasible solution, we see the following corollary.

Corollary 5.5.6. *ASU is in Q^{trac}.*

We now restate Algorithm 2 for the problem ASU in Algorithm 4 below. We use a shortest-path routing in N as initial routing.

Algorithm 4 Iterative heuristic for ASU

Require: Instance $I = (N, \mathcal{OD}, \alpha)$ of ASU. Maximal number of iterations s.
Ensure: Solution $(a, \mathcal{R})$ to I
1: Find a shortest-path routing $\mathcal{R}^0$ for $\mathcal{OD}$ in N.
2: Calculate arc weights $w_a^0 := \sum_{P_{(u,v)} \in \mathcal{R}^0, P_{uv} \ni a} w_{uv}$.
3: Choose $a^0 \in \operatorname{argmax}_{a \in A} \{w(\mathcal{R})_a L_a\}$.
4: **for** $i = 1, \ldots, s$ **do**
5:     Find a shortest-path $\mathcal{R}^i$ routing in $N(a^{i-1})$.
6:     Calculate arc weights $w_a^i := \sum_{P_{(u,v)} \in \mathcal{R}^i, P_{uv} \ni a} w_{uv}$.
7:     Choose $a^i \in \operatorname{argmax}_{a \in A} \{w_a^i L_a\}$.
8:     **if** $a^i = a^{i-1}$ **then**
9:         Stop. **return** $(a^i, \mathcal{R}^i)$
10:    **end if**
11: **end for**
12: Stop. **return** $(a^s, \mathcal{R}^s)$

For ASU, we obtain a polynomial overall running time of the iterative heuristic:

Lemma 5.5.7. *Let $I = (N, \mathcal{OD}, \alpha)$ be an instance of ASU with n nodes for which we apply Algorithm 4. Then*

- *every iteration of Algorithm 4 is in $O(n^3)$, and*
- *the overall running time of Algorithm 4 is $O(n^5)$.*

Proof. According to Lemma 5.5.4 and 5.5.5, every iteration of Algorithm 4 is in $O(n^3)$. Due to the stopping criterion in step 8, the algorithm stops after at most $|A|$ iterations, independently of the choice of s. □

5.5.3 Solving Arc Speed-Up Exactly

The price of sequentiality compares the solution found by the iterative heuristic with an exact solution. In ASU, exact solutions can be found in polynomial time, which facilitates such a comparison. An exact algorithm for ASU is stated in this section.

Let $I = (N, \mathcal{OD}, \alpha)$ be an instance of ASU and let n denote the number of nodes in N. I can be solved in time $O(n^5)$ using an all-pairs shortest path algorithm to find a routing in $N(a')$ for every choice of $a' \in \mathcal{A}$. However, since the networks N and $N(a')$ differ only in the length of arc a', this approach can be speeded up:

The following Algorithm 5 solves ASU to optimality by finding an optimal routing in N first and then enumerating all arcs and updating shortest path lengths in the speeded-up network by a label-correcting step.

Algorithm 5 Exact algorithm for ASU

Require: Instance $I = (N, \mathcal{OD}, \alpha)$ of ASU.
Ensure: Arc a^* such that $(a^*, \mathcal{R}(a^*))$ is an optimal solution to I for any $\mathcal{R}(a^*) \in \text{OPT}(a^*)$ and objective value $c^* = (a^*, \mathcal{R}(a^*))$.
1: Determine the shortest path distance $d(i, j)$ for every pair of nodes $(i, j) \in V \times V$.
2: Set $c^* := \sum_{(u,v) \in \mathcal{OD}} w_{uv} d(u, v)$.
3: **for** $a' = (i, j) \in \mathcal{A}$ **do**
4:     **for** $(u, v) \in \mathcal{OD}$ **do**
5:         Set $d'(u, v) := \min\{d(u, v), d(u, i) + \alpha L_{a'} + d(j, v)\}$.
6:     **end for**
7:     Set $c' := \sum_{(u,v) \in \mathcal{OD}} w_{uv} d'(u, v)$.
8:     **if** $c' < c^*$ **then**
9:         Set $c^* := c'$, and $a^* := a'$.
10:     **end if**
11: **end for**
12: **return** c^* and a^*.

Theorem 5.5.8. *Algorithm 5 finds an optimal solution to ASU with routing in time $O(n^4)$, where n is the number of nodes in the considered network.*

Proof. For $i, j \in V$, like in Algorithm 5 denote by $d(i, j)$ the shortest path distance from i to j. Let P_{uv}^N be a shortest path from u to v in N and $a' = (i, j) \in \mathcal{A}$. Then P_{uv}^N is a shortest path from u to v in N if and only if $d(u, v) \leq d(v, i) + \alpha L_{a'} + d(j, v)$. Otherwise, the path $P_{uv}^{N(a')}$, consisting of a shortest path from u to i in N, $a' = (i, j)$ and a shortest path from j to v in N, is a shortest path from u to v in $N(a')$ with length $d(v, i) + \alpha L_{a'} + d(j, v)$.

The minimum $d'(u, v)$ calculated in step 5 of Algorithm 5 thus is the length of a shortest path from u to v in $N(a')$. For every arc $a' \in \mathcal{A}$, based on these numbers $d'(u, v)$, the objective value is computed and compared. Algorithm 5 then outputs an arc speed-up for which the objective value is minimal; thus it is correct.

To find the shortest path distance between all pairs of nodes in step 1, we can apply an all-pair shortest path algorithm, e.g., the Floyd-Warshall algorithm in $O(n^3)$.

For every arc and every OD-pair, step 5 is done in constant time, and for every arc, steps 7–10 are performed in $O(|\mathcal{OD}|)$; thus, steps 3–11 are done in $O(|A|\cdot|\mathcal{OD}|)$. Since step 2 can also be done in $O(|\mathcal{OD}|)$, the overall running time of Algorithm 5 is $O(n^3+|A|\cdot|\mathcal{OD}|)$ which in terms of n is $O(n^4)$, because $|A| \in O(n^2)$ and $|\mathcal{OD}| \in O(n^2)$. □

Hence, the running time of Algorithm 5 is even lower than the worst-case running time of the iterative approach specified in Algorithm 2.

However, the goal of this section is not to speed up the calculation of optimal solutions to ASU, but to analyze the price of sequentiality for this problem. This is done in the following sections.

5.5.4 The Price of Sequentiality for Arc Speed-Up

Let $I = (N, \mathcal{OD}, \alpha)$ be an instance of ASU. Analogous to Notation 5.3.1, we denote by $S^0(I)$ the set of all solutions $(\mathcal{S}^0, \mathcal{R}^0)$ that may be returned at the end of step 3 of Algorithm 4 and by $S^i(I)$ for $i = 1,\ldots,s$ the set of solutions $(\mathcal{S}^i, \mathcal{R}^i)$ that may be returned at the end of iteration i of step 7 of Algorithm 4.

For the sake of a compact representation, in this section, we furthermore use the following notations.

- $\hat{c}(P_{uv}) := \sum_{a \in P_{uv}} L_a$ for a path P_{uv} in N.
- $\hat{c}(I) := \sum_{P_{uv} \in \mathcal{R}} w_{uv} \sum_{a \in P_{uv}} L_a$ for a shortest-path routing $\mathcal{R}$ in N.
- $c^*(I) := c(a^*, \mathcal{R}^*)$ for an optimal solution $(a^*, \mathcal{R}^*)$ to I.
- $c^i(I) := \max_{(a^i, \mathcal{R}^i) \in S^i(I)} c(a^i, \mathcal{R}^i)$ for $i = 0,\ldots,|A|-1$.
- $c(I) := \lim_{i \to \infty} c(I)$ (for $s := \infty$ and ignoring the stopping criterion defined in steps 8–10 of Algorithm 4).

Since $c^*(I) \geq \alpha \hat{c}(I)$, the natural bound on the improvement that speeding up an arc can achieve for the overall travel time is $\frac{\hat{c}(I)}{c^*(I)} \geq \frac{1}{\alpha}$. This provides the first bound on the price of sequentiality.

Lemma 5.5.9. *For any instance I and every $i \in \mathbb{N}$ of ASU we have*

$$\frac{c^i(I)}{c^*(I)} \leq \frac{\hat{c}(I)}{c^*(I)} \leq \frac{1}{\alpha}.$$

We now develop bounds on $c^*(I)$ and $c^0(I)$ in terms of $\hat{c}(I)$. To simplify notation we use the following abbreviations: Given an instance $I = (N, \mathcal{OD}, \alpha)$ of ASU with arc set A we define

5.5 Analysis of the Iterative Heuristic for Arc Speed-Up

- $L^{\max}(I) := \max_{a \in A} L_a$,
- $W^{\text{sum}}(I) := \sum_{(u,v) \in \mathcal{OD}} w_{uv}$, and
- $W^{\max}(I) := \max_{(u,v) \in \mathcal{OD}} w_{uv}$.

Lemma 5.5.10. *Let $I = (N, \mathcal{OD}, \alpha)$ be an instance of ASU. The optimal objective value $c^*(I)$ of I is bounded from below by*

$$c^*(I) \geq \hat{c}(I) - L^{\max}(I)(1 - \alpha) W^{\text{sum}}(I). \tag{5.3}$$

and

$$c^*(I) \geq \alpha[\hat{c}(I) - W^{\text{sum}}(I) + W^{\max}(I)] + W^{\text{sum}}(I) - W^{\max}(I). \tag{5.4}$$

Proof. It follows from (5.1) that

$$c(a^*, P_{uv}) \geq \hat{c}(P_{uv}) - L^{\max}(I)(1 - \alpha).$$

Summarizing over all OD-pairs we obtain (5.3).

If there are several different OD-pairs, at most the path of one OD-pair (u', v') can entirely be covered by the arc speed-up. For this path $P_{u'v'}$ it holds that

$$c(a^*, P_{u'v'}) \geq \alpha \hat{c}(P_{u'v'})$$

For every $(u, v) \in \mathcal{OD} \setminus \{(u', v')\}$ the corresponding path P_{uv} contains at least one arc $a \neq a^*$. Since $L_a \in \mathbb{N}$ for all $a \in A$ we have

$$c(a^*, P_{uv}) \geq \alpha[\hat{c}(P_{uv}) - 1] + 1 \quad \forall (u, v) \in \mathcal{OD} \setminus \{(u', v')\}.$$

Summarizing over all OD-pairs we obtain

$$c^*(I) \geq \alpha\left[\hat{c}(I) - \left(\sum_{(u,v) \in \mathcal{OD}} w_{uv} - w_{u'v'}\right)\right] + \sum_{(u,v) \in \mathcal{OD}} w_{uv} - w_{u'v'}$$

which is minimal if we choose (u', v') such that $w_{u'v'}$ is maximal. This implies (5.4). □

Comparing the bounds (5.3) and (5.4) from Lemma 5.5.10, we obtain the following result:

Lemma 5.5.11. *Let $I = (N, \mathcal{OD}, \alpha)$ be an instance of ASU. Then*

$$\hat{c}(I) - L^{\max}(I)(1 - \alpha) W^{\text{sum}}(I) \leq \alpha[\hat{c}(I) - W^{\text{sum}}(I) + W^{\max}(I)]$$
$$+ W^{\text{sum}}(I) - W^{\max}(I)$$

$$\Leftrightarrow \hat{c}(I) \leq W^{\text{sum}} L^{\max}(I) + W^{\text{sum}} - W^{\max}.$$

Lemma 5.5.12. *For an instance* $I = (N, \mathcal{OD}, \alpha)$, *the objective value* $c(I)$ *determined using the iterative heuristic 4 is bounded from above by*

$$c(I) \leq c^i(I) \leq c^0(I) \leq \hat{c}(I) - (1-\alpha)W^{max}(I) \quad \forall i \in \mathbb{N}. \tag{5.5}$$

Proof. The first two inequalities follow from Lemma 5.2.2. To see the last inequality, we note that in step 1 Algorithm 4 chooses a shortest-path routing $\mathcal{R}^0$ in N and in step 3 an arc $a^0 \in \operatorname{argmax}_{a \in A}\{w(\mathcal{R})_a L_a\}$. We obtain

$$c^0(I) = \hat{c}(I) - (1-\alpha)\max_{a \in A} w(\mathcal{R})_a L_a \leq \hat{c}(I) - (1-\alpha)\max_{(u,v) \in \mathcal{OD}} w_{uv},$$

since $\max_{a \in A} L_a w(\mathcal{R}^0)_a \geq \max_{(u,v) \in \mathcal{OD}} w_{uv}$. □

In the subsequent sections we use the bounds from Lemma 5.5.10 and 5.5.12 to deduce bounds on whole classes of instances.

Let $\mathcal{I}_{AR}$ denote the class of all instances of ASU. As we will see later in Lemma 5.5.23, $PoS(\mathcal{I}_{AR}) = \infty$. We hence impose restrictions on the classes of instances of ASU.

Definition 5.5.13. We classify instances $I = (N, \mathcal{OD}, \alpha)$ with $N = (V, A)$ of ASU according to the following characteristics:

1. CGN: $n := |V|$ and $L^{max} := \max_{a \in A} L_a$.
2. OD-pairs: $k := |\mathcal{OD}|$, $W^{sum} := \sum_{(u,v) \in \mathcal{OD}} w_{uv}$ and $W^{max} := \max_{(u,v) \in \mathcal{OD}} w_{uv}$.
3. Speed-up factor: $\alpha' := \alpha$.

For fixed $(n, L^{max}, k, W^{sum}, W^{max}, \alpha') \in \mathbb{N}^5 \times [0, 1)$ let $\mathcal{I}[(n, L^{max}), (k, W^{sum}, W^{max}), \alpha']$ denote the set of instances having the described characteristics (1.)–(3.). When we consider a class of instances where a certain parameter is not restricted, we use the entry "-."

Not every combination $(n, L^{max}, k, W^{sum}, W^{max}, \alpha') \in \mathbb{N}^5 \times [0, 1)$ corresponds to instances of ASU. We say that a combination $(k, W^{sum}, W^{max}) \in \mathbb{N}^3$ is *feasible* if there exist numbers $\{w_1, \ldots, w_k\}$ such that $W^{max} = \max_{j=1,\ldots,k} w_j$ and $W^{sum} = \sum_{j=1,\ldots,k} w_k$.

Furthermore, for the sake of a shorter notation, we use the following abbreviations:

$$PoS^i[(n, L^{max}), (k, W^{sum}, W^{max}), \alpha] := PoS^i(\mathcal{I}[(n, L^{max}), (k, W^{sum}, W^{max}), \alpha]),$$

$$PoS[(n, L^{max}), (k, W^{sum}, W^{max}), \alpha] := PoS(\mathcal{I}[(n, L^{max}), (k, W^{sum}, W^{max}), \alpha]).$$

5.5.4.1 Price of Sequentiality for Networks of Arbitrary Size and Fixed α

In this section we derive the price of sequentiality for some classes of instances $\mathcal{I}$ which have in common that the speed-up factor α is fixed and the size n is not fixed.

We start in Lemma 5.5.14, fixing all values except n.

5.5 Analysis of the Iterative Heuristic for Arc Speed-Up

Lemma 5.5.14. *Let $L^{max} \in \mathbb{N}$, $\alpha \in [0, 1)$ and a feasible $(k, W^{sum}, W^{max}) \in \mathbb{N}^3$ be given. Then it holds that*

$$PoS^0[(-, L^{max}), (k, W^{sum}, W^{max}), \alpha] \leq \frac{(L^{max} + 1)W^{sum} - W^{max} - (1 - \alpha)W^{max}}{\alpha L^{max} W^{sum} + W^{sum} - W^{max}}.$$

Proof. We define a function $g : \mathbb{R}_0^+ \to \mathbb{R}$ as the ratio between the bounds on the objective value found using Algorithm 4 and on the optimal objective value.

$$g(s) = \begin{cases} \frac{s - (1-\alpha)W^{max}}{(s - (W^{sum} - W^{max}))\alpha + W^{sum} - W^{max}} & \text{if } s \leq L^{max} W^{sum} + W^{sum} - W^{max} \\ \frac{s - (1-\alpha)W^{max}}{s - L^{max}(1-\alpha)W^{sum}} & \text{if } s > L^{max} W^{sum} + W^{sum} - W^{max}. \end{cases}$$

Note that then for every instance $I \in \mathcal{I}[(-, L^{max}), (k, W^{sum}, W^{max}), \alpha]$, it holds that

$$PoS^0(I) = \sup_{(\mathcal{S}^0, \mathcal{R}^0) \in S^0(I)} \frac{c(\mathcal{S}^0, \mathcal{R}^0)}{c(\mathcal{S}^*, \mathcal{R}^*)} \leq g(\hat{c}(I)). \tag{5.6}$$

Hence

$$PoS^0[(-, L^{max}), (k, W^{sum}, W^{max}), \alpha] \leq \sup_{I \in \mathcal{I}[(-, L^{max}), (k, W^{sum}, W^{max}), \alpha]} g(\hat{c}(I))$$

$$\leq \sup_{s \in \mathbb{R}_0^+} g(s). \tag{5.7}$$

We now calculate the supremum of g on $\mathbb{R}_0^+$.

- Note that for the break point $s = L^{max} W^{sum} + W^{sum} - W^{max}$, we have

$$g(L^{max} W^{sum} + W^{sum} - W^{max})$$

$$= \frac{L^{max} W^{sum} + W^{sum} - W^{max} - (1 - \alpha)W^{max}}{(L^{max} W^{sum} + W^{sum} - W^{max} - (W^{sum} - W^{max}))\alpha + W^{sum} - W^{max}}$$

$$= \frac{L^{max} W^{sum} + W^{sum} - W^{max} - (1 - \alpha)W^{max}}{L^{max} W^{sum} + W^{sum} - W^{max} - L^{max}(1-\alpha)W^{sum}}$$

$$= \lim_{s \searrow (L^{max} W^{sum} + W^{sum} - W^{max})} g(s).$$

- We observe that $g(s)$ is monotonously increasing on the interval $[0, L^{max} W^{sum} + W^{sum} - W^{max}]$, since

$$\frac{d}{d(s)} \frac{s - (1-\alpha)W^{\max}}{(s - (W^{\text{sum}} - W^{\max}))\alpha + W^{\text{sum}} - W^{\max}}$$

$$= \frac{(1-\alpha)(W^{\text{sum}} - W^{\max} + \alpha W^{\max})}{[(s - (W^{\text{sum}} - W^{\max}))\alpha + W^{\text{sum}} - W^{\max}]^2} > 0.$$

- Furthermore, $g(s)$ decreases monotonously on $[L^{\max}W^{\text{sum}} + W^{\text{sum}} - W^{\max}, \infty)$, since

$$\frac{d}{d(s)} \frac{s - (1-\alpha)W^{\max}}{s - L^{\max}(1-\alpha)W^{\text{sum}}} = -\frac{(1-\alpha)(L^{\max}W^{\text{sum}} - W^{\max})}{[s - L^{\max}(1-\alpha)W^{\text{sum}}]^2} < 0$$

Thus, g takes its maximum in $s = L^{\max}W^{\text{sum}} + W^{\text{sum}} - W^{\max}$. Due to (5.7) it follows that

$$PoS^0(I) \leq \frac{(L^{\max} + 1)W^{\text{sum}} - W^{\max} - (1-\alpha)W^{\max}}{\alpha L^{\max}W^{\text{sum}} + W^{\text{sum}} - W^{\max}}$$

for every $I \in \mathcal{I}[(-, L^{\max}), (k, W^{\text{sum}}, W^{\max}), \alpha]$. □

We can show that the bound given in Lemma 5.5.14 is tight by constructing a worst-case instance.

Lemma 5.5.15. *Let $L^{\max} \in \mathbb{N}$, $\alpha \in [0,1)$ and a feasible $(k, W^{\text{sum}}, W^{\max}) \in \mathbb{N}^3$ be given. Then*

$$PoS[(-, L^{\max}), (k, W^{\text{sum}}, W^{\max}), \alpha] = PoS^0[(-, L^{\max}), (k, W^{\text{sum}}, W^{\max}), \alpha]$$

$$= \frac{(L^{\max} + 1)W^{\text{sum}} - W^{\max} - (1-\alpha)W^{\max}}{\alpha L^{\max}W^{\text{sum}} + W^{\text{sum}} - W^{\max}}. \tag{5.8}$$

Proof. In Lemma 5.5.14 it is shown that

$$PoS^0[(-, L^{\max}), (k, W^{\text{sum}}, W^{\max}), \alpha] \leq \frac{(L^{\max}+1)W^{\text{sum}} - W^{\max} - (1-\alpha)W^{\max}}{\alpha L^{\max}W^{\text{sum}} + W^{\text{sum}} - W^{\max}}.$$

Hence, since

$$PoS[(-, L^{\max}), (k, W^{\text{sum}}, W^{\max}), \alpha] \leq PoS^0[(-, L^{\max}), (k, W^{\text{sum}}, W^{\max}), \alpha]$$

it remains to show that there is an instance $\hat{I} \in \mathcal{I}[(-, L^{\max}), (k, W^{\text{sum}}, W^{\max}), \alpha]$ such that

$$PoS(\hat{I}) = \frac{(L^{\max} + 1)W^{\text{sum}} - W^{\max} - (1-\alpha)W^{\max}}{\alpha L^{\max}W^{\text{sum}} + W^{\text{sum}} - W^{\max}}.$$

5.5 Analysis of the Iterative Heuristic for Arc Speed-Up

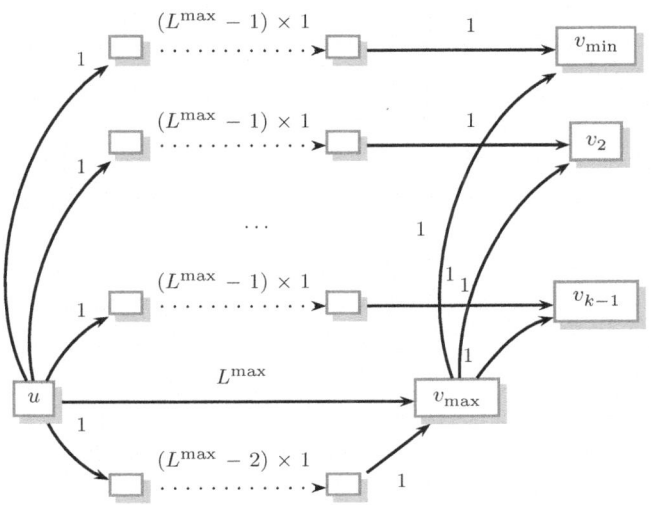

Fig. 5.3 Network N in the worst-case instance $\hat{I}$ for $PoS[(-, L^{\max}), (k, W^{\text{sum}}, W^{\max}), \alpha]$

We now construct such an instance $\hat{I}$.

Consider the network N with $(L^{\max} + 1)k$ nodes depicted in Fig. 5.3. The first $k - 1$ dotted arcs represent a sequence of $(L^{\max} - 1)$ arcs of length 1, directed in the direction of the dotted arcs. The dotted arc at the bottom represents a sequence of $(L^{\max} - 2)$ arcs of length 1, directed in the direction of the dotted arcs. The numbers written at the other arcs are the arc lengths. That is, there is one arc a^* of maximal length $L_{a^*} := L^{\max}$, all other arcs a have length $L_a := 1$.

We define a set of OD-pairs $\mathcal{OD} = \{(u, v_k) : k \in K\}$, i.e., all OD-pairs have the same origin u but different destinations v_k with arbitrary weights w_k for $(u, v_k') \in \mathcal{OD}$ such that $\sum_{(u, v_k) \in \mathcal{OD}} w_k = W^{\text{sum}}$ and $\max_{(u, v_k) \in \mathcal{OD}} w_k = W^{\max}$. Without loss of generality we assume that $w_1 \leq w_2 \leq \ldots, \leq w_k$. Therefore, we write $v_{\max} := v_k$ and $v_{\min} := v_1$.

This defines our instance $\hat{I} := (N, \mathcal{OD}, \alpha)$.

For every $(u, v_k) \in \mathcal{OD}$ there are two paths from u to v_i, namely, a path P_k' containing the arc of length $L^{\max}$ and a path P_k which contains only arcs of length 1. Note that $c(\alpha, P_k') = c(\alpha, P_k)$ for every $(u, v_k) \in \mathcal{OD}$.

Assume that the routing $\mathcal{R}^0 := \{P_k : k \in K\}$ is chosen in step 1 of Algorithm 4. Then $\max_{a \in A}\{w(\mathcal{R})_a L_a\} = \max_{(u_i, v_i) \in \mathcal{OD}} w_{u_i v_i} = w_{u_k v_k}$ and in step 3 an arc a^0 of the lowermost path is chosen for speed-up. Hence

$$c^0(\hat{I}) = (L^{\max} + 1)W^{\text{sum}} - W^{\max} - (1 - \alpha)W^{\max}. \tag{5.9}$$

Then $\mathcal{R}^0 \in \mathrm{OPT}(a^0)$, hence we can assume that $\mathcal{R}^1 = \mathcal{R}^0$ and $a^1 = a^0$ in steps 5–7 of the first iteration which causes Algorithm 4 to stop in step 9 after the first iteration with solution $(\mathcal{R}^1, a^1) = (\mathcal{R}^0, a^0)$.

Clearly, the optimal solution in this example is $(a^*, \mathcal{R}')$ with $\mathcal{R}' := \{P'_k : k \in K\}$. Hence

$$c^*(\hat{I}) = c(a^*, \mathcal{R}') = \alpha L^{\max} W^{\mathrm{sum}} + W^{\mathrm{sum}} - W^{\max}. \tag{5.10}$$

Combining (5.9) and (5.10) we obtain

$$PoS(\hat{I}) = \frac{(L^{\max} + 1)W^{\mathrm{sum}} - W^{\max} - (1-\alpha)W^{\max}}{\alpha L^{\max} W^{\mathrm{sum}} + W^{\mathrm{sum}} - W^{\max}}.$$

□

We now consider more general classes of instances, obtaining a higher price of sequentiality. To this end, we regard $PoS[(-, L^{\max}), (k, W^{\mathrm{sum}}, W^{\max}), \alpha]$ as a function in its parameters $L^{\max}$, $(k, W^{\mathrm{sum}}, W^{\max})$ and α and maximize this function. Since

$$PoS^0[(-, L^{\max}), (k, W^{\mathrm{sum}}, W^{\max}), \alpha] = PoS[(-, L^{\max}), (k, W^{\mathrm{sum}}, W^{\max}), \alpha]$$

for every $L^{\max} \in \mathbb{N}$, feasible $(k, W^{\mathrm{sum}}, W^{\max}) \in \mathbb{N}^3$ and $\alpha \in [0, 1)$ as shown in Lemma 5.5.15, the equality $PoS^0[\mathcal{I}] = PoS[\mathcal{I}]$ also holds for all classes of instances for which we derive the price of sequentiality in the following sections. However, to avoid a lengthy notation, we omit the price of sequentiality after iteration 0 from the formulation of the statements given in the following.

Lemma 5.5.16. *For every $\alpha \in [0, 1)$ and every feasible $(k, W^{sum}, W^{max}) \in \mathbb{N}^3$, it holds that*

$$PoS[-, (k, W^{sum}, W^{max}), \alpha] = \frac{1}{\alpha}.$$

Proof. According to Lemma 5.5.15, for fixed $L^{\max}$, the price of sequentiality is given as

$$PoS[(-, L^{\max}), (k, W^{\mathrm{sum}}, W^{\max}), \alpha] = \frac{(L^{\max} + 1)W^{\mathrm{sum}} - W^{\max} - (1-\alpha)W^{\max}}{\alpha L^{\max} W^{\mathrm{sum}} + W^{\mathrm{sum}} - W^{\max}}.$$

We now regard the price of sequentiality as a function g in $L^{\max}$. In order to find the maximum of this function, we consider the first derivative. We have

$$\frac{d}{dL^{\max}} g(L^{\max}) = \frac{d}{dL^{\max}} \frac{(L^{\max} + 1)W^{\mathrm{sum}} - W^{\max} - (1-\alpha)W^{\max}}{\alpha L^{\max} W^{\mathrm{sum}} + W^{\mathrm{sum}} - W^{\max}}$$

5.5 Analysis of the Iterative Heuristic for Arc Speed-Up

$$= \frac{(1-\alpha)(W^{\text{sum}}(W^{\text{sum}} - W^{\text{max}}) + \alpha W^{\text{max}} W^{\text{sum}})}{[\alpha L^{\text{max}} W^{\text{sum}} + W^{\text{sum}} - W^{\text{max}}]^2}$$

$$> 0,$$

hence

$$PoS[-, (k, W^{\text{sum}}, W^{\text{max}}), \alpha] = \sup_{L^{\text{max}} \in \mathbb{N}} PoS[(-, L^{\text{max}}), (k, W^{\text{sum}}, W^{\text{max}}), \alpha]$$

$$= \lim_{L^{\text{max}} \to \infty} PoS[(-, L^{\text{max}}), (k, W^{\text{sum}}, W^{\text{max}}), \alpha]$$

$$= \lim_{L^{\text{max}} \to \infty} \frac{(L^{\text{max}} + 1)W^{\text{sum}} - W^{\text{max}} - (1-\alpha)W^{\text{max}}}{\alpha L^{\text{max}} W^{\text{sum}} + W^{\text{sum}} - W^{\text{max}}}$$

$$= \frac{1}{\alpha}.$$

□

In Lemma 5.5.9 we saw that for every instance I $\frac{\hat{c}(I)}{c^*(I)} \leq \frac{1}{\alpha}$. Thus, Lemma 5.5.16 reveals that if we do not impose any restrictions on the networks, i.e., restrict the arc lengths or the number of nodes, in the considered classes of instances, the worst-case behavior of Algorithm 4 is asymptotically not better than not choosing any arc for speed-up. This, of course, is due to the fact that for increasing path lengths, the impact on the overall travel time of speeding up a single arc of length 1 (as done in the worst-case instance constructed in the proof of Lemma 5.5.15) decreases.

We hence keep L^{max} fixed in the following and consider the subclasses of instances with less restrictions on the set of OD-pairs.

Since (5.8) does not depend on k, the following corollary to Lemma 5.5.15 follows directly.

Corollary 5.5.17. *Let $L^{\text{max}} \in \mathbb{N}$, $\alpha \in [0, 1)$ and $(W^{\text{sum}}, W^{\text{max}}) \in \mathbb{N}^2$ with $W^{\text{sum}} \geq W^{\text{max}}$ be given. Then*

$$PoS[(-, L^{\text{max}}), (-, W^{\text{sum}}, W^{\text{max}}), \alpha] = \frac{(L^{\text{max}} + 1)W^{\text{sum}} - W^{\text{max}} - (1-\alpha)W^{\text{max}}}{\alpha L^{\text{max}} W^{\text{sum}} + W^{\text{sum}} - W^{\text{max}}}. \tag{5.11}$$

Lemma 5.5.18. *For every $L^{\text{max}} \in \mathbb{N}$, $(W^{\text{sum}}, k) \in \mathbb{N}^2$ and $\alpha \in [0, 1)$, we have*

$$PoS[(-, L^{\text{max}}), (k, W^{\text{sum}}, -), \alpha]$$

$$= \begin{cases} \frac{(L^{\text{max}}+1)W^{\text{sum}} - \lceil \frac{W^{\text{sum}}}{k} \rceil - (1-\alpha)\lceil \frac{W^{\text{sum}}}{k} \rceil}{\alpha L^{\text{max}} W^{\text{sum}} + W^{\text{sum}} - \lceil \frac{W^{\text{sum}}}{k} \rceil} & \text{if } (1-\alpha)L^{\text{max}} < 1 \\ \frac{W^{\text{sum}}(L^{\text{max}} - (1-\alpha)) + (2-\alpha)(k-1)}{\alpha L^{\text{max}} W^{\text{sum}} + (k-1)} & \text{otherwise.} \end{cases} \tag{5.12}$$

Proof. For fixed $W^{\max}$,

$$PoS[(-, L^{\max}), (k, W^{\text{sum}}, W^{\max}), \alpha] = \frac{(L^{\max} + 1)W^{\text{sum}} - W^{\max} - (1-\alpha)W^{\max}}{\alpha L^{\max} W^{\text{sum}} + W^{\text{sum}} - W^{\max}}.$$

It holds that $W^{\max} \geq \frac{W^{\text{sum}}}{k}$ and $W^{\max} \leq W^{\text{sum}} - k + 1$. Hence

$$PoS[(-, L^{\max}), (k, W^{\text{sum}}, -), \alpha]$$
$$= \sup_{W^{\max} \in [\frac{W^{\text{sum}}}{k}, W^{\text{sum}} - k + 1] \cap \mathbb{N}} \frac{(L^{\max} + 1)W^{\text{sum}} - W^{\max} - (1-\alpha)W^{\max}}{\alpha L^{\max} W^{\text{sum}} + W^{\text{sum}} - W^{\max}}.$$

We now calculate the maximum of

$$g(W^{\max}) := \frac{(L^{\max} + 1)W^{\text{sum}} - W^{\max} - (1-\alpha)W^{\max}}{\alpha L^{\max} W^{\text{sum}} + W^{\text{sum}} - W^{\max}}$$

on $[\frac{W^{\text{sum}}}{k}, W^{\text{sum}} - k + 1]$.

$$\frac{dg}{dW^{\max}} = \frac{(1-\alpha)[(1-\alpha)L^{\max} - 1]W^{\text{sum}}}{[\alpha L^{\max} W^{\text{sum}} + W^{\text{sum}} - W^{\max}]^2} = \begin{cases} < 0 & \text{if } (1-\alpha)L^{\max} < 1 \\ \geq 0 & \text{otherwise.} \end{cases} \quad (5.13)$$

Hence, g is monotonously decreasing and takes its maximum in $\frac{W^{\text{sum}}}{k}$ if $(1-\alpha)L^{\max} < 1$ and is monotonously increasing and takes the maximum in $W^{\text{sum}} - (k+1)$ otherwise. Since we assume that $W^{\max} \in \mathbb{N}$ it follows that

$$PoS[(-, L^{\max}), (k, W^{\text{sum}}, W^{\max}), \alpha]$$
$$= \begin{cases} g(\lceil \frac{W^{\text{sum}}}{k} \rceil) & \text{if } (1-\alpha)L^{\max} < 1 \\ g(W^{\text{sum}} - k + 1) & \text{otherwise.} \end{cases}$$
$$= \begin{cases} \frac{(L^{\max}+1)W^{\text{sum}} - \lceil \frac{W^{\text{sum}}}{k} \rceil - (1-\alpha)\lceil \frac{W^{\text{sum}}}{k} \rceil}{\alpha L^{\max} W^{\text{sum}} + W^{\text{sum}} - \lceil \frac{W^{\text{sum}}}{k} \rceil} & \text{if } (1-\alpha)L^{\max} < 1 \\ \frac{W^{\text{sum}}(L^{\max} - (1-\alpha)) + (2-\alpha)(k-1)}{\alpha L^{\max} W^{\text{sum}} + (k-1)} & \text{otherwise.} \end{cases}$$

$\square$

Lemma 5.5.19. *For every $L^{\max} \in \mathbb{N}$, $k \in \mathbb{N}$ and $\alpha \in [0, 1)$ we have*

$$PoS[(-, L^{\max}), (k, -, -), \alpha] = \begin{cases} \frac{L^{\max}k + k - 2 + \alpha}{\alpha L^{\max} k + k - 1} & \text{if } (1-\alpha)L^{\max} < 1 \\ \frac{L^{\max} - (1-\alpha)}{\alpha L^{\max}} & \text{otherwise.} \end{cases} \quad (5.14)$$

5.5 Analysis of the Iterative Heuristic for Arc Speed-Up

Proof. • Suppose that $(1 - \alpha)L^{\max} < 1$. Since the function g, defined in the proof of Lemma 5.5.18, is monotonously decreasing, for fixed W^{sum}, we have

$$PoS[(-, L^{\max}), (k, W^{\text{sum}}, -), \alpha] = g\left(\left\lceil \frac{W^{\text{sum}}}{k} \right\rceil\right) \leq g\left(\frac{W^{\text{sum}}}{k}\right) = \frac{L^{\max}k + k - 2 + \alpha}{\alpha L^{\max}k + k - 1}.$$

Equality holds whenever W^{sum} is divisible by k. Hence,

$$PoS[(-, L^{\max}), (k, -, -), \alpha] = \sup_{W^{\text{sum}} \in \mathbb{N}} PoS[(-, L^{\max}), (k, W^{\text{sum}}, -), \alpha]$$

$$= \frac{L^{\max}k + k - 2 + \alpha}{\alpha L^{\max}k + k - 1}.$$

• Now suppose that $(1 - \alpha)L^{\max} \geq 1$. For fixed W^{sum} we have

$$PoS[(-, L^{\max}), (k, W^{\text{sum}}, -), \alpha] = \frac{W^{\text{sum}}(L^{\max} - (1 - \alpha)) + (2 - \alpha)(k - 1)}{\alpha L^{\max} W^{\text{sum}} + (k - 1)}$$

due to Lemma 5.5.18. We consider the price of sequentiality as a function g in W^{sum}:

$$g(W^{\text{sum}}) := \frac{W^{\text{sum}}(L^{\max} - (1 - \alpha)) + (2 - \alpha)(k - 1)}{\alpha L^{\max} W^{\text{sum}} + (k - 1)}.$$

We have

$$\frac{dg}{dW^{\text{sum}}} := \frac{(k - 1)(1 - \alpha)(L^{\max}(1 - \alpha) - 1)}{[\alpha L^{\max} W^{\text{sum}} + (k - 1)]^2} > 0,$$

hence g is monotonously increasing. Furthermore,

$$\lim_{W^{\text{sum}} \to \infty} g(W^{\text{sum}}) := \lim_{W^{\text{sum}} \to \infty} \frac{W^{\text{sum}}(L^{\max} - (1 - \alpha)) + (2 - \alpha)(k - 1)}{\alpha L^{\max} W^{\text{sum}} + (k - 1)}$$

$$= \frac{L^{\max} - (1 - \alpha)}{\alpha L^{\max}}.$$

It follows that

$$PoS[(-, L^{\max}), (k, -, -), \alpha] = \sup_{W^{\text{sum}} \in \mathbb{N}} PoS[(-, L^{\max}), (k, W^{\text{sum}}, -), \alpha]$$

$$= \lim_{W^{\text{sum}} \to \infty} PoS[(-, L^{\max}), (k, W^{\text{sum}}, -), \alpha]$$

$$= \frac{L^{\max} - (1 - \alpha)}{\alpha L^{\max}}.$$

□

If we do not impose any restrictions on the set of OD-pairs, that is, we are looking at the class of instances $\mathcal{I}[(-, L^{\max}), -, \alpha]$ with maximal arc length $L^{\max}$ and speed-up factor α, we have

Lemma 5.5.20. *For every $L^{max} \in \mathbb{N}$ it holds that*

$$PoS[(-, L^{max}), -, \alpha] = \begin{cases} \frac{L^{max}+1}{\alpha L^{max}+1} & \text{if } (1-\alpha)L^{max} < 1 \\ \frac{L^{max}-(1-\alpha)}{\alpha L^{max}} & \text{otherwise.} \end{cases}$$

Proof. For $(1 - \alpha)L^{\max} \geq 1$ there is nothing to show. We hence assume that

$$(1 - \alpha)L^{\max} < 1.$$

Due to Lemma 5.5.19, we have $PoS[(-, L^{\max}), (k, -, -), \alpha] = \frac{L^{\max}k+k-2+\alpha}{\alpha L^{\max}k+k-1}$. Regarding the price of sequentiality as a function g of k we obtain

$$\frac{d}{dk}g(k) = \frac{d}{dk} \frac{L^{\max}k+k-2+\alpha}{\alpha L^{\max}k+k-1} = \frac{(1-\alpha)(1-(1-\alpha)L^{\max})}{[\alpha L^{\max}k+k-1]^2} > 0$$

Hence, g increases monotonously and asymptotically behaves like $\frac{L^{\max}+1}{\alpha L^{\max}+1}$. Hence, in this case

$$PoS[(-, L^{\max}), -, \alpha] = \sup_{k \in \mathbb{N}} PoS[(-, L^{\max}), (k, -, -), \alpha]$$

$$= \lim_{k \to \infty} PoS[(-, L^{\max}), (k, -, -), \alpha]$$

$$= \lim_{k \to \infty} \frac{L^{\max}k+k-2+\alpha}{\alpha L^{\max}k+k-1}$$

$$= \frac{L^{\max}+1}{\alpha L^{\max}+1}.$$

□

The results of this section are summarized in Table 5.1.

5.5.4.2 Price of Sequentiality for Networks of Arbitrary Size and Unrestricted α

Looking once more at Lemma 5.5.15, we observe that we can say something about the class of instances $\mathcal{I}[(-, L^{\max}), (k, W^{\text{sum}}, W^{\max}), -]$ where the speed-up factor $\alpha \in [0, 1)$ is not fixed.

Lemma 5.5.21. *For every $L^{max} \in \mathbb{N}$ and every feasible $(k, W^{sum}, W^{max}) \in \mathbb{N}^3$, it holds that*

5.5 Analysis of the Iterative Heuristic for Arc Speed-Up

Table 5.1 Price of sequentiality for the classes of instances considered in Sect. 5.5.4.1

	$\alpha > 1 - \frac{1}{L^{\max}}$	$\alpha \leq 1 - \frac{1}{L^{\max}}$
$PoS[(-, L^{\max}), (k, W^{\text{sum}}, W^{\max}), \alpha]$ $= PoS[(-, L^{\max}), (-, W^{\text{sum}}, W^{\max}), \alpha]$	$\dfrac{(L^{\max}+1)W^{\text{sum}} - W^{\max} - (1-\alpha)W^{\max}}{\alpha L^{\max} W^{\text{sum}} + W^{\text{sum}} - W^{\max}}$	$\dfrac{(L^{\max}+1)W^{\text{sum}} - W^{\max} - (1-\alpha)W^{\max}}{\alpha L^{\max} W^{\text{sum}} + W^{\text{sum}} - W^{\max}}$
$PoS[(-, L^{\max}), (k, W^{\text{sum}}, -), \alpha]$	$\dfrac{(L^{\max}+1)W^{\text{sum}} - \left\lceil \frac{W^{\text{sum}}}{k} \right\rceil - (1-\alpha)\left\lceil \frac{W^{\text{sum}}}{k} \right\rceil}{\alpha L^{\max} W^{\text{sum}} + W^{\text{sum}} - \left\lceil \frac{W^{\text{sum}}}{k} \right\rceil}$	$\dfrac{W^{\text{sum}}(L^{\max} - (1-\alpha)) + (2-\alpha)(k-1)}{\alpha L^{\max} W^{\text{sum}} + (k-1)}$
$PoS[(-, L^{\max}), (k, -, -), \alpha]$	$\dfrac{L^{\max} k + k - 2 + \alpha}{\alpha L^{\max} k + k - 1}$	$\dfrac{L^{\max} - (1-\alpha)}{\alpha L^{\max}}$
$PoS[(-, L^{\max}), -, \alpha]$	$\dfrac{L^{\max}+1}{\alpha L^{\max}+1}$	$\dfrac{L^{\max} - (1-\alpha)}{\alpha L^{\max}}$
$PoS[-, (k, W^{\text{sum}}, W^{\max}), \alpha]$	$\dfrac{1}{\alpha}$	$\dfrac{1}{\alpha}$

$$PoS[(-, L^{max}), (k, W^{sum}, W^{max}), -] = \frac{(L^{max} + 1)W^{sum} - 2W^{max}}{W^{sum} - W^{max}}.$$

Proof. For fixed α the price of sequentiality is given as

$$PoS[(-, L^{max}), (k, W^{sum}, W^{max}), \alpha] = \frac{(L^{max} + 1)W^{sum} - W^{max} - (1-\alpha)W^{max}}{\alpha L^{max} W^{sum} + W^{sum} - W^{max}}$$

due to Lemma 5.5.15. We regard the price of sequentiality as a function g in α. We have

$$\frac{d}{d\alpha} g(\alpha)$$

$$= \frac{d}{d\alpha} \frac{(L^{max} + 1)W^{sum} - W^{max} - (1-\alpha)W^{max}}{\alpha L^{max} W^{sum} + W^{sum} - W^{max}}$$

$$= -\frac{2L^{max} W^{sum}(W^{sum} - W^{max}) + W^{sum}(L^{max}(L^{max} - 1)W^{sum} - W^{max}) + (W^{max})^2}{[\alpha L^{max} W^{sum} + W^{sum} - W^{max}]^2}$$

(5.15)

We show that (5.15) is negative:

1. If $L^{max} \geq 2$, (5.15) is obviously negative.
2. If $L^{max} = 1$ we have

$$-\frac{2L^{max} W^{sum}(W^{sum} - W^{max}) + W^{sum}(L^{max}(L^{max} - 1)W^{sum} - W^{max}) + (W^{max})^2}{[\alpha L^{max} W^{sum} + W^{sum} - W^{max}]^2}$$

$$= -\frac{(W^{sum} - W^{max})^2 + W^{sum}(W^{sum} - W^{max})}{[\alpha W^{sum} + W^{sum} - W^{max}]^2}$$

$$< 0$$

Hence, the supremum of $g(\alpha)$ is attained for $\alpha = 0$. We obtain

$$PoS[(-, L^{max}), (k, W^{sum}, W^{max}), -] = \sup_{\alpha \in [0,1)} PoS[(-, L^{max}), (k, W^{sum}, W^{max}), \alpha]$$

$$= \sup_{\alpha \in [0,1)} \frac{(L^{max} + 1)W^{sum} - W^{max} - (1-\alpha)W^{max}}{\alpha L^{max} W^{sum} + W^{sum} - W^{max}}$$

$$= \frac{(L^{max} + 1)W^{sum} - W^{max} - W^{max}}{W^{sum} - W^{max}}.$$

□

Like in Sect. 5.5.4.1 we can derive the price of sequentiality for more general classes of instances.

5.5 Analysis of the Iterative Heuristic for Arc Speed-Up

Lemma 5.5.22. *For every $L^{max} \in \mathbb{N}$ and $(k, W^{sum}) \in \mathbb{N}^2$, we have*

$$PoS[(-, L^{max}), (k, W^{sum}, -), -] = \frac{(L^{max} - 1)W^{sum} + 2(k - 1)}{k - 1}$$

Proof. Due to Lemma 5.5.21, we have for fixed W^{max}

$$PoS[(-, L^{max}), (k, W^{sum}, W^{max}), -] = \frac{(L^{max} + 1)W^{sum} - 2W^{max}}{W^{sum} - W^{max}}.$$

Analogously to the calculations of Lemma 5.5.18 (with $\alpha = 0$) we see that the price of sequentiality increases monotonously in W^{max} and hence takes its maximum in $W^{sum} - (k - 1)$. It follows that

$$PoS[(-, L^{max}), (k, W^{sum}, -), -] = \frac{(L^{max} + 1)W^{sum} - 2(W^{sum} - (k - 1))}{W^{sum} - (W^{sum} - (k - 1))}$$
$$= \frac{(L^{max} - 1)W^{sum} + 2(k - 1)}{k - 1}.$$

$\square$

Corollary 5.5.23. *For every $L^{max} \in \mathbb{N}$ and $k \in \mathbb{N}$, we have*

$$PoS[(-, L^{max}), (k, -, -), -] = \infty.$$

Proof. This follows directly from Lemma 5.5.22, since

$$\lim_{W^{sum} \to \infty} \frac{(L^{max} - 1)W^{sum} + 2(k - 1)}{k - 1} = \infty.$$

$\square$

Hence, we can continue Table 5.2 for unrestricted α as shown in Table 5.1.

5.5.4.3 Price of Sequentiality for One OD-Pair and Networks of Fixed Size

As we have seen in Lemma 5.5.16,

$$\lim_{L^{max} \to \infty} PoS[(-, L^{max}), (k, W^{sum}, W^{max}), \alpha] = \frac{1}{\alpha},$$

which is no improvement compared to not choosing any arc for speed-up at all. However, having a closer look at the proof of Theorem 5.5.15, the worst-case example constructed there needs $n = (L^{max} + 1)k$ nodes.

Table 5.2 Price of sequentiality for the classes of instances considered in Sect. 5.5.4.2

$PoS[(-, L^{\max}), (k, W^{\text{sum}}, W^{\max}), -]$	
$= PoS[(-, L^{\max}), (k, W^{\text{sum}}, W^{\max}), -]$	$\frac{(L^{\max}+1)W^{\text{sum}} - 2W^{\max}}{W^{\text{sum}} - W^{\max}}$
$PoS[(-, L^{\max}), (k, W^{\text{sum}}, -), -]$	$\frac{(L^{\max}-1)W^{\text{sum}} + 2(k-1)}{k-1}$
$PoS[(-, L^{\max}), (k, -, -), -]$	∞
$PoS[(-, L^{\max}), -, -]$	∞
$PoS[-, -, -]$	∞

In this section we analyze whether we can find bounds on the price of sequentiality also for classes of instances with unbounded $L^{\max}$ if we bound the number of nodes n. For the sake of simplicity, we consider only instances with $|\mathcal{OD}| = 1$.

Lemma 5.5.24. *Let $I = (N, \mathcal{OD}, \alpha)$ be an instance of ASU with $|\mathcal{OD}| = 1$. Without loss of generality we can assume that $w_{uv} = 1$.*

Proof. Let $I = (N, \mathcal{OD}, \alpha)$ be an instance of ASU with $|\mathcal{OD}| = 1$ and $w_{uv} > 1$ for the OD-pair $(u, v) \in \mathcal{OD}$. We define $I' := (N, \mathcal{OD}', \alpha)$ with $\mathcal{OD}' = \{(u, v)\}$, and $w_{uv} = 1$.

Let $(a^*, \mathcal{R}^*)$ denote an optimal solution to I and $(a^i, \mathcal{R}^i)$ be a solution of Algorithm 4 after step 3 for $i = 0$ or after step 7 if $i > 0$. $\mathcal{R}^*$ consists of one path P^* and $\mathcal{R}^i$ consists of one path P^i. Hence

$$PoS(I) = \frac{\hat{c}(I)}{c^*(I)} = \frac{w_{uv} \sum_{a \in P^i} L_a^{a^i}}{w_{uv} \sum_{a \in P^*} L_a^{a^*}} = \frac{\sum_{a \in P^i} L_a^{a^i}}{\sum_{a \in P^*} L_a^{a^*}} = PoS(I').$$

□

We now improve the bound given in Lemma 5.5.12 for instances in $I \in \mathcal{I}[(n, -), (1, -, -), \alpha]$.

Lemma 5.5.25. *For every instance $I \in \mathcal{I}[(n, -), (1, -, -), \alpha]$, it holds that*

$$c(I) \leq c^i(I) \leq c^0(I) \leq \hat{c}(I) - (1 - \alpha) \left\lceil \frac{\hat{c}(I)}{n-1} \right\rceil \quad \forall i \in \mathbb{N}.$$

Proof. Let $I = (N, \mathcal{OD}, \alpha) \in \mathcal{I}[(n, -), (1, -, -), \alpha]$. Without loss of generality we assume that $w_{uv} = 1$ for the OD-pair $(u, v) \in \mathcal{OD}$ (see Lemma 5.5.24).

For the OD-pair $(u, v) \in \mathcal{OD}$ let P^0 denote the shortest path from u to v in N which is found in step 1 of Algorithm 4. Then $\hat{c}(I) = \sum_{a \in P^0} L_a$.

5.5 Analysis of the Iterative Heuristic for Arc Speed-Up

Since P^0 contains every node at most once, it contains at most $n-1$ arcs. This implies

$$\max_{a \in P^0} L_a \geq \frac{\hat{c}(I)}{n-1}.$$

Hence, for a^0 chosen in step 3 of Algorithm 4, it holds that $L_{a^0} \geq \frac{\hat{c}(I)}{n-1}$. Thus

$$c^0(I) \leq \hat{c}(I) - (1-\alpha)\left\lceil \frac{\hat{c}(I)}{n-1} \right\rceil.$$

$\square$

In the following Lemma 5.5.26 we see that we can indeed find better bounds on the price of sequentiality if the number of nodes n is bounded.

Lemma 5.5.26. *For every $n \in \mathbb{N}$, $L^{max} \geq n-1$ and $\alpha \in [0, 1)$, we have*

$$PoS^0[(n, L^{max}), (1, -, -), \alpha] \leq \frac{(n-1)-(1-\alpha)}{(n-1)\alpha}.$$

Proof. Consider an instance $I = (N, \mathcal{OD}, \alpha) \in \mathcal{I}[(n, L^{max}), (1, -, -), \alpha]$ with $L^{max} \geq n-1$. Without loss of generality we assume that $w_{uv} = 1$ for the OD-pair $(u, v) \in \mathcal{OD}$ (see Lemma 5.5.24). We define a function $g : [1, (n-1)L^{max}] \to \mathbb{R}$ as the ratio between the bounds on the objective value found using Algorithm 4 and on the optimal objective value.

$$g(s) = \begin{cases} \dfrac{s - (1-\alpha)\left\lceil \frac{s}{n-1} \right\rceil}{s\alpha} & \text{if } s \leq L^{max} \\[2ex] \dfrac{s - (1-\alpha)\left\lceil \frac{s}{n-1} \right\rceil}{s - L^{max}(1-\alpha)} & \text{if } L^{max} < s \leq (n-1)L^{max}. \end{cases}$$

Due to Lemma 5.5.25 and 5.5.10 and since every simple path in N contains at most $n-1$ arcs, we see that

$$PoS^0[(n, L^{max}), (1, -, -), \alpha] \leq \max_{s \in [1, (n-1)L^{max}]} g_{n, L^{max}, \alpha}(s).$$

Note that for the break point $s = L^{max}$, we have

$$g(L^{max}) = \frac{L^{max} - (1-\alpha)\left\lceil \frac{L^{max}}{n-1} \right\rceil}{L^{max}\alpha} = \frac{L^{max} - (1-\alpha)\left\lceil \frac{L^{max}}{n-1} \right\rceil}{L^{max} - L^{max}(1-\alpha)} = \lim_{s \searrow L^{max}} g(s). \tag{5.16}$$

We now show that

$$\max_{s\in[1,(n-1)L^{\max}]} g(s) = \frac{(n-1)-(1-\alpha)}{(n-1)\alpha}.$$

To this end, we maximize $g(s)$ both on $[1, L^{\max}]$ and $(L^{\max}, (n-1)L^{\max}]$ and compare the results.

1. We consider the function g on $[1, L^{\max}]$. We note that g is continuous on intervals for which $\lceil \frac{s}{n-1} \rceil$ is constant. Consider g on such an interval $((b-1)(n-1), b(n-1)]$, where $\lceil \frac{s}{n-1} \rceil = b$. On this interval, g is given as

$$g(s) = \frac{s-(1-\alpha)b}{s\alpha} = \frac{1}{\alpha} - \frac{(1-\alpha)b}{s\alpha}.$$

Since $\alpha \in [0, 1)$, this function is monotonously increasing in s and thus takes its maximum on the right end of the interval $((b-1)(n-1), b(n-1)]$, i.e., for $s = b(n-1)$.

In order to find the maximum of g on $[1, L^{\max}]$, we compare the maximum values for all such intervals. We note that for every $b \in \mathbb{N}$

$$g(b(n-1)) = \frac{b(n-1)-(1-\alpha)b}{b(n-1)\alpha} = \frac{(n-1)-(1-\alpha)}{(n-1)\alpha}$$

and thus does not depend on b.

Hence,

$$\sup_{s\in[1,L^{\max}]} \frac{s-(1-\alpha)b}{s\alpha} = \sup_{b\in\mathbb{Z}\cap[1,\frac{L^{\max}}{n-1}]} \frac{(n-1)b-(1-\alpha)b}{(n-1)b\alpha} = \frac{(n-1)-(1-\alpha)}{(n-1)\alpha}.$$

(We remark that $[1, \frac{L^{\max}}{n-1}] \neq \emptyset$ since we claimed that $L^{\max} \geq n-1$.)

2. Now we consider the function g on $(L^{\max}, (n-1)L^{\max}]$. Also here, g is continuous on intervals for which $\lceil \frac{s}{n-1} \rceil$ is constant and again we first maximize g on such an interval $((b-1)(n-1), b(n-1)]$ with $\lceil \frac{s}{n-1} \rceil = b$ for a fixed $b \in [\lfloor \frac{L^{\max}}{n-1} \rfloor, L^{\max}] \cap \mathbb{Z}$.

Since

$$\frac{d}{ds} \frac{s-(1-\alpha)b}{s-L^{\max}(1-\alpha)} = -\frac{(1-\alpha)(L^{\max}-b)}{[s-L^{\max}(1-\alpha)]^2} \begin{cases} = 0 \text{ if } b = L^{\max}, \\ < 0 \text{ otherwise} \end{cases},$$

g is monotonously decreasing on every interval $((b-1)(n-1), b(n-1)]$ for

$$b \in [\lfloor \frac{L^{\max}}{n-1} \rfloor, L^{\max} - 1] \cap \mathbb{Z}$$

5.5 Analysis of the Iterative Heuristic for Arc Speed-Up

while it is constant on $((L^{\max} - 1)(n - 1), (n - 1)L^{\max}]$, i.e., in the case of $b = L^{\max}$.

Thus,

$$g(s) \leq \lim_{s \searrow b(n-1)} g(s) = \frac{b(n-1)-(1-\alpha)b}{b(n-1)-L^{\max}(1-\alpha)} \quad \forall s \in ((b-1)(n-1), b(n-1)].$$

Hence,

$$\sup_{s \in (L^{\max},(n-1)L^{\max}]} g(s) = \sup_{b \in [\lfloor \frac{L^{\max}}{n-1} \rfloor, L^{\max}] \cap \mathbb{Z}} \frac{b(n-1)-(1-\alpha)b}{b(n-1)-L^{\max}(1-\alpha)}$$

We now maximize $h(b) := \frac{b(n-1)-(1-\alpha)b}{b(n-1)-L^{\max}(1-\alpha)}$ on $[\lceil \frac{L^{\max}}{n-1} \rceil, L^{\max}]$.
Since we can assume that $n > 2$

$$\frac{d}{db}h(b) = \frac{d}{db}\frac{b(n-1)-(1-\alpha)b}{b(n-1)-L^{\max}(1-\alpha)} = -\frac{L^{\max}(1-\alpha)(n-2+\alpha)}{[b(n-1)-L^{\max}(1-\alpha)]^2} < 0, \quad (5.17)$$

i.e., $h(b)$ is monotonously decreasing.

Since $\frac{L^{\max}}{n-1}$ may be fractional, we have two candidates for the supremum of g on

$$(L^{\max}, (n-1)L^{\max}],$$

namely

$$z_1 := \lim_{s \searrow L^{\max}} g(s) \text{ and } z_2 := h\left(\left\lceil \frac{L^{\max}}{n-1} \right\rceil\right) = g\left(\left\lceil \frac{L^{\max}}{n-1} \right\rceil (n-1)\right).$$

According to (5.16) and (1.)

$$\lim_{s \searrow L^{\max}} g(s) = g(L^{\max}) \leq \frac{(n-1)-(1-\alpha)}{(n-1)\alpha}. \quad (5.18)$$

3. Thus, as a last step, we compare the result from (1.) to

$$h\left(\left\lceil \frac{L^{\max}}{n-1} \right\rceil\right) = \frac{\left\lceil \frac{L^{\max}}{n-1} \right\rceil (n-2+\alpha)}{\left\lceil \frac{L^{\max}}{n-1} \right\rceil (n-1) - L^{\max}(1-\alpha)}.$$

We show that

$$\frac{(n-1)-(1-\alpha)}{(n-1)\alpha} \geq \frac{\left\lceil \frac{L^{\max}}{n-1} \right\rceil (n-2+\alpha)}{\left\lceil \frac{L^{\max}}{n-1} \right\rceil (n-1) - L^{\max}(1-\alpha)}.$$

Assume that

$$\frac{(n-1)-(1-\alpha)}{(n-1)\alpha} < \frac{\left\lceil\frac{L^{\max}}{n-1}\right\rceil(n-2+\alpha)}{\left\lceil\frac{L^{\max}}{n-1}\right\rceil(n-1) - L^{\max}(1-\alpha)}.$$

Dividing by $(n-2+\alpha)$ and taking the inverse, we obtain

$$\frac{\left\lceil\frac{L^{\max}}{n-1}\right\rceil(n-1) - L^{\max}(1-\alpha)}{\left\lceil\frac{L^{\max}}{n-1}\right\rceil} < (n-1)\alpha.$$

We subtract $\left[\frac{L^{\max}}{\left\lceil\frac{L^{\max}}{n-1}\right\rceil}\alpha\right]$ on both sides and obtain the equivalent inequality

$$\frac{\left\lceil\frac{L^{\max}}{n-1}\right\rceil(n-1) - L^{\max}}{\left\lceil\frac{L^{\max}}{n-1}\right\rceil} < (n-1 - \frac{L^{\max}}{\left\lceil\frac{L^{\max}}{n-1}\right\rceil})\alpha.$$

Since

$$n-1 - \frac{L^{\max}}{\left\lceil\frac{L^{\max}}{n-1}\right\rceil} = \frac{L^{\max}}{\left\lceil\frac{L^{\max}}{n-1}\right\rceil} - \frac{L^{\max}}{\left\lceil\frac{L^{\max}}{n-1}\right\rceil} > 0,$$

dividing by $(n-1 - \frac{L^{\max}}{\left\lceil\frac{L^{\max}}{n-1}\right\rceil})$, we obtain the equivalent inequality

$$\alpha > \frac{(n-1)\left\lceil\frac{L^{\max}}{n-1}\right\rceil - L^{\max}}{(n-1)\left\lceil\frac{L^{\max}}{n-1}\right\rceil - L^{\max}} = 1,$$

which is a contradiction to the choice of $\alpha \in [0, 1)$.
□

In Lemma 5.5.26 we saw a bound on the price of sequentiality for instances in the class $\mathcal{I}[(n, L^{\max}), (1, -, -), \alpha]$ with $L^{\max} \geq n - 1$. The following lemma shows that this bound holds in general for all instances in $\mathcal{I}[(n, -), (1, -, -), \alpha]$.

Lemma 5.5.27. *For every $n \in \mathbb{N}$ and $\alpha \in [0, 1)$*

$$PoS^0[(n, -), (1, -, -), \alpha] \leq \frac{(n-1)-(1-\alpha)}{(n-1)\alpha}.$$

Proof. As we have seen in Lemma 5.5.19, for every $L^{\max} \in \mathbb{N}$

$$PoS[(-, L^{\max}), (1, -, -), \alpha] = \frac{L^{\max} - 1 + \alpha}{\alpha L^{\max}} = \frac{1}{\alpha} - \frac{1-\alpha}{\alpha L^{\max}}. \quad (5.19)$$

This function is monotonously increasing in $L^{\max}$; hence

5.5 Analysis of the Iterative Heuristic for Arc Speed-Up

$$PoS^0[(n,-),(1,-,-),\alpha]$$
$$= \sup_{L^{\max} \in \mathbb{N}} PoS[(n, L^{\max}), (1, -, -), \alpha]$$
$$= \sup\{ \sup_{L^{\max} \in [1, n-2]} PoS[(n, L^{\max}), (1, -, -), \alpha],$$
$$\sup_{L^{\max} \in [n-1, \infty)} PoS[(n, L^{\max}), (1, -, -), \alpha]\}$$
$$\leq \sup\{\frac{(n-2)-(1-\alpha)}{\alpha(n-2)}, \frac{(n-1)-(1-\alpha)}{\alpha(n-1)}\}$$
$$= \frac{(n-1)-(1-\alpha)}{\alpha(n-1)}.$$

□

We show that the bound provided by Lemma 5.5.27 is tight.

Lemma 5.5.28. *For every* $n \in \mathbb{N}$ *and* $\alpha \in [0, 1)$, *we have*

$$PoS[(n,-),(1,-,-),\alpha] = \frac{(n-1)-(1-\alpha)}{(n-1)\alpha}.$$

Proof. In Lemma 5.5.27 it is shown that

$$PoS[(n,-),(1,-,-),\alpha] \leq \frac{(n-1)-(1-\alpha)}{(n-1)\alpha}.$$

We now construct an instance $\hat{I} \in \mathcal{I}[(n,-),(1,-,-),\alpha]$ such that

$$PoS(\hat{I}) = \frac{(n-1)-(1-\alpha)}{(n-1)\alpha}.$$

Consider a network N with n nodes and n arcs as depicted in Fig. 5.4. The dotted arcs there represent a sequence of $(n-3)$ arcs of length $L_a := 1$, directed in the direction of the dotted arcs. The other arc, a^*, has arc length $L_{a^*} := n - 1$.

The considered OD-pair is (u, v). Due to Lemma 5.5.24, without loss of generality, we can assume that $w_{uv} = 1$.

Note that there are two paths P and P' from u to v. P' consists of a^* and P consists of the short arcs. We have $\hat{c}(P) = \hat{c}(P') = n - 1$. Assume that in step 1 of Algorithm 4 $\mathcal{R}^0 := \{P\}$ is chosen. Then in step 3, an arc a with length $L_a = 1$ is chosen and in steps 5, and 7 of the first iteration the solution remains unchanged. Hence, $c(I) = n - 1 - (1 - \alpha)$.

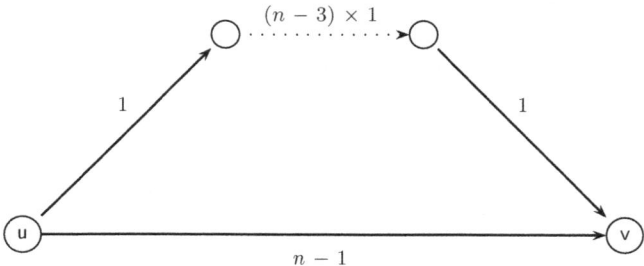

Fig. 5.4 Worst-case instance for $PoS[(n, L^{\max}), (1, -, -), \alpha']$ for $L^{\max} > n-1$

We observe that the optimal solution is $(a^*, \{P'\})$ with objective value $c(a^*, \{P'\}) = \alpha(n-1)$.

Thus,

$$PoS(\hat{I}) = \frac{(n-1) - (1-\alpha)}{(n-1)\alpha}.$$

□

Chapter 6
Conclusions and Outlook

In this book we investigated the problem of integrating routing decisions in network problems, in particular in the problems line planning, timetabling, and delay management. We mainly considered the following aspects, as stated in the introduction:

- How can existing models for line planning, timetabling, and delay management be extended to allow a free choice of the routing?
- What is the computational complexity of determining a line concept/timetable/ disposition timetable and a routing, simultaneously?
- To what extent can existing solution methods for line planning, timetabling, and delay management support the simultaneous calculation of both a line plan/timetable/disposition timetable and a routing?

With respect to the first question, we found that after some small modifications, the networks CGN, LN, and EAN used in preceding works on line planning, timetabling, or delay management are highly suitable for the formulation of the problems of uLPwR, scLPwR, cLPwR, TTwR, and DMwR, respectively, since routings can be represented as collections of paths in the corresponding networks.

The second aspect investigated in this book was the computational complexity of the considered problems. Table 6.1 summarizes the obtained results.

We found that after fixing either the routing or the line concept/timetable/ disposition timetable, a solution to uLPwR, TTwR, or DMwR which is optimal with regard to the fixed input can be found in polynomial time. This is not the case in scLPwR and cLPwR: finding a feasible routing for a given line concept, due to the added capacity restrictions, is a multi-commodity flow problem instead of a shortest path problem.

If only one OD-pair is considered, TTwR can be solved in polynomial time by solving a series of linear programs. In contrast, uLPwR and DMwR are already strongly NP-hard for only one OD-pair if no additional assumptions on the network

Table 6.1 Summary of complexity results

	uLPwR	scLPwR/cLPwR	TTwR	DMwR
Find corresponding network component for given routing	Trivial	Trivial	Solve one LP	Critical path method
Find corresponding routing for given network component	Shortest path problem in CGN	Multi-commodity flow problem (strongly NP-hard)	Shortest path problem in EAN	Shortest path problem in EAN and critical path method
Find feasible solution	Strongly NP-hard	Strongly NP-hard	Solve one LP and shortest path problem in EAN	Shortest path problem in EAN
Find optimal solution (1 OD-pair)	Strongly NP-hard, pseudopolynomially solvable by modified RCSP algorithm for instances with NLT property	Strongly NP-hard, even for instances with NLT property	Solve series of LPs	Strongly NP-hard, polynomially with NLT property path algorithm for instances with NTT property
Find optimal solution (same origin)	Strongly NP-hard, even for instances with OD-NLT property	Strongly NP-hard, even for instances with OD-NLT property	Strongly NP-hard	Strongly NP-hard

Special cases	See Table 2.1 for complexity analysis of uLPwR in linear PTNs	Strongly NP-hard under strongest considered restrictions	In P for arbitrary but fixed number of OD-pairs, or in case of only one departure and arrival event per OD-pair	—
Polynomial-time approximability (assuming $P \neq NP$)	Not possible	Not possible	No constant-factor appr. algorithm in general, trivial $(\max_{a \in A_{op}} \frac{u_a}{l_a})$-appr. algorithm	No constant-factor appr. algorithm in general, 2-appr. algorithm for one OD-pair trivial $(1+M)$-appr. algorithm

structure are made. However, for both problems we identified a restriction on the class of instances for which an optimal solution can be found using modified shortest path algorithms.

For all problems, it was not possible to extend the results of polynomial- or pseudopolynomial-time solvability to instances with more OD-pairs. Even if all OD-pairs have the same origin (and under the above-mentioned restrictions on the instances of uLPwR and DMwR), the problems uLPwR, TTwR, and DMwR are strongly NP-hard.

As to approximation algorithms, things do not look much easier. We saw that for uLPwR it is already strongly NP-hard to find a feasible solution; hence, we cannot hope for polynomial- or pseudopolynomial-time algorithms unless $P = NP$. Furthermore, we showed that (unless $P = NP$) there is no constant-factor approximation algorithm for TTwR (even if all OD-pairs have the same origin) or DMwR (even under the restriction that there are at most two OD-pairs). However, it is still an open question whether there exists an approximation algorithm for DMwR if we restrict the class of instances such that all OD-pairs have the same origin or if we the require that no OD-pair can enter a train more than once.

For uLPwR in linear networks, we extensively studied how the complexity of the problem depends on restrictions of the different input parameters. We were able to classify all defined instance classes to be solvable in either polynomial time or NP-hard; however, for some of the problems, a pseudopolynomial-time algorithm neither could be found nor could its existence be disproved. Anyway, when adding capacity restrictions, even under the strongest restrictions on the instances, the problems scLPwR and cLPwR turned out to be strongly NP-hard.

Although the problems uLPwR, scLPwR, cLPwR, TTwR, and DMwR were proven to be strongly NP-hard even under strong restrictions, there may be hidden parameters which, when fixed, allow polynomial-time solution methods. It may even be possible to find algorithms whose running time is superpolynomial only in some input parameters. The identification of such parameters and proofs of existence (or nonexistence) of such parameterized algorithms could further refine the classification in tractable and intractable problems made in this book.

In this book we focused on the exact optimization algorithms and only touched on approximation aspects. The development of approximation algorithms or the proof of nonexistence of such algorithms should be investigated in more detail.

As a third aspect we investigated whether existing solution methods for line planning, timetabling, and delay management can be modified to solve the problems when the passenger routing is integrated. Apart from specialized algorithms developed for restricted classes of instances, we applied two different general strategies to solve network problems with routing.

On the one hand, for all of the considered problems, existing integer programming formulations could be adapted to solve the problems with integrated routing. IP formulations for TTwR and DMwR could be obtained by the addition of multi-commodity flow constraints to IP formulations used in timetabling and delay management problems with fixed routing.

6 Conclusions and Outlook

Also the IP formulation for scLPwR from [SS06] uses this type of constraints. To overcome the "unrealistic" behavior of this model in presence of capacity constraints, in cLPwR we added implicit shortest path constraints to the IP formulation from [SS06], implying that a line concept can only be chosen as a solution if it allows all passengers to travel on shortest paths in the corresponding routing network. A comparison of the solution and the running time of scLPwR and cLPwR would be interesting, but is beyond the scope of this book.

In TTwR, we made use of the fact that a passenger's travel time depends only on the timetable and the chosen virtual activity, i.e., the trains chosen to start and end the journey at origin and destination which led to a different integer programming approach, based on virtual activities instead of flows.

Having a closer look at the structure of the constraint matrices of the flow-based formulations for uLPwR, scLPwR, TTwR, and DMwR, we observe that they have almost a block diagonal structure. For example, in the flow-based IP formulation for TTwR in Sect. 3.7, we have a submatrix of timetabling variables and timetable feasibility constraints, $|\mathcal{OD}|$ submatrices of routing variables and routing constraints, and only $2|\mathcal{OD}|$-constraints connecting these blocks. Hence, using the Dantzig–Wolfe decomposition method could lead to a better running time.

Also for the other IP formulations developed in this book, it should be investigated whether decomposition techniques can be applied or whether there is some other structure that can be exploited when solving these problems.

As a second solution approach, we proposed an iterative heuristic that alternatingly calculates the routing and the network component required as a solution to general network problems with routing. The use of this heuristic is reasonable if an optimal solution to the considered network problem with respect to a given routing or a given network component can be found much faster than an optimal solution for the actual problem and if a feasible initial solution is at hand. The performance of the heuristic was theoretically analyzed for TTwR and the arc speed-up problem (ASU). For many classes of instances of ASU, a tight bound on the price of sequentiality could be proven. However, for networks with restricted size but unrestricted arc lengths, the price of sequentiality was only identified for instances with only one OD-pair and is still open in the general case.

In [Anh12] the iterative heuristic was tested and compared to the integer programming formulations for TTwR where it showed good performance. Of course, it would be interesting to test it also on line planning, delay management, and other network problems with routing.

The integration of the passenger routing in the optimization process leads to a more realistic estimation of passengers' behavior and hence to better solutions. However, compared to real-world situations, the public transportation models considered in this book are highly simplified, neglecting, e.g., aspects like the capacity of tracks and stations or periodicity requirement on timetables. Hence, many extensions to the considered problems are possible and would make the models more applicable in practice.

In this book we assumed that passengers aim at minimizing their travel time. However, other aspects like the number of transfers, the costs, and the reliability of the chosen route play an important role in practice and should be taken into account in further work. When restrictions on the train capacity are taken into account, new models for route choice and more sophisticated solution techniques are required. A first attempt to include selfish routing in line planning is made by investigating cLPwR in Sect. 2.6; however, alternative models and methods should be taken into account. Furthermore, in practice most passengers would only travel by public transportation if the offered routes are fast and convenient enough. Hence, public transportation models which include not only passengers' route choice but also variable demand would be an interesting extension to the questions investigated in this book.

In this book, we regarded the problems of line planning, timetabling, and delay management independently. However, when these planning steps are operated one after another, suboptimal solutions are obtained. For example, in line planning, the line concepts were evaluated using only rough estimations of the changing times at stations. It was neglected that the "real" changing times depend on the timetable. Hence, it is possible that a line concept which is optimal with respect to the objective function specified in line planning leads to a worse timetable than a "suboptimal" line concept. Furthermore, timetables that are good with respect to the objective functions defined in timetabling tend to be vulnerable to delays. Hence, it could be helpful to have the next planning steps in mind when doing line planning or delay management or even to integrate some of the planning steps.

Frequently Used Notation

General Notation

	$\mathbb{N}$	Natural numbers, not including 0
	$\mathbb{N}_0$	Natural numbers, including 0
	$\mathbb{Z}$	Integer numbers
	$\mathbb{Q}$	Rational numbers
	$\mathbb{R}$	Real numbers
	$\mathbb{R}_0^+$	Nonnegative real numbers
	$\{i,j\}$	Undirected edge in a graph
	(i,j)	Directed edge in a digraph
	L_{ij}	Length of an undirected or directed edge in a graph
	I	Instance of a problem
	$\mathcal{I}$	Class of instances of a problem

Public Transportation Concepts

$\mathcal{PTN} = (S, E)$		Public transportation network (PTN) with set of stations S and set of tracks or edges E
	$\mathcal{L}$	Line pool with elements l
$\mathcal{L}(s)/\mathcal{L}(e)$		Lines at station s/edge e
$S(l)/E(l)$		Stations/edges visited by line l
	α_l	Line driving time function of line l
	$p_s^{ll'}$	Transfer penalty for a transfer from line l to line l' at station s
	β_l	Cost of line l
	b_l	Per-train cost of line l
	B	Budget
	f_l	Frequency of line l
	$\mathcal{L}'$	Line concept
	TR	Set of trains with elements tr
	$l(\text{tr})$	Line corresponding to train tr
	Cap	Capacity of the trains
	$\mathcal{C}$	Set of planned connections with elements $(\text{tr}_1, \text{tr}_2, s)$

Frequently Used Notation

	π	Timetable
	d	Source delays
	x	Disposition timetable

Change-and-Go Networks

	$N = (V, A)$	Change-and-go network (CGN)
	V_{travel}	Set of travel nodes
	$V_{\text{org}}/V_{\text{dest}}$	Set of origin nodes with elements $u^{\text{org}}/v^{\text{dest}}$
	$A_{\text{drive}}/A_{\text{trans}}/A_{\text{org}}/A_{\text{dest}}$	Set of driving/transfer/origin/destination
	$l(a)$	Line corresponding to arc a
	c_a	Length of arc $a \in A$
	n_N/m_N	Number of nodes/arcs in the CGN
	$N(\mathcal{L}') = (V(\mathcal{L}), A(\mathcal{L}'))$	Routing network for line concept $\mathcal{L}'$
	$N(\mathcal{R}) = (V(\mathcal{R}), A(\mathcal{R}))$	Routing network for routing $\mathcal{R}$

Line Networks

	$N_L = (V_L, A_L)$	Line network
	$A_{\text{trans}}/A_{\text{org}}/A_{\text{dest}}$	Set of transfer/origin/destination arcs
	$c_{ll'}$	Length of arc $(l, l') \in A_L$
	n_{N_L}/m_{N_L}	Number of nodes/arcs in the line network
	$N_L(\mathcal{L}') = (V_L(\mathcal{L}), A_L(\mathcal{L}'))$	Line-routing network for line concept $\mathcal{L}'$
	$N_L(\mathcal{R}) = (V_L(\mathcal{R}), A_L(\mathcal{R}))$	Line-routing network for routing $\mathcal{R}$
	$l(a)$	Line corresponding to arc a

Event-Activity Networks

	$\mathcal{N} = (\mathcal{E}, \mathcal{A})$	Event-activity network (EAN) consisting of events $\mathcal{E}$ and activities $\mathcal{A}$
	$\mathcal{E}_{\text{dep}}/\mathcal{E}_{\text{arr}}$	Set of departure/arrival events
	$\mathcal{E}_{\text{dep}}^k/\mathcal{E}_{\text{arr}}^k$	Set of departure/arrival events for OD-pair $(u_k, v_k) \in \mathcal{OD}$
	$\mathcal{A}_{\text{drive}}/\mathcal{A}_{\text{wait}}/\mathcal{A}_{\text{change}}$	Driving, waiting, changing activities
	$\mathcal{N}_{\text{op}} = (\mathcal{E}_{\text{op}}, \mathcal{A}_{\text{op}})$	Operational network with set of operational events $\mathcal{E}_{\text{op}}$ and set of operational activities $\mathcal{A}_{\text{op}}$
	l_a/u_a	Lower/upper bound on operational activity a
	s_a	Slack time on operational activity a
	$\mathcal{E}_{\text{org}}$	Set of origin events (in TTwR) with elements u^{org} or u_k^{org}
	$\mathcal{E}'_{\text{org}}$	Set of origin events (in DMwR) with elements $u^{\text{org}}(\sigma)$ or $u_k^{\text{org}}(\sigma_k)$
	$\mathcal{E}_{\text{dest}}$	Set of destination events with elements v^{dest} or v_k^{dest}
	$\mathcal{A}_{\text{org}}$	Set of origin activities (in TTwR)
	$\mathcal{A}'_{\text{org}}$	Set of origin activities (in DMwR)
	$\mathcal{A}_{\text{dest}}$	Set of destination activities
	$\mathcal{A}_{\text{fix}}$	Set of maintained changing activities
	$\text{tr}(e)/\text{tr}(a)$	Train to which event e/activity a belongs
	$\mathcal{A}_{\text{virt}}$	Virtual activities
	$l_k^{\text{down}}/l_k^{\text{upp}}$	Length of shortest path from u^{org} to v^{dest} according to lower/upper bounds
	$\mathcal{N}(\pi)$	Routing network for timetable π
	$\mathcal{N}(x)$	Routing network for disposition timetable x
	$P_{ij}^{\mathcal{A}_{\text{fix}}}$	Path from node i to node j defined by a set of maintained changing activities $\mathcal{A}_{\text{fix}}$

Frequently Used Notation

Routings and Functions

$\mathcal{OD}$ — Set of OD-pairs with elements (u,v) or $(u_k, v_k) \in S \times S$ with weight w_{uv}/w_k (in uLPwR/scLPwR/cLPwR/TTwR) or elements (u, v, σ) or $(u_k, v_k, \sigma_k) \in S \times S \times \mathbb{Z}$ with weight $w_{uv\sigma}/w_k$ (in DMwR)

P_{uv}/P_k — Path or line-route/train-route for OD-pair $(u,v)/(u_k, v_k)$ (in uLPwR/scLPwR/cLPwR/TTwR)

$P_{uv\sigma}/P_k$ — Path or ST-train-route for OD-pair $(u, v, \sigma)/(u_k, v_k)$ (in DMwR)

$\mathcal{R}_{uv}$ — Routing for OD-pair (u,v) with path weight w_{uv}^P for $P_{uv} \in \mathcal{R}_{uv}$ (in scLPwR/cLPwR)

$\mathcal{R}$ — Routing or line-routing/train-routing/ST-train-routing

$c(\mathcal{S}, P)$ — Travel time along path or line-route/train-route/ST-train-route P for network component $\mathcal{S}$ (e.g., line concept, timetable, disposition timetable)

$c(\mathcal{S}, \mathcal{R})$ — Travel time for line-routing/train-routing/ST-train-routing $\mathcal{R}$ for network component $\mathcal{S}$ (e.g., line concept, timetable, disposition timetable)

$\mathrm{OPT}(\mathcal{R})$ — Set of network components (e.g., line concept, timetable, disposition timetable) for routing $\mathcal{R}$

$\mathrm{OPT}(\mathcal{S})$ — Set of routings for network component $\mathcal{S}$ (e.g., line concept, timetable, disposition timetable)

Arc Speed-up

α — Speed-up factor
$\mathcal{R}$ — Routing
$N(a')$ — Speed-up network for arc a'

Price of Sequentiality

$PoS^i(I)/PoS(I)$ — Price of sequentiality (in iteration i) for instance I

$PoS_X^i(I)/PoS_X^{\mathrm{stop}}(I)$ — (in iteration i) for instance I using strategy X

$\widehat{PoS}_X^i(I)/\widehat{PoS}_X^{\mathrm{stop}}(I)$ — (in iteration i) for instance I using strategy X

$PoS^i(\mathcal{I})/PoS(\mathcal{I})$ — Price of sequentiality (in iteration i) for the class of instances $\mathcal{I}$

References

[ADOGT99] B. Adenso-Díaz, M. Oliva González, and P. González-Torre. On-line timetable rescheduling in regional train services. *Transportation Research Part B*, 33:387–398, 1999.

[AMO93] R. K. Ahuja, T. L. Magnanti, and J. B. Orlin. *Network flows*. Prentice Hall, Inc., 1993.

[AMS06] N. Alon, D. Moshkovitz, and S. Safra. Algorithmic construction of sets for k-restrictions. *ACM Trans. Algorithms*, 2(2):153–177, 2006.

[Anh12] J. Anhalt. Eine iterative Heuristik für aperiodische Fahrplangestaltung mit OD-Paaren. Master's thesis, Georg-August-Universität Göttingen, 2012. (in German).

[APW02] L. Anderegg, P. Penna, and P. Widmayer. Online train disposition: to wait or not to wait? *Electronic Notes in Theoretical Computer Science*, 66(6), 2002.

[Bau10] R. Bauer. *Theory and Engineering for Shortest Paths and Delay Management*. PhD thesis, Karlsruher Institut für Technologie, 2010.

[BC89] J. E. Beasley and N. Christofides. An algorithm for the resource constrained shortest path problem. *Networks*, 19(4):379–394, 1989.

[BGLR93] M. Bellare, S. Goldwasser, C. Lund, and A. Russel. Efficient probabilistically checkable proofs and applications to approximation. In *STOC '93 Proceedings of the twenty-fifth annual ACM symposium on Theory of computing*, pages 294–304, New York, USA, 1993. ACM.

[BGP07] R. Borndörfer, M. Grötschel, and M. E. Pfetsch. A column-generation approach to line planning in public transport. *Transportation Science*, 41(1):123–132, 2007.

[BGP08] R. Borndörfer, M. Grötschel, and M. E. Pfetsch. Models for line planning in public transport. In M. Hickman, P. Mirchandani, and S. Voß, editors, *Computer-aided Systems in Public Transport*, volume 600 of *Lecture Notes in Economics and Mathematical Systems*, pages 363–378. Springer Berlin Heidelberg, 2008.

[BHLS07] A. Berger, R. Hoffmann, U. Lorenz, and S. Stiller. Online delay management: PSPACE hardness and simulation. Technical Report ARRIVAL-TR-0097, ARRIVAL Project, 2007.

[BK10] R. Burdett and E. Kozan. A sequencing approach for train timetabling. *OR Spectrum*, 32(1):163–193, 2010.

[BK12] R. Borndörfer and M. Karbstein. A Direct Connection Approach to Integrated Line Planning and Passenger Routing. In D. Delling and L. Liberti, editors, *12th Workshop on Algorithmic Approaches for Transportation Modelling, Optimization, and Systems*, volume 25 of *OpenAccess Series in Informatics (OASIcs)*, pages 47–57, Dagstuhl, Germany, 2012. Schloss Dagstuhl–Leibniz-Zentrum fuer Informatik.

[BKZ97] M. R. Bussieck, P. Kreuzer, and U. T. Zimmermann. Optimal lines for railway systems. *European Journal of Operational Research*, 96(1):54–63, 1997.

[BLL04] M. R. Bussieck, T. Lindner, and M. E. Lübbecke. A fast algorithm for near cost optimal line plans. *Mathematical Methods of Operations Research*, 59:205–220, 2004.

[BLNN98] U. Brännlund, P. O. Lindberg, A. Nou, and J. E. Nilsson. Railway timetabling using lagrangian relaxation. *Transportation Science*, 32(4):358, 1998.

[BLZ97] M. R. Bussieck, T. Lindner, and U. T. Zimmermann. Discrete optimization in public rail transport. *Mathematical Programming*, 79:415–444, 1997.

[BN10] R. Borndörfer and M. Neumann. Models for line planning with transfers. Technical Report 10-11, ZIB, Takustr.7, 14195 Berlin, 2010.

[BNP09] R. Borndörfer, M. Neumann, and M. E. Pfetsch. The line connectivity problem. In B. Fleischmann, K.-H. Borgwardt, R. Klein, and A. Tuma, editors, *Operations Research Proceedings 2008*, pages 557–562. Springer Berlin Heidelberg, 2009.

[Bus98] M. Bussieck. *Optimal Lines in Public Rail Transport*. PhD thesis, Technische Universität Braunschweig, 1998.

[CCT10] V. Cacchiani, A. Caprara, and P. Toth. Non-cyclic train timetabling and comparability graphs. *Operations Research Letters*, 38(3):179–184, 2010.

[CDD$^+$07] S. Cicerone, G. D'Angelo, G. Di Stefano, D. Frigioni, and A. Navarra. On the interaction between robust timetable planning and delay management. Technical Report ARRIVAL-TR-0116, ARRIVAL project, 2007.

[CDPP12] F. Corman, A. D'Ariano, D. Pacciarelli, and M. Pranzo. Bi-objective conflict detection and resolution in railway traffic management. *Transportation Research Part C*, 20(1):79–94, 2012.

[CDS$^+$09] S. Cicerone, G. D'Angelo, G. Di Stefano, D. Frigioni, A. Navarra, M. Schachtebeck, and A. Schöbel. Recoverable robustness in shunting and timetabling. In R. K. Ahuja, R. H. Möhring, and C.D. Zaroliagis, editors, *Robust and Online Large-Scale Optimization*, volume 5868 of *Lecture Notes in Computer Science*, pages 28–60. Springer, Heidelberg, 2009.

[CF95] I. Constantin and M. Florian. Optimizing frequencies in a transit network: a nonlinear bi-level programming approach. *International Transactions in Operational Research*, 2(2):149, 1995.

[CFT02] A. Caprara, M. Fischetti, and P. Toth. Modeling and solving the train timetabling problem. *Operations Research*, 50(5):851–861, 2002.

[Chu84] T. A. Chua. The planning of urban bus routes and frequencies: A survey. *Transportation*, 12:147–172, 1984.

[CKT10] A. Caprara, L. Kroon, and P. Toth. Optimization problems in passenger railway systems. In J. J. Cochran, L. A. Cox, P. Keskinocak, J. P. Kharoufeh, and J. C. Smith, editors, *Wiley Encyclopedia of Operations Research and Management Science*. John Wiley & Sons, Inc., 2010.

[CR11] R. Cordone and F. Redaelli. Optimizing the demand captured by a railway system with a regular timetable. *Transportation Research Part B: Methodological*, 45(2):430–446, 2011.

[CS07] C. Conte and A. Schöbel. Identifying dependencies among delays. In *proceedings of IAROR 2007*, 2007. ISBN 978-90-78271-02-4.

[CT12] V. Cacchiani and P. Toth. Nominal and robust train timetabling problems. *European Journal of Operational Research*, 219(3):727–737, 2012.

[CTV98] J.-F. Cordeau, P. Toth, and D. Vigo. A survey of optimization models for train routing and scheduling. *Transportation Science*, 32(4):380–404, November 1998.

[CvDZ98] M. T. Claessens, N. M. van Dijk, and P. J. Zwaneveld. Cost optimal allocation of rail passenger lines. *European Journal of Operational Research*, 110:474–489, 1998.

[DCDH12] T. Dollevoet, F. Corman, A. D'Ariano, and D. Huisman. An iterative optimization framework for delay management and train scheduling. Technical report, Econometric Institute Report EI2012-10,Erasmus University Rotterdam, 2012.

References

[dDD98] R. de Vries, B. De Schutter, and B. De Moor. On max-algebraic models for transportation networks. In *Proceedings of the International Workshop on Discrete Event Systems*, pages 457–462, Cagliari, Italy, 1998.

[DH07] G. Desaulniers and M. D. Hickmann. Public transit. In *Handbook in OR & MS*, volume 14, pages 69–127. Elsevier, 2007.

[DH11] T. Dollevoet and D. Huisman. Fast heuristics for delay management with passenger rerouting. Technical report, Econometric Institute Report EI2011-35,Erasmus University Rotterdam, 2011.

[DHSS09] T. Dollevoet, D. Huisman, M. Schmidt, and A. Schoebel. Delay Management with Re-Routing of Passengers. In J. Clausen and G. Di Stefano, editors, *9th Workshop on Algorithmic Approaches for Transportation Modeling, Optimization, and Systems (ATMOS'09)*, volume 12 of *OpenAccess Series in Informatics (OASIcs)*, Dagstuhl, Germany, 2009. Schloss Dagstuhl–Leibniz-Zentrum fuer Informatik.

[DHSS12] T. Dollevoet, D. Huisman, M. Schmidt, and A. Schöbel. Delay management with rerouting of passengers. *Transportation Science*, 46(1):74–89, February 2012.

[Die78] H. Dienst. *Linienplanung im spurgeführten Personenverkehr mit Hilfe eines heuristischen Verfahrens*. PhD thesis, Technische Universität Braunschweig, 1978. (in German).

[Die06] R. Diestel. *Graph Theory*. Springer, 3rd edition, 2006.

[dMI06] L. dell'Olio, J. L. Moura, and A. Ibeas. Bi-level mathematical programming model for locating bus stops and optimizing frequencies. *Transportation Research Record: Journal of the Transportation Research Board*, 1971, 2006.

[DPP07] A. D'Ariano, D. Pacciarelli, and M. Pranzo. A branch and bound algorithm for scheduling trains in a railway network. *European Journal of Operational Research*, 183(2):643–657, 2007.

[DSS11] T. Dollevoet, M. Schmidt, and A. Schöbel. Delay management including capacities of stations. In A. Caprara and S. Kontogiannis, editors, *11th Workshop on Algorithmic Approaches for Transportation Modelling, Optimization, and Systems (ATMOS)*, volume 20 of *OASIcs*, pages 88–99, Dagstuhl, Germany, 2011. Schloss Dagstuhl–Leibniz-Zentrum für Informatik.

[DV95] J. R. Daduna and S. Voß. Practical experiences in schedule synchronization. In J. R. Daduna, I. Branco, and J. M. Pinto Paixão, editors, *Computer-aided transit scheduling: proceedings of the Sixth International Workshop on Computer-aided Scheduling of Public Transport*, volume 430 of *Lecture Notes in Economics and Mathematical Systems*, 1995.

[Dv01] B. De Schutter and T. van den Boom. Model predictive control for railway networks. In *Proceedings of the 2001 IEEE/ASME International Conference on Advanced Intelligent Mechatronics, Como, Italy*, pages 105–110, 2001.

[Fei98] U. Feige. A threshold of ln n for approximating set cover. *Journal of the ACM*, 45:634–652, July 1998.

[FM09] M. Fischetti and M. Monaci. Light robustness. In R. K. Ahuja, R.H. Möhring, and C.D. Zaroliagis, editors, *Robust and online large-scale optimization*, volume 5868 of *Lecture Note on Computer Science*, pages 61–84. Springer, 2009.

[FSZ09] M. Fischetti, S. Salvagnin, and A. Zanette. Fast approaches to improve the robustness of a railway timetable. *Transportation Science*, 43:321–335, 2009.

[FT87] M. L. Fredman and R. E. Tarjan. Fibonacci heaps and their uses in improved network optimization algorithms. *Journal of the ACM*, 34:596–615, 1987.

[Fuh08] R. Fuhse. *Heuristiken zur Erstellung von Linienkonzepten*. Master's thesis, Georg-August-Universität Göttingen, 2008. (in German).

[Gat07] M. Gatto. *On the Impact of Uncertainty on Some Optimization Problems: Combinatorial Aspects of Delay Management and Robust Online Scheduling*. PhD thesis, ETH Zürich, 2007.

[GGJ+04] M. Gatto, B. Glaus, R. Jacob, L. Peeters, and P. Widmayer. Railway delay management: Exploring its algorithmic complexity. In *Proc. 9th Scandinavian Workshop on Algorithm Theory (SWAT)*, volume 3111 of *Lecture Notes in Computer Science*, pages 199–211, 2004.

[GH08] V. Guihaire and J.-T. Hao. Transit network design and scheduling: A global review. *Transportation Research Part E*, 42:1251–1273, 2008.

[GHMH+13] M. Goerigk, S. Heße, M. Müller-Hannemann, M. Schmidt, and A. Schöbel. Recoverable robust timetable information. In 13th Workshop on Algorithmic Approaches for Transportation Modelling, Optimization, and Systems (ATMOS), volume 33 of OpenAccess Series in Informatics (OASIcs), pages 1–14. Schloss Dagstuhl–Leibniz-Zentrum fuer Informatik, 2013.

[Gib85] A. Gibbons. *Algorithmic graph theory*. Cambridge University Press, 1985.

[GJ79] M. R. Garey and D. S. Johnson. *Computers and Intractability—A Guide to the Theory of NP-Completeness*. Freeman, San Francisco, 1979.

[GJP+04] M. Gatto, R. Jacob, L. Peeters, B. Weber, and P. Widmayer. Theory on the tracks: A selection of railway optimization problems. *Bulletin of the EATCS*, 84:41–70, 2004.

[GJPS05] M. Gatto, R. Jacob, L. Peeters, and A. Schöbel. The computational complexity of delay management. In D. Kratsch, editor, *Graph-Theoretic Concepts in Computer Science: 31st International Workshop (WG 2005)*, volume 3787 of *Lecture Notes in Computer Science*, 2005.

[GJPW07] M. Gatto, R. Jacob, L. Peeters, and P. Widmayer. Online delay management on a single train line. In *Algorithmic Methods for Railway Optimization*, number 4359 in Lecture Notes in Computer Science, pages 306–320. Springer, 2007.

[GKMH+13] M. Goerigk, M. Knoth, M. Müller-Hannemann, M. Schmidt, and A. Schöbel. The price of robustness in timetable information. *Transportation Science*, 2013. available online before print.

[Gov05] R. M. P. Goverde. *Punctuality of railway operations and timetable stability analysis*. PhD thesis, TRAIL Research School, 2005.

[GS07] A. Ginkel and A. Schöbel. To wait or not to wait? The bicriteria delay management problem in public transportation. *Transportation Science*, 41(4):527–538, 2007.

[GS10] M. Goerigk and A. Schöbel. An empirical analysis of robustness concepts for timetabling. In T. Erlebach and M. Lübbecke, editors, *Proceedings of the 10th Workshop on Algorithmic Approaches for Transportation Modelling, Optimization, and Systems (ATMOS)*, volume 14 of *OASIcs*, pages 100–113, Dagstuhl, Germany, 2010. Schloss Dagstuhl–Leibniz-Zentrum für Informatik.

[GS11] M. Goerigk and A. Schöbel. Engineering the modulo network simplex heuristic for the periodic timetabling problem. In P. M. Pardalos and S. Rebennack, editors, *Experimental Algorithms*, volume 6630 of *Lecture Notes in Computer Science*, pages 181–192. Springer Berlin Heidelberg, 2011.

[GS13] M. Goerigk and A. Schöbel. Improving the modulo simplex algorithm for large-scale periodic timetabling. *Computers & Operations Research*, 40(5):1363–1370, 2013.

[GSS04] Z. Gao, H. Sun, and L. L. Shan. A continuous equilibrium network design model and algorithm for transit systems. *Transportation Research Part B: Methodological*, 38(3):235–250, 2004.

[GvHK04] J.-W. Goossens, S. van Hoesel, and L. Kroon. A branch-and-cut approach for solving railway line planning problems. *Transportation Science*, 38(3):379–393, 2004.

[GvHK06] J.-W. Goossens, S. van Hoesel, and L. Kroon. On solving multi-type railway line planning problems. *European Journal of Operational Research*, 168(2):403–424, 2006. Feature Cluster on Mathematical Finance and Risk Management.

References

[GYW06] J. F. Guan, Hai Yang, and S.C. Wirasinghe. Simultaneous optimization of transit line configuration and passenger line assignment. *Transportation Research Part B*, 40(10):885–902, 2006.

[Hd01] B. Heidergott and R. de Vries. Towards a control theory for transportation networks. *Discrete Event Dynamic Systems*, 11:371–398, 2001.

[HGL08] G. Heilporn, L. De Giovanni, and M. Labbé. Optimization models for the single delay management problem in public transportation. *European Journal of Operational Research*, 189(3):762–774, 2008.

[HKF97] A. Higgins, E. Kozan, and L. Ferreira. Heuristic techniques for single line train scheduling. *Journal of Heuristics*, 3(1):43–62, 1997.

[HKLV05] D. Huisman, L. G. Kroon, R. M. Lentink, and M. J. C. M. Vromans. Operations research in passenger railway transportation. *Statistica Neerlandica*, 59:467–497, 2005.

[HPB13] M. Heydar, M. E. H. Petering, and D. R. Bergmann. Mixed integer programming for minimizing the period of a cyclic railway timetable for a single track with two train types. *Computers & Industrial Engineering*, 66(1):171–185, 2013.

[IC95] Y. Israeli and A. Ceder. Transit route design using scheduling and multiobjective programming techniques. In J. R. Daduna, I. Branco, and J. M. Pinto Paixão, editors, *Computer-aided transit scheduling*, volume 430 of *Lecture Notes in Economics and Mathematical Systems*, pages 56–75, Berlin, 1995. Springer.

[Kas10] M. Kaspi. Service oriented train timetabling. Master's thesis, Tel Aviv University, 2010.

[Kin08] M. Kinder. Models for periodic timetabling. Master's thesis, Technische Universität Berlin, 2008.

[KMH$^+$08] L. Kroon, G. Maróti, M. R. Helmrich, M. Vromans, and R. Dekker. Stochastic improvement of cyclic railway timetables. *Transportation Research Part B: Methodological*, 42(6):553–570, 2008.

[KP03] L. G. Kroon and L. W. P. Peeters. A variable trip time model for cyclic railway timetabling. *Transportation Science*, 37:198–212, 2003.

[KS11] N. Kliewer and L. Suhl. A note on the online nature of the railway delay management problem. *Networks*, 57, 2011.

[KTZ11] S. O. Krumke, C. Thielen, and C. Zeck. Extensions to online delay management on a single train line: new bounds for delay minimization and profit maximization. *Mathematical Methods of Operations Research*, 74(1):53–75, 2011.

[Lie05] C. Liebchen. A cut-based heuristic to produce almost feasible periodic railway timetables. In S. E. Nikoletseas, editor, *Experimental and Efficient Algorithms*, volume 3503 of *Lecture Notes in Computer Science*, pages 354–366. Springer Berlin Heidelberg, 2005.

[Lie06] C. Liebchen. *Periodic Timetable Optimization in Public Transport*. PhD thesis, Technische Universität Berlin, 2006. published by dissertation.de.

[Lin00] T. Lindner. *Train Schedule Optimization in Public Rail Transport*. PhD thesis, Technische Universität Braunschweig, 2000.

[LLER11] R. Lusby, J. Larsen, M. Ehrgott, and D. Ryan. Railway track allocation: models and methods. *OR Spectrum*, 33:843–883, 2011.

[LLMS09] C. Liebchen, M. Lübbecke, R. Möhring, and S. Stiller. The concept of recoverable robustness, linear programming recovery, and railway applications. In R. K. Ahuja, R. H. Möhring, and C. D. Zaroliagis, editors, *Robust and online large-scale optimization*, volume 5868, pages 1–27. Springer, 2009.

[LM07a] C. Liebchen and R. Möhring. The modeling power of the periodic event scheduling problem: Railway timetables and beyond. In F. Geraets, L. Kroon, A. Schoebel, D. Wagner, and C. Zaroliagis, editors, *Algorithmic Methods for Railway Optimization*, volume 4359 of *Lecture Notes in Computer Science*, pages 3–40. Springer Berlin Heidelberg, 2007.

[LM07b] C. Liebchen and R. H. Möhring. The modeling power of the periodic event scheduling problem: Railway timetables– and beyond. In F. Geraets, L. Kroon, A. Schöbel, D. Wagner, and C. D. Zaroliagis, editors, *Algorithmic Methods for Railway Optimization*, volume 4359 of *Lecture Notes in Computer Science*, pages 3–40. Springer Berlin Heidelberg, 2007.

[LPW08] C. Liebchen, M. Proksch, and F. H. Wagner. Performance of algorithms for periodic timetable optimization. In G. Fandel, W. Trockel, M. Hickman, P. Mirchandani, and S. Voß, editors, *Computer-aided Systems in Public Transport*, volume 600 of *Lecture Notes in Economics and Mathematical Systems*, pages 151–180. Springer Berlin Heidelberg, 2008.

[LS67] W. Lampkin and P. D. Saalmans. The design of routes, service frequencies, and schedules for a municipal bus undertaking: a case study. *Operational Research Quartely*, 18:375–397, 1967.

[LSS+10] C. Liebchen, M. Schachtebeck, A. Schöbel, S. Stiller, and A. Prigge. Computing delay resistant railway timetables. *Computers & Operations Research*, 37(5):857–868, 2010.

[Lüb09] J. Lübbe. Passagierrouting und Taktfahrplanoptimierung. Master's thesis, Technische Universität Berlin, 2009. (in German).

[LZ05] T. Lindner and U. T. Zimmermann. Cost optimal periodic train scheduling. *Mathematical Methods of Operations Research*, 62(2):281–295, 2005.

[Man80] C. E. Mandl. Evaluation and optimization of urban public transportation networks. *European Journal of Operational Research*, 5(6):396–404, 1980.

[Nac98] K. Nachtigall. *Periodic Network Optimization and Fixed Interval Timetables*. Deutsches Zentrum für Luft– und Raumfahrt, Institut für Flugführung, Braunschweig, 1998. Habilitationsschrift.

[NJ08] K. Nachtigall and K. Jerosch. Simultaneous network line planning and traffic assignment. In M. Fischetti and P. Widmayer, editors, *8th Workshop on Algorithmic Approaches for Transportation Modeling, Optimization, and Systems*, 2008.

[NO08] K. Nachtigall and J. Opitz. Solving periodic timetable optimisation problems by modulo simplex calculations. In M. Fischetti and P. Widmayer, editors, *ATMOS 2008 - 8th Workshop on Algorithmic Approaches for Transportation Modeling, Optimization, and Systems*, Dagstuhl, Germany, 2008. Schloss Dagstuhl - Leibniz-Zentrum für Informatik, Germany.

[NV96] K. Nachtigall and S. Voget. A genetic algorithm approach to periodic railway synchronization. *Computers & Operations Research*, 23(5):453–463, 1996.

[NW88] G. L. Nemhauser and L. A. Wolsey. *Integer and Combinatorial Optimization*. Wiley, 1988.

[Odi96] M. A. Odijk. A constraint generation algorithm for the construction of periodic railway timetables. *Transportation Research Part B*, 30(6):455–464, 1996.

[Odi98] M. A. Odijk. *Railway Timetable Generation*. PhD thesis, Technische Universiteit Delft, 1998.

[ORv06] M. A. Odijk, H. E. Romeijn, and H. van Maaren. Generation of classes of robust periodic railway timetables. *Computers & Operations Research*, 33(8):2283–2299, 2006.

[PB06] M. Pfetsch and R. Borndörfer. Routing in line planning for public transport. In H.-D. Haasis, H. Kopfer, and J. Schönberger, editors, *Operations Research Proceedings 2005*, volume 2005 of *Operations Research Proceedings*, pages 405–410. Springer Berlin Heidelberg, 2006.

[Pee03] L. Peeters. *Cyclic Railway Timetable Optimization*. PhD thesis, Erasmus University Rotterdam, 2003.

[Phi93] C. A. Phillips. The network inhibition problem. In *Proceedings of the twenty-fifth annual ACM symposium on Theory of computing*, pages 776–785, New York, 1993. ACM.

[Roc84] R. T. Rockafellar. *Network flows and monotropic optimization.* John Wiley & Sons, Inc., 1984.

[RS97] R. Raz and S. Safra. A sub-constant error-probability low-degree test, and a sub-constant error-probability PCP characterization of NP. In *Proceedings of the twenty-ninth annual ACM symposium on Theory of computing*, STOC '97, pages 475–484, New York, NY, USA, 1997. ACM.

[SBK01] L. Suhl, C. Biederbick, and N. Kliewer. Design of customer-oriented dispatching support for railways. In Stefan Voß and Joachim R. Daduna, editors, *Computer-Aided Scheduling of Public Transport*, volume 505 of *Lecture Notes in Economics and Mathematical Systems*, pages 365–386. Springer Berlin Heidelberg, 2001.

[SBP74] L. A. Silman, Z. Barzily, and U. Passy. Planning the route system for urban busses. *Computers & Operations research*, 1:201–211, 1974.

[Sch01] A. Schöbel. A model for the delay management problem based on mixed-integer programming. *Electronic Notes in Theoretical Computer Science*, 50(1), 2001.

[Sch05] S. Scholl. *Customer-Oriented Line Planning*. PhD thesis, Technische Universität Kaiserslautern, 2005. published by dissertation.de.

[Sch06] A. Schöbel. *Optimization in public transportation. Stop location, delay management and tariff planning from a customer-oriented point of view.* Optimization and Its Applications. Springer, New York, 2006.

[Sch07] A. Schöbel. Integer programming approaches for solving the delay management problem. In *Algorithmic Methods for Railway Optimization*, number 4359 in Lecture Notes in Computer Science, pages 145–170. Springer, 2007.

[Sch09] A. Schöbel. Capacity constraints in delay management. *Public Transport*, 1(2):135–154, 2009.

[Sch10] M. Schachtebeck. *Delay Management in Public Transportation: Capacities, Robustness, and Integration*. PhD thesis, Georg-August-Universität Göttingen, 2010.

[Sch11] A. Schöbel. Line planning in public transportation: models and methods. *OR Spectrum*, pages 1–20, 2011. 10.1007/s00291-011-0251-6.

[Sch12] M. Schmidt. Line planning with equilibrium routing. Technical report, Preprint series, Institute for Numerical and Applied Mathematics, Georg-August-University Göttingen, 2012. submitted.

[Sch13] M. Schmidt. Simultaneous optimization of delay management decisions and passenger routes. *Public Transport*, 5(1):125–147, 2013.

[SG13] M. Siebert and M. Goerigk. An experimental comparison of periodic timetabling models. *Computers & Operations Research*, 40(10):2251–2259, 2013.

[SK09] A. Schöbel and A. Kratz. A bicriteria approach for robust timetabling. In R. K. Ahuja, R. H. Möhring, and C. D. Zaroliagis, editors, *Robust and Online Large-Scale Optimization*, volume 5868 of *Lecture Notes in Computer Science*, pages 119–144. Springer Berlin Heidelberg, 2009.

[SM99] L. Suhl and T. Mellouli. Requirements for, and design of, an operations control system for railways. In *Computer-Aided Transit Scheduling*. Springer, 1999.

[SMBG01] L. Suhl, T. Mellouli, C. Biederbick, and J. Goecke. Managing and preventing delays in railway traffic by simulation and optimization. In *Mathematical methods on Optimization in Transportation Systems*, pages 3–16. Kluwer, 2001.

[Son79] H. Sonntag. Ein heuristisches Verfahren zum Entwurf nachfrageorientierter Linienführung im öffentlichen Personennahverkehr. *ZOR - Zeitschrift für Operations-Research*, 23:B15, 1979. (in German).

[SS06] A. Schöbel and S. Scholl. Line planning with minimal transfers. In *5th Workshop on Algorithmic Methods and Models for Optimization of Railways*, number 06901 in Dagstuhl Seminar Proceedings, 2006.

[SS08] M. Schachtebeck and A. Schöbel. IP-based techniques for delay management with priority decisions. In M. Fischetti and P. Widmayer, editors, *ATMOS 2008 - 8th Workshop on Algorithmic Approaches for Transportation Modeling, Optimization, and Systems*, Dagstuhl Seminar proceedings, 2008.

[SS10a] M. Schachtebeck and A. Schöbel. To wait or not to wait and who goes first? Delay management with priority decisions. *Transportation Science*, 44(3):307–321, 2010.

[SS10b] M. Schmidt and A. Schöbel. The complexity of integrating routing decisions in public transportation models. In T. Erlebach and M. Lübbecke, editors, *Proceedings of the 10th Workshop on Algorithmic Approaches for Transportation Modelling, Optimization, and Systems*, volume 14 of *OASIcs*, pages 156–169, Dagstuhl, Germany, 2010. Schloss Dagstuhl–Leibniz-Zentrum für Informatik.

[SS12a] M. Schmidt and A. Schöbel. The complexity of integrating routing decisions in public transportation models. Technical report, Preprint series, Institute for Numerical and Applied Mathematics, Georg-August-University Göttingen, 2012. submitted.

[SS12b] M. Schmidt and A. Schöbel. Timetabling with passenger routing. Accepted for publication in OR Spectrum. 2013.

[SU89] P. Serafini and W. Ukovich. A mathematical model for periodic scheduling problems. *SIAM Journal on Discrete Mathematics*, 2:550–581, 1989.

[TTBP11] L. M. Torres, R. Torres, R. Borndörfer, and M. E. Pfetsch. Line planning on tree networks with applications to the quito trolebús system. *International Transactions in Operational Research*, 18:455–472, 2011.

[WC96] Z. Wang and J. Crowcroft. Quality of service routing for supporting multimedia applications. *IEEE Journal on Selected areas in communications*, 14(7):1228–1234, 1996.

[WYFL08] R. C. W. Wong, T. W. Y. Yuen, K. W. Fung, and J. M. Y. Leung. Optimizing timetable synchronization for rail mass transit. *Transportation Science*, 42(1):57–69, 2008.

Index

A
activity
  artificial, 102
  changing, 83
  destination, 83
  driving, 83
  maintained changing, 124
  maintained changing activities
    corresponding to a routing, 129
  operational, 83
  origin, 83, 123
  reasonable virtual, 100
  virtual, 98
  waiting, 83
activity weight, 86
acyclic, 7
all-wait strategy, 161
arc, 6
  destination, 22, 27
  driving, 21
  origin, 22, 27
  transfer, 21, 27
arc speed-up, 181, *see* ASU
ASU, 182

B
budget, 11

C
capacitated line planning, 10
capacitated line planning with routing, *see* cLPwR
capacity, 15
CGN, 22
change-and-go network, *see* CGN

cLPwR, 19, 25
connected, 7
connected component, 7
cost
  along a path, 16, 23, 29
  of a line, *see* line cost
critical path method, 127

D
delay
  in a node, 125
  of a solution, 125
delay management, 113
delay management with fixed connections,
    see DMwFC
delay management with routing,
    see DMwR
digraph, 6
direct traveler approach, 11
Directed Two-Commodity Integral Flow, 32
disposition timetable, 115, 120, 124
  corresponding to a routing, 129
  corresponding to a set of maintained
    activities, 128
DMwFC, 127
DMwR, 122, 125
driving time, 12, 16, 23
  between two stations, 79
  of a line, 16

E
EAN, 83
  operational, 83, 124
edge, 5
  directed, 6

event
    arrival, 82
    departure, 82
    destination, 82
    operational, 82
    origin, 82, 123
event-activity network, *see* EAN

F
feasible, 74
    for a disposition timetable, 121, 122, 124
    for a line concept, 17, 24, 29
    for a line-routing, 17
    for a network component, 170
    for a routing, 24, 124, 170
    for an ST-train-routing, 122
    line concept, 15
    line-route, 16
    timetable, 79, 85
frequency, 9, 15
frequency setting, 9

G
graph
    directed, 6
    undirected, 5

H
Hamiltonian Path, 131
Hitting Set, 36

L
length
    of a path, 7
line, 9, 15
    bi-directional, 15
    one-directional, 15
line concept, 15
    corresponding to a routing, 30, 31
line connectivity problem, 35
line cost, 10
line driving time function, 10
line network, *see* LN
line planning, 9
    cost-oriented, 11
    passenger-oriented, 11
line pool, 9
line routing network, 29
line-route, 16
line-routing, 17
    for an OD-pair, 17
linear network, 6
LN, 28

N
nested, 50
network
    directed, 6
    undirected, 6
NLT property, 38, 61
no-line-twice property, *see* NLT property
no-train-twice property, *see* NTT property
node, 5
    destination, 21
    origin, 21
    travel, 21
node-OD-pair, 167
NTT property
    for a path, 137
    for an instance, 146
number of passengers, 16, 121
    using a line-route, 17
    using a path, 24, 168

O
OD-NLT-property, 57
OD-pair, 16, 121
    operational, 97
OD-pairs, 11
origin-destination pair, *see* OD-pair

P
Partition, 41
passenger weight, *see* number of passengers
path, 6
    directed, 6
    for a node-OD-pair, 168
    for an operational OD-pair, 97
    undirected, 6
planned connections, 79
predecessor, 7
    direct, 7
price of sequentiality, 175
PTN, 8
public transportation network, *see* PTN

R
RCSP, *see* Resource-Constrained Shortest Path
Resource-Constrained Shortest Path, 39
routing, 24, 29, 85, 97, 124, 168, 181

Index

corresponding to a disposition timetable, 127
corresponding to a line concept, 30
corresponding to a set of maintained changing activities, 128
corresponding to a timetable, 88
for a node-OD-pair, 168
for an OD-pair, 24
routing network, 24, 84, 125

S
scLPwR, 18, 25
Set Cover, 88
shortest path line-routing
   for a line concept, 19
shortest path problem, 7
   all pairs, 7
shortest-path routing, 7
   for a line concept, 24
simple, 6
simple capacitated line planning with routing, *see* scLPwR, *see* scLPwR
simplified timetabling problem, *see* sTT
slack time, 74, 84
solution, 5
   corresponding to a disposition timetable, 127
   corresponding to a line concept, 30
   corresponding to a routing, 30, 31, 87, 129
   corresponding to a set of maintained changing activities, 128
   corresponding to a timetable, 88
source delay, 114, 120
speed-up factor, 181
speed-up network, 182
ST-train-route, 121
ST-train-routing, 122
start time, 116, 121
Steiner Tree, 61
sTT, 86
subgraph, 6
   induced, 6
subnetwork, 6
   induced, 6

successor, 7
   direct, 7

T
timetable, 74, 79
   corresponding to a routing, 87
timetabling, 73
   aperiodic, 73
   periodic, 73
timetabling with routing, *see* TTwR
timetabling with routing between events, *see* TTwRE
train, 10, 15, 79
train-route, 80
train-routing, 81
transfer penalty, 12, 16
   station-independent, 27
transfer time, 12, 16, 23, 26
   along a path, 28
transportation mode, 10
travel time, 12, 17, 23, 25, 81, 85, 86, 122, 124
   along a line-route, 16
   along a path, 80, 85, 97, 121
   nominal, 125
tree, 8
TTwR, 81, 85, 100
TTwRE, 97

U
uLPwR, 17, 25, 29
uncapacitated line planning, 10
uncapacitated line planning with routing, *see* uLPwR

V
vertex, 5

W
wait-depart decision, 113, 122
waiting time
   at a station, 79

The manufacturer's authorised representative in the EU is Springer Nature Customer Service Centre GmbH, Europaplatz 3, 69115 Heidelberg, Germany. If you have any concerns regarding our products, please contact ProductSafety@springernature.com

Printed and bound by CPI Group (UK) Ltd, Croydon, CR0 4YY

26/03/2026

02078916-0003